T0282433

Microbial Resources

Microbial Resources

From Functional Existence in Nature to Applications

Edited by

Ipek Kurtböke

GeneCology Research Centre and Faculty of Science, Health, Education and Engineering, University of the Sunshine Coast, Maroochydore DC, QLD 4558, Australia

Academic Press is an imprint of Elsevier
125 London Wall, London EC2Y 5AS, United Kingdom
525 B Street, Suite 1800, San Diego, CA 92101-4495, United States
50 Hampshire Street, 5th Floor, Cambridge, MA 02139, United States
The Boulevard, Langford Lane, Kidlington, Oxford OX5 1GB, United Kingdom

Library of Congress Cataloging-in-Publication Data
A catalog record for this book is available from the Library of Congress

British Library Cataloguing-in-Publication Data
A catalogue record for this book is available from the British Library

ISBN: 978-0-12-804765-1

For information on all Academic Press publications visit our website at
https://www.elsevier.com/books-and-journals

Publisher: Sara Tenney
Acquisition Editor: Linda Versteeg-Buschman
Editorial Project Manager: Timothy Bennett
Production Project Manager: Edward Taylor
Designer: Mark Rogers

Typeset by Thomson Digital

Contents

2. A Flavor of Prokaryotic Taxonomy: Systematics Revisited

Paul De Vos, Fabiano Thompson, Cristiane Thompson and Jean Swings

3. Bioactive Actinomycetes: Reaching Rarity Through Sound Understanding of Selective Culture and Molecular Diversity

İpek Kurtböke

4. Microbial Resources for Global Sustainability

Jim Philp and Ronald Atlas

10. The Nagoya Protocol Applied to Microbial Genetic Resources

Philippe Desmeth

11. Fungal Genetic Resources for Biotechnology

Kevin McCluskey

12. Industrial Culture Collections: Gateways from Microbial Diversity to Applications

Olga Genilloud

Contributors

Ronald Atlas, University of Louisville, Louisville, KY, United States

C. Benjamin Naman, Center for Marine Biotechnology and Biomedicine, Scripps Institution of Oceanography, University of California, San Diego, CA, United States

Vera Bussas, Leibniz Institute DSMZ-German Collection of Microorganisms and Cell Cultures, Braunschweig, Germany

Paul De Vos, Ghent University, Gent, Belgium

Arnold L. Demain, RISE Institute, Drew University, Madison, NJ, United States

Philippe Desmeth, Belgian Science Policy Office, Brussels, Belgium

William Fenical, Center for Marine Biotechnology and Biomedicine, Scripps Institution of Oceanography, University of California, La Jolla, CA, United States

John A. Fuerst, The University of Queensland, St. Lucia, QLD, Australia

Olga Genilloud, Fundación MEDINA, Health Sciences Technology Park, Granada, Spain

William H. Gerwick, Center for Marine Biotechnology and Biomedicine, Scripps Institution of Oceanography, University of California, San Diego, CA, United States

Irena B. Ivshina, Institute of Ecology and Genetics of Microorganisms, Ural Branch of the Russian Academy of Sciences, Perm State University, Perm, Russia

Anastasiya V. Krivoruchko, Institute of Ecology and Genetics of Microorganisms, Ural Branch of the Russian Academy of Sciences, Perm State University, Perm, Russia

İpek Kurtböke, Genecology Research Centre and Faculty of Science, Health, Education and Engineering, University of the Sunshine Coast, Maroochydore DC, QLD 4558, Australia

Maria S. Kuyukina, Institute of Ecology and Genetics of Microorganisms, Ural Branch of the Russian Academy of Sciences, Perm State University, Perm, Russia

Christopher A. Leber, Center for Marine Biotechnology and Biomedicine, Scripps Institution of Oceanography, University of California, San Diego, CA, United States

Evan Martens, Cempra, Inc., Chapel Hill, NC, United States

Kevin McCluskey, Kansas State University, Manhattan, KS, United States

Kozo Ochi, Hiroshima Institute of Technology, Saeki-ku, Hiroshima, Japan

Lynette Bueno Pérez, Center for Marine Biotechnology and Biomedicine, Scripps Institution of Oceanography, University of California, La Jolla, CA, United States

Jim Philp, Science and Technology Policy Division, Paris, France

Avinash Sharma, Microbial Culture Collection, National Centre for Cell Science, Sutarwadi, Pashan Pune, Maharashtra, India

Yogesh Shouche, Microbial Culture Collection, National Centre for Cell Science, Sutarwadi, Pashan Pune, Maharashtra, India

Ken-ichiro Suzuki, Tokyo University of Agriculture, Setagaya-ku, Tokyo, Japan

Jean Swings, Ghent University, Gent, Belgium; SAGE-COPPE-UFRJ, Rio de Janeiro, Brazil

Cristiane Thompson, Federal University of Rio de Janeiro (UFRJ); SAGE-COPPE-UFRJ, Rio de Janeiro, Brazil

Fabiano Thompson, SAGE-COPPE-UFRJ, Rio de Janeiro, Brazil

Author Biographies

Ronald M. Atlas is Professor of Biology at the University of Louisville. He is a fellow in the *American Academy of Microbiology*, has served as President of *American Society for Microbiology*, as a member of the NIH Recombinant Advisory committee, as chair of NASA's Planetary Protection Subcommittee, as chair of the Wellcome Trust Pathogens, Immunology and Population Health Strategy Committee, and as chair of the Board of Directors of the One Health Commission. He is the author of 300 manuscripts and 20 books. He chairs the American Society for Microbiology Public and Scientific Affairs Board and regularly advices the US Government on issues of environment and infectious diseases.

Dr. Vera Bussas is a microbiologist. Since 1990 she has been committed to the activities of the patent depositary of the Leibniz-Institut DSMZ-*Deutsche Sammlung von Mikroorganismen und Zellkulturen* GmbH, Braunschweig, Germany. As an International Depositary Authority (IDA) representative,) she is the responsible curator of biological material deposited for patent purposes according to the *Budapest Treaty on The International Recognition of the Deposit of Microorganisms for the Purposes of Patent Procedure*. The handling of the biological material implicates additional qualifications as biological safety officer and as trained shipper according to IATA regulations. Since 2007 Dr. Bussas is member of the WFCC Executive Board.

Prof. Arnold L. Demain is a pioneer in industrial microbiology. He was formerly the Professor of Industrial Microbiology in the Department of Biology at the Massachusetts Institute of Technology (MIT) and advancing fermentation biology for more than 60 years. He contributed to the elucidation and regulation of the biosynthetic pathways leading to penicillins and cephalosporins, improved microbial production of cholesterol-lowering drugs, immunosuppressives, antitumor agents, and antifungal drugs. He has over 560 publications, has coauthored or coedited 14 books, holds 21 US patents, and is a member of the National

Academies of Sciences of the United States, Mexico, and Hungary. He is currently at The Charles A. Dana Research Institute for Scientists Emeriti (RISE) at Drew University in Madison, New Jersey.

Dr. Philippe Desmeth is bioengineer and environmental advisor. He gained experience in agro-industrial production and training farmers in West Africa and Southeast Asia. He worked in private companies before joining the Belgian Coordinated Collections of Micro-organisms (BCCM) in 1996, as international cooperation manager. At the Belgian Science Policy Office, he was also involved in the management of the Belgian Biodiversity Platform. He follows the Access and Benefit Sharing problematic and has coordinated EU-funded projects such as MOSAICC, MOSAICS and BRIO. He co-launched and coordinates the TRUST project to implement the Nagoya Protocol in microbiology. He was reelected President of the *World Federation for Culture Collections* (WFCC) in October 2013.

Paul De Vos is Emeritus Professor in Microbiology since October 2015 at the Ghent University where he was full professor until 2000. He was director of the BCCM/LMG bacteria collection between 2005 and 2015. He is member of the Bergey Trust. His general research interests are bacterial diversity, evolution, taxonomy and identification. Research projects are linked with bacteria involved in the nitrogen cycle, methanotrophs, the genera *Bacillus* and *Pseudomonas*; plant pathogens (*Pseudomonas*, *Xanthomonas*, *Clavibacter* and *Enterobacteriaceae*). He is author/coauthor of over 280 peer reviewed papers in biodiversity, taxonomy, and bacterial fermentation.

Prof. William (Bill) Fenical joined the Scripps Institution of Oceanography (SIO), UC-San Diego, in 1973. He is currently Distinguished Professor of Oceanography and Pharmaceutical Science, and Founding Director of SIO's Center for Marine Biotechnology and Biomedicine. His research interests have focused on the field of marine microbiology and the utilization of marine microorganisms as a source for new drug discovery. He has authored more than 460 papers in this field, and was awarded the NCI's Merit Award (2003), the Ernest Guenther Award from the American Chemical Society (2006), the Lifetime Achievement Award from the American Society of Pharmacognosy (2006). In 2008, he was elected Fellow of the American Association of Science (AAAS).

John Fuerst is an Emeritus Professor at the School of Chemistry and Molecular Biosciences, The University of Queensland, Australia. His research interests are microbial evolution including planctomycetes, early origins of eukaryotes, and the actinomycetes of marine sponge bacteria and their natural products. In 2008 he was Visiting Professor of the Royal Netherlands Academy of Arts and Sciences at Radboud University of Nijmegen collaborating on anammox planctomycete structure research. He is author of reviews in *Annual Reviews of Microbiology* and *Nature Reviews Microbiology* on planctomycetes and their cell compartments, summarizing research on cell structure in phylum *Planctomycetes* and their evolutionary significance.

Dr. Olga Genilloud is a biochemist with an extended career in applied microbiology focusing on the biosynthesis of bioactive secondary metabolites. She has more than 20 years research experience in bacterial natural products drug discovery in the pharma sector, as Group Leader in the Basic Research Center at Merck Sharp & Dohme in Spain. In 2008, she took on the leadership position to establish Fundación MEDINA from the MSD-Spain R&D Center. Currently she is Scientific Director at Fundación MEDINA and Head of the Microbiology department. She is a Fellow of the Royal Society of Chemistry, the Society of Industrial Microbiology and Biotechnology, the Spanish Society of Microbiology, and the Spanish Society of Biotechnology.

Prof. William H. Gerwick received his BS in Biochemistry at UC Davis (1976) and PhD in Oceanography at the Scripps Institution of Oceanography (Bill Fenical, 1981), pursued postdoctoral studies at the University of Connecticut (Steven Gould, 1981–82), was Professor of Chemistry at the University of Puerto Rico (1982–84) and then moved to the College of Pharmacy, Oregon State University (1984–2005) as Professor of Pharmaceutical Sciences. In 2005 he moved to become Professor of Oceanography and Pharmaceutical Sciences at Scripps Institution of Oceanography and the Skaggs School of Pharmacy and Pharmaceutical Sciences, UC San Diego, and in 2011 was promoted to Distinguished Professor. His research focuses on discovery of novel natural products from marine cyanobacteria, their biomedical applications, and investigations of their biosynthesis using orthogonal approaches.

Prof. Dr Irena Ivshina specializes in environmental microbiology and industrial biotechnology. She leads the Laboratory of Alkanotrophic Microorganisms, Institute of Ecology and Genetics of Microorganisms (IEGM), Urals Branch RAS, and teaches at Microbiology and Immunology department, Perm State National Research University (PSNRU). For over 30 years she has been researching into the applications of microorganisms for rehabilitation of terrestrial ecosystems contaminated by petroleum hydrocarbons, heavy metals, and pharma pollutants. She has authored over 200 articles and 15 patents. She is a Laureate of the Russian Federation Government award for the development of a complex remediation biotechnology for disturbed and hydrocarbon-contaminated northern biogeocenoses.

Dr Anastasiya Krivoruchko is a young researcher at the Laboratory of Alkanotrophic Microorganisms, IEGM. She focuses on the adhesion properties of actinobacteria and their application in biodegradation of recalcitrant petroleum hydrocarbons. She is an assistant professor at the Microbiology and Immunology department, PSNRU. She is a Laureate of the Perm Krai Young Scientist award for the most promising research project in biology and agriculture.

Dr. Ipek Kurtböke has worked in the field of biodiscovery in Turkey, Italy, UK and Australia since 1982. She currently conducts research and teaches in the field of applied microbiology and biotechnology and is a senior lecturer at the University of the Sunshine Coast (USC), Queensland, Australia. She has established a bioactive actinomycete library for the purposes of research and teaching at the USC in partnership with regional, national and international collaborators for discovery of new therapeutic agents, agrobiologicals, enzymes and environmentally friendly biotechnological innovations. She has been one of the founding members of the *Australian Microbial Resources Research Network* (AMRIN) as well as an active member of the *World Federation of Culture Collections* (WFCC) and also served as the Vic-President of the Federation (2010–2013).

Prof. Dr. Maria Kuyukina has been working for the Laboratory of Alkanotrophic Microorganisms, IEGM in Perm, Russian Federation for 25 years. She is a senior researcher and focuses on the application of molecular biological methods in bioremediation, and is also involved in developing polyfunctional biocatalysts for bioremediation of oil-contaminated environments. She is currently with the Microbiology and Immunology Department, PSNRU. She is a Laureate of the Russian Federation Government award for the development of a complex remediation biotechnology for disturbed and hydrocarbon-contaminated northern biogeocenoses.

Mr. Christopher A. Leber graduated from UCLA in 2015, earning his BS in Marine Biology along with a minor in Atmospheric and Oceanic Science. He then began his pursuit of a PhD in Marine Chemical Biology at the Scripps Institution of Oceanography, under the guidance of Professor William H. Gerwick. Chris's research is aimed at investigating the ecological roles and biomedical potential of novel secondary metabolites from marine organisms.

Mr. Evan Martens attended Drew University in Madison, NJ from 2007–11 majoring in biology/premedicine with a minor in public health. After completing his junior year, he did microbiology research under the supervision of Prof. Arnold Demain, at the Research Institute for Scientists Emeriti (R.I.S.E) program and published several papers. Upon graduating from Drew in 2011, Evan was employed by Emergency Medical Associates (EMA) in Parsippany, NJ as a medical scribe/clinical information manager. Evan is currently working as a drug development biologist and scientific editor for Cempra, Inc. in Chapel Hill, which is a clinical-stage pharmaceutical company focused on developing antibiotics to meet critical medical needs in the treatment of bacterial infectious diseases.

Prof. McCluskey obtained his Doctorate in Botany and Plant Pathology at Oregon State University and after a postdoctoral fellowship at the University of Arizona, he accepted the position of Curator of the Fungal Genetics Stock Center in 1995. His activities on behalf of culture collections have led to election onto the Executive Board of the World Federation for Culture Collections and to a leadership role in the nascent US Culture Collection Network. He is a member of the *American Phytopathological Society Public Policy Board* and has represented the WFCC at the meetings of the Subsidiary Body on Scientific, Technical, and Technological Advice to the Convention on Biological Diversity. He is now a Research Professor at the FGSC, Kansas State University, and has published extensively on genome biology of fungi.

Dr. C. Benjamin Naman in 2006 received his BS in Chemistry and Psychology from the Mellon College of Science, Carnegie Mellon University, and worked from 2006–10 in the Natural Products Discovery Group at Givaudan Flavors Corporation. Ben completed his PhD in Pharmaceutical

Sciences with a specialization in Medicinal Chemistry and Pharmacognosy at the College of Pharmacy, The Ohio State University in 2015 (A. Douglas Kinghorn, Thesis Advisor), before going on to a postdoctoral research position at Scripps Institution of Oceanography, University of California, San Diego (William H. Gerwick, Principal Investigator). His research focuses on the discovery and human health application of natural products from terrestrial plant and marine organisms, with particular emphasis on drug discovery for infectious disease and cancer chemotherapy.

Prof. Kozo Ochi is a microbiologist who has worked at Fujisawa Pharmaceutical Company (1982–91), National Food Research Institute (1992–2010), and Hiroshima Institute of Technology (2011–present). His research interests have been bacterial stress response, stringent response, sporulation, antibiotic production, and strain improvement. He has developed a new method, ribosome engineering, to activate bacterial silent genes, which is useful for screening of novel secondary metabolites, in addition to productivity enhancement.

Dr. Lynette Bueno Pérez is currently an Assistant Professor in the Department of Pharmaceutical Sciences of the School of Pharmacy at the University of Puerto Rico Medical Sciences Campus, San Juan. She holds degrees in microbiology from the University of Puerto Rico Mayagüez Campus (BS 2001), pharmaceutical sciences from the University of Puerto Rico Medical Sciences Campus (MS 2008), and medicinal chemistry and pharmacognosy from The Ohio State University (PhD 2014). She was a postdoctoral scientist at Scripps Institution of Oceanography, University of California, San Diego (2014–16). Her research interests have focused in the discovery of new natural product compounds from marine bacteria and higher plants with potential antibacterial and anticancer activity.

Dr. Jim Philp is a microbiologist who has worked as a policy analyst since 2011 at the OECD, specializing in industrial biotechnology. He has been an academic, researching, and teaching environmental and industrial biotechnology. He spent 8.5 years working for Saudi Aramco in Saudi Arabia as an oil biotechnologist. He has authored over 300 articles. In 2015 he was inducted into *Who's Who*. He is a Fellow of the Royal Society of Chemistry, and an Associate Fellow of the Institution of Chemical Engineers.

Dr. Avinash Sharma is an environmental microbiologist and scientist at Microbial Culture Collection, National Centre for Cell Science, Pune, India. He is Incharge of International Depositary Authority at MCC, Pune under the Budapest Treaty on International Recognition of Deposit of Microorganisms for the Purposes of Patent Procedure. Dr. Sharma obtained his PhD in 2013 and his research is focused on the microbial taxonomy and metagenomics approaches to explore microbial community structure of extremophilic organisms.

Dr. Yogesh Shouche started his career as scientist at National Centre for Cell Science (NCCS) in Pune, India. He has worked on the bacterial diversity of various ecological niches, his current area of research relates to understanding of the succession of microbial communities in human gut. He has been working in the field of microbial ecology, microbial molecular taxonomy, and biodiversity over the last 25 years; published more than 150 publications in peer reviewed journals. He is on the editorial board of prestigious journals, such as the *Current Science, European Journal of Soil Biology, Plos One*, and *Scientific Reports*. In 2009, he was given the responsibility of establishing the Microbial Culture Collection (MCC), which is the recognized International Depositary Authority (IDA). Dr. Yogesh currently leads a large group of researchers at MCC including 30 Scientists and Technicians.

Ken-ichiro Suzuki is a professor of Department of Fermentation Sciences of Tokyo University of Agriculture since April 2016. He obtained his PhD in 1982 with the taxonomic study of coryneform bacteria, especially focusing on their chemotaxonomy under supervision of Prof. Kazuo Komagata at the University of Tokyo. He has been working in culture collections in Japan, JCM (1982–2001) and NBRC (2001–16) for their management and taxonomic studies of aerobic bacteria especially of actinobacteria. He has more than 140 original papers, 20 chapters of books, and 35 reviews in microbial taxonomy and preservation methods. He has been an executive board member of the World Federation of Culture Collections since 2000 and an associate editor of International Journal of Systematic and Evolutionary Microbiology since 2012. He has been contributing to establishment of culture collections and the network in Asian region.

Jean Swings is Emeritus professor in Microbiology at Ghent University (Belgium) and visiting professor at the Federal University of Rio de Janeiro (UFRJ). He is the former director of LMG Bacteria Collection and served as president of the World Federation for Culture Collections (WFCC) from 2000 to 2004. He coauthored over 470 peer reviewed articles in the field of prokaryotic taxonomy.

Cristiane Thompson is Associate Professor in the Institute of Biology of the Federal University of Rio de Janeiro (UFRJ). She leads the Microbiology laboratory in the same institute. She has developed research on microbial genomic taxonomy and marine systems. She is a visiting professor of the College of Sciences of the San Diego State University. She has published over 70 peer reviewed papers and serves as a coeditor for *Peer J*.

Fabiano Thompson is full professor at COPPE-UFRJ, University of Rio de Janeiro in the field of marine microbial biodiversity and biotechnology. He published over 150 peer reviewed papers. He served as an associate editor of The Prokaryotes (4th Ed.) and is a coeditor for *Microbial Ecology, BMC Genomics, Peer J,* and for *Plos One*. He coordinates the Brazilian network on marine biotechnology.

Preface

Microorganisms are the most diverse and ubiquitous biological entities, which have successfully colonized diverse ecological niches of the planet. Clear fossil evidence of microbial life was found to exist in ancient stromatolites, such as the fossilized filamentous prokaryote of 3.5 billion years of age in Western Australia. Although Archaea have remained close to the characteristics of the universal ancestor and still occupy extreme habitats with similar characteristics to the ones in existence in the early days of the Earth, bacteria continued to evolve into impressive morphological diversity and occupy a broad range of habitats. They also provided the forerunners of today's mitochondria and chloroplast, leading to the establishment and evolution of Eukaryotic organisms about 1.7 billion years ago. Bacteria, Archaea, and Eukarya throughout the evolution of the Earth established their diverse ecological niches and specialized in diverse functions to maintain ecosystem health, thanks to their metabolic diversity and genetic adaptabilities. Through the application of molecular techniques, it has become clear that microbial diversity is greater than that which their cultured representatives displayed. Metagenomics combined with improved sampling techniques is now taking our understanding further and revealing the existence of novel microorganisms in marine, extreme environments, as well as in the previously explored environments, such as the terrigenous ones. Most importantly, these novel molecular tools are now revealing the functional existence of microorganisms in different niches, providing support for the sustainable existence of higher organisms in these ecosystems.

Biodiscovery and microbial biotechnology are based on the search for exploitable and diverse biological resources. In this search, the screening of microbial natural products still continues to represent an important route to the discovery of novel chemicals for the development of new therapeutic agents. The evaluation of the biosynthetic potential of lesser known and/or new microbial taxa is thus of increasing interest. However, selection of novel bioactive compounds producing microorganisms from nature requires a sound microbial taxonomical knowledge and a better understanding of microbial ecology, physiology, and metabolism. Therefore, taxonomic expertise combined with effective preservation of microbial genetic resources will provide a stronger platform for novel discoveries. Furthermore, emerging new technologies associated with bioinformatics and biogeography are fundamentally changing our capability of applying target-based approaches to selectively culture industrially important microorganisms.

Microbial Resources: From Functional Existence in Nature to Industrial Applications will provide the reader with an exciting interdisciplinary journey covering sequential aspects of microbiology, ranging from functional existence of microorganisms in nature to the transformation of this knowledge at industrial and application levels. The book covers and provides:

- the aspects of foundational information related to the microbial ecology, taxonomy, and preservation of diversity, as well as providing information on microbial genetics, genomics, and cryptic genes with importance to industry;
- information on the sustainable utilization of microbial resources for biodiscovery and biotechnology with significant added value due to the inclusion of comprehensive coverage of the culture collections and biological resource centers;
- legislative issues related to the Convention on Biological Diversity (CBD) and its Nagoya Protocol that identify benefit-sharing deriving from the use of microbial resources; and
- information on the IP, biosafety, and transport of microbiological material.

I thank all authors for their valuable contributions. This book brings pioneers of the field and early career scientists together, which is no doubt an indication of the commitment toward the advancement of microbial biotechnology and biodiscovery, to be continued across the generations to come.

We will embrace emerging molecular advancements, while treasuring the traditions and passions passed to us by the earlier generations, when working with talented microorganisms that will continue to contribute to the benefit of mankind.

Ipek Kurtböke

Chapter 1

Planctomycetes—New Models for Microbial Cells and Activities

John A. Fuerst
The University of Queensland, St. Lucia, QLD, Australia

INTRODUCTION

Recent advances in genomics and metagenomics have revealed the exceptional breadth of the diversity of microorganisms, including those of domains Bacteria and Archaea, and within the Bacteria, there is a universe of species beyond the more familiar *E. coli*, *Bacillus subtilis*, and *Streptomyces griseus* to explore regarding new phenotypes and new genotypes. This implies many new opportunities for discovery in basic science and new types of application for biotechnology and industry. Within the Bacteria and its more than 29 extant phyla with cultured type strains (not to mention the probable thousand or more phyla from uncultured sequence data) (Yarza et al., 2014), the divergent phylum Planctomycetes is a good example of the resources both scientific and technological to be explored in a new region of Bacterial geography.

WHAT ARE PLANCTOMYCETES AND WHAT PROPERTIES MARK THEM OUT AS UNUSUAL?

The planctomycetes comprise a distinct separate phylum of Domain Bacteria. Members of this phylum appear to be related more closely to members of several other Bacterial phyla than to members of other phyla, forming a PVC superphylum including phylum Verrucomicrobia, Chlamydiae, Lentisphaerae, and probably several other so far uncultured bacterial phyla, such as Candidatus Omnitrophica (OP3) as well as the Planctomycetes (Devos et al., 2013; Devos and Ward, 2014; Fuerst, 2013). They were originally described on the basis of rosette-forming budding morphotypes in freshwater habitats and enrichments [e.g., *Planctomyces bekefii* (Gimesi, 1924), the still uncultured type strain of genus *Planctomyces*], and were originally mistaken for fungi due to formation of noncellular stalks by some types which were interpreted as cellular mycelia before electron microscopy could be applied (for the history of

Microbial Resources. http://dx.doi.org/10.1016/B978-0-12-804765-1.00001-1

their discovery see Fuerst, 1995; Jenkins and Staley, 2013). Budding reproduction and rosette formation are common in many cultured species, and some also produce stalks. There are now at least 23 genera with members in pure culture as well as 6 genera with Candidatus status so far not in pure culture but observed in either cocultures or from mixed culture bioreactors. Some genera with members in pure culture, such as *Gemmata*, *Rhodopirellula*, and *Thermogutta* now have two or more species, but some past genera with more than one species have been genetically heterogeneous enough to support proposals for new genera organized from those species. At least two Classes and Orders have been proposed within phylum *Planctomycetes* including strains in pure culture, the *Phycisphaerae* and *Phycisphaerales* (Fukunaga et al., 2009) part of the previously informally recognized WPS-1 group with many uncultured members but also the cultured marine *Phycisphera*, *Algisphaera*, and *Tepidisphaera* species (Elshahed et al., 2007; Fukunaga et al., 2009; Kovaleva et al., 2015; Yoon et al., 2014) and the class *Planctomycetia*, family *Planctomycetaceae* (Schlesner and Hirsch, 1986), and order *Planctomycetales* (Ward, 2015), while the anaerobic anammox planctomycetes have been classified as a separate order *Brocadiales* within the class *Planctomycetia* (Jetten et al., 2015). However, an analysis of 16S rRNA gene sequences derived from marine water column microbial communities alone has indicated a much greater taxonomic and genetic diversity than this (Yilmaz et al., 2015), suggesting the phylum should be divided into at least 10 classes, 16 orders, and 43 families, and this implies a much greater potential for microbial resource exploration than suspected so far. Members of PVC superphylum lead diverse lifestyles, but they seem to share phylogenetic relationship and a shared signature protein for the superphylum has been identified (Lagkouvardos et al., 2014). For several decades, the marker for Bacterial cell walls, the mechanical strength-conferring polymer peptidoglycan, was thought on the basis of chemical analysis to be absent from planctomycetes (Liesack et al., 1986) and chlamydia, but recent reexamination has revealed peptidoglycan in both some planctomycetes (Jeske et al., 2015; van Teeseling et al., 2015) and in chlamydia species (Pilhofer et al., 2013), though in the case of planctomycetes it is not yet clear where it is localized or what role it plays in the walls of planctomycete species, previously reported to be predominantly composed of protein (Liesack et al., 1986; Stackebrandt et al., 1986). Planctomycetes have been reported to be quite deeply branching within the Bacteria tree—both 16S rRNA phylogenetic analysis (Brochier and Philippe, 2002) and genomic analysis of whole proteomes (Jun et al., 2010) have indicated this, though phylogenetics of branch order of Bacterial phyla is a complex topic.

SUMMARY OF RELEVANT PLANCTOMYCETE PHYSIOLOGY

Many planctomycetes, such as *Pirellula*, *Blastopirellula*, *Rhodopirellula*, *Planctopirus*, *Gimesia*, and *Gemmata* species are chemoheterotrophic aerobes, albeit sometimes oligotrophic (e.g., *Isosphaera pallida*) and slow-growing (e.g., *G. obscuriglobus* with an 11 h generation time, *Gimesia maris* 13–100 h depending on medium, and the anammox planctomycetes with typical generation times of 2 weeks) and in the case of acid peat bog species like *Singulisphaera* and *Zavarzinella*, acidophilic. Many may well have a central role in plant and algal heteropolysaccharide degradation including exopolysaccharides produced by other bacteria. Some planctomycetes reported from habitats, such as sulfur-rich hot springs, for example, the anaerobic, sulfide-saturated sediments of the Zodletone, Oklahoma mesophilic spring, may have the ability to reduce sulfur to sulfide, and perhaps sulfur respiration as such (Elshahed et al., 2007). A Zodletone organism in the *Blastopirellula* group has been isolated and grows anaerobically with sulfur probably via carbohydrate fermentation (Elshahed et al., 2007). Most planctomycetes described in pure culture are mesophiles, but some thermophile species exist (Giovannoni et al., 1987; Kovaleva et al., 2015; Slobodkina et al., 2016). *Isosphaera pallida* from a hot spring was the first of these isolated, and is a moderate thermophile with 55°C maximum growth temperature (Giovannoni et al., 1987). The marine thermophilic anaerobic and microaerobic planctomycete *Thermostilla marina* from a submarine hydrothermal vent habitat can definitely use elemental sulfur as an electron acceptor generating sulfide as well as being able to respire with nitrate, using mono-, di-, or polysaccharides as electron donors (Slobodkina et al., 2016). One of the unusual properties of planctomycetes is the possession by at least one species of a number of enzymes concerned with single-carbon compound transformations linked to methanopterin and methanofuran (Chistoserdova et al., 2004). This is thought to be part of a formaldehyde detoxification mechanism, rather than part of any metabolism of, for example, methane or methanol as carbon sources, but this has still to be confirmed experimentally with pure cultures (Kalyuzhnaya et al., 2005). Phylogenetically the C1 transfer enzymes of planctomycetes seem to fall between domain Bacteria and domain Archaea representative homologs (Chistoserdova et al., 2004), consistent with the deep-branching of planctomycetes noted by Brochier and Philippe (2002) but also possibly due to horizontal gene transfer between domains at an early stage in domain formation (Bauer et al., 2004). In any case, C1 transfer enzymes of planctomycetes are important for understanding origins of methylotrophy and methanogenesis pathways (Chistoserdova, 2013), and might conceivably be useful for their genetic modification biotechnologically.

Species of so-called anammox planctomycetes are remarkable among the Bacteria for possessing a unique ability to oxidize ammonium anaerobically and autotrophically (Kartal et al., 2012; van Niftrik and Jetten, 2012). These anammox planctomycetes have all so far been grown only in effectively mixed culture bioreactors rather than in pure culture, so are of Candidatus status nomenclaturally, for example, "*Candidatus* Kuenenia stuttgartiensis," "*Candidatus* Brocadia anammoxidans" etc. but metagenomic approaches have yielded much knowledge about the mechanisms underlying their metabolism, and they have unique ultrastructural properties related to this metabolism.

The existence of a much wider diversity of marine planctomycete clades than recorded on the basis of cultured planctomycetes (Yilmaz et al., 2015) suggests a much wider physiological diversity of planctomycetes may be explored and applied in the future.

The isolation of planctomycetes in pure culture has concentrated on chemoheterotrophs, and methods for growing anammox planctomycetes in pure culture have not yet been developed. Isolation has been achieved using carbohydrate carbon sources like *N*-acetylglucosamine, for example, together with antibiotics, such as ampicillin (Schlesner, 1994) and streptomycin (Wang et al., 2002), to which planctomycetes are considerably more resistant than many other bacteria (Wang et al., 2002). However other approaches, such as direct micromanipulation of cells and growth of microcolonies on "ecological" media, such as low-nutrient lake water agar (Franzmann and Skerman, 1984), and the use of Gelrite or gellan to solidify culture media instead of agar (Storesund and Ovreas, 2013), combined with use of unusual polysaccharide carbon sources, such as a mix of alginate, xylan, and pectin, have also been successful (Davis et al., 2005). In the case of *Telmatocola*, strains do not grow on agar media but will grow on media containing Phytagel, an agar substitute produced from exopolysaccharide of *Sphingomonas elodea* (Kulichevskaya et al., 2009). *Bythopirellula goksoyri* behaves in a similar way (Storesund and Ovreas, 2013). So, for tapping greater diversity of planctomycetes, one way forward is via the use of alternative culture medium-solidifying polymers. In the case of some oligotrophic strains, planctomycete growth may be inhibited by threshold concentrations of carbon substrate—for example, *Telmatocola* strains inhibited by $>0.01\%$ *N*-acetylglucosamine (Kulichevskaya et al., 2012) and *Isosphaera pallida* by $\geq 0.05\%$ glucose (Giovannoni et al., 1987). A great variety of aquatic habitats and geographical regions have yielded isolates, including both marine, hypersaline, hot springs, and freshwater, and regions in or close to Europe, Australia, North and South America, and Antarctica (Bauld and Staley, 1976; Schlesner, 1989, 1994; Wang et al., 2002) and metagenomic and molecular ecology studies indicate they are ubiquitous in aquatic and soil habitats. There have been several isolations from hot springs, for example, *Thermogutta terrifontis* (Slobodkina et al., 2016), *Isosphaera pallida* (Giovannoni et al., 1987), and submarine hydrothermal vents (*Thermostilla marina*—see Slobodkina et al., 2016), but most strains isolated from other habitats are not thermophilic. Acidophilic strains of several different genera have been isolated from acidic wetlands (e.g., *Sphagnum* peat bogs) in Russia (Dedysh and Kulichevskaya, 2013; Kulichevskaya et al., 2007, 2008). Marine microalgae have emerged as significant habitats for planctomycetes and sources of inoculum for isolation (Bengtsson and Øvreås, 2010; Lage and Bondoso, 2014).

No attempts appear to have been made to isolate iron- or manganese-oxidizing planctomycetes, anaerobic photosynthetic planctomycetes, or anaerobes using unusual electron acceptors, such as ferric iron or manganese oxides, but such approaches may be worthwhile to explore a potential wider diversity given the apparent existence of sulfur respirers and anaerobic ammonium oxidizers, and the evidence for ferrous iron as an electron donor and ferric iron as an electron acceptor for at least one anammox planctomycete species (Strous et al., 2006). The anammox planctomycete "*Candidatus* Kuenenia stuttgartiensis" can respire anaerobically using both iron and manganese oxides as electron acceptors when formate is supplied as an electron donor and ferrous iron as electron donor can also oxidized with nitrate as electron acceptor. This is consistent with the presence of a branched respiratory chain which would enable the respiration of different energy sources with different electron acceptors. It is in the ultrastructure of their cells that planctomycetes display some of their most remarkable properties, and these are so significant that they require separate detailed treatment.

COMPARTMENTALIZATION OF PLANCTOMYCETE CELLS

All planctomycete species so far examined using appropriate electron microscopy techniques possess internal membranes dividing the cell into at least two major compartments and in some cases apparently more than two depending on the genus (Fig. 1.1). The exact nature and number of these additional compartments has been subject of controversy and varying interpretation of ultrastructural data. A basic cell plan shared by all planctomycetes is that involving a major compartment, the pirellulosome, containing a highly condensed bundle of genomic DNA and all the cell's ultrastructurally visible ribosomes (Fuerst, 2005; Fuerst and Sagulenko, 2011). This compartment is surrounded by a simple bilayer intracytoplasmic membrane (ICM), and between this membrane and the cytoplasmic membrane lies a ribosome-free region, termed the paryphoplasm. The composition of this region is not known, but thin section staining indicates it is not empty. Species in the genus *Pirellula*, *Blastopirellula*, and *Rhodopirellula* display this simple form of cell plan, but other planctomycete species possess at least the two major compartments and the ICM. In two further types of planctomycete, pure-cultured species in the *Gemmata* group and the Candidatus species not yet in pure culture within the anammox planctomycetes, such as "*Candidatus* Kuenenia stuttgartiensis," there is a further compartment within the ribosome-rich pirellulosome region.

In *Gemmata obscuriglobus*, a major experimental model for complex compartmentalization in planctomycetes, there are regions of closely apposed double membrane which can appear to surround the nucleoid DNA in transmission electron micrographs (TEMs) of sectioned cells prepared via the cryosubstitution process (Fig. 1.2), while in anammox planctomycetes there is a ribosome-free region surrounded by a single membrane, a region containing anammox metabolic enzymes (de Almeida et al., 2015).

This anammoxosome (Fig. 1.3) is remarkable for containing several enzymes central to the anaerobic ammonium-oxidizing metabolism unique to anammox planctomycetes (de Almeida et al., 2015), as well as harboring cytochrome *c*, and even more remarkable for being surrounded by a membrane containing ATP

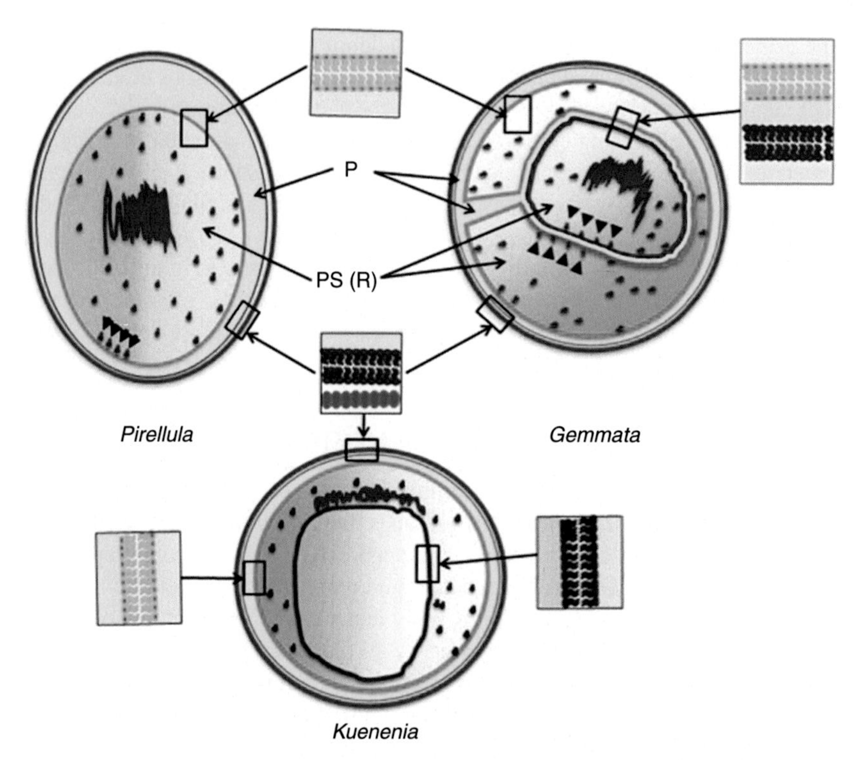

FIGURE 1.1 **Schematic diagrams of cell plans of planctomycetes.** Figure displays shared features as well as features varying in different genera (*Pirellula*, Gemmata and "*Candidatus* Kuenenia stuttgartiensis"), but several different genera share each cell plan seen in *Pirellula* and "*Candidatus* Kuenenia stuttgartiensis." *P*, paryphoplasm compartment (shown in *yellow*); *PS*, pirellulosome compartment (shown in *blue*); *R*, riboplasm. Nucleoid is shown in *red*. *Black arrows* indicate linear grouping of ribosomes bound to internal membranes. Exploded insets show membrane conformation in detail at the relevant location at compartment boundaries or at the boundary of the whole cell. *(Modified from Fig. 1 of Fuerst, J.A., Sagulenko, E., 2013. Nested bacterial boxes: nuclear and other intracellular compartments in planctomycetes. J. Mol. Microbiol. Biotechnol. 23 (1–2), 95–103, author owns copyright.)*

synthase (van Niftrik et al., 2010). This is consistent with extant biochemical models for this metabolism involving generation by anammox reaction electron transfers of a protonmotive force across the anammoxosome membrane, making the anammoxosome as an energy-generating membrane-bounded internal organelle analogous in some ways to a eukaryotic mitochondrion (Jogler, 2014; Neumann et al., 2011). The unique nature of anammox plantomycetes and the anammoxosome is indicated by the presence of ladderane membrane lipids unique within any organism, containing a "ladder" hydrocarbon chain of concatenated cyclobutane rings (Sinninghe Damste et al., 2002).

Cryoelectron tomography of the organelle indicates that it is clearly a separate entity separated from the rest of the cell, distributes to daughter cells during

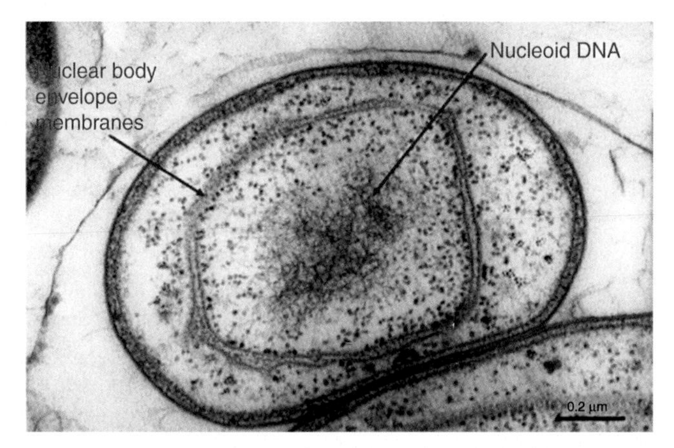

FIGURE 1.2 **Transmission electron micrograph of sectioned cell of *Gemmata obscuriglobus* prepared via cryosubstitution displaying a ribosome-containing pirellulosome compartment containing the fibrillar nucleoid.** Bar: 0.2 μm. *(Modified from Fig. 6C of Lindsay, M.R., Webb, R.I., et al., 2001. Cell compartmentalisation in planctomycetes: novel types of structural organisation for the bacterial cell. Arch. Microbiol. 175(6), 413–429, author owns copyright.)*

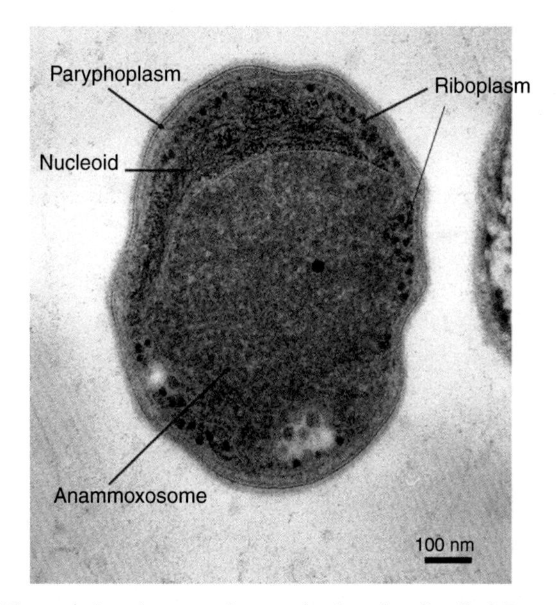

FIGURE 1.3 **Transmission electron micrograph of sectioned cell of *Brocadia anammoxidans* prepared via cryosubstitution.** An anammoxosome without ribosome but filled with tubule-like structures occupies much of the cell, surrounded by a single anammoxosome membrane. The roughly crescent-shaped fibrillar nucleoid is closelyapposed to this membrane, and lies within the ribosome-containing pirellulosome, which is bounded by an intracellular membrane. Between the intracellular membrane and the cytoplasmic membrane and cell wall is a ribosome-less space, the paryphoplasm. Bar, 100 nm, author owns copyright.

division (van Niftrik et al., 2009), and can be isolated in pure form from lysed cells (Neumann et al., 2014). However, such methods have not yet been able to resolve a controversy concerning the nucleoid-associated membranes of *G. obscuriglobus*, which has in the past been postulated to possess a double-membrane-bounded nuclear body containing the nucleoid of the cell and separating it from the rest of the cell and the remaining contents of the pirellulosome (Fuerst and Webb, 1991; Lindsay et al., 2001). One study (Santarella-Mellwig et al., 2013) proposes from its data that the nucleoid-associated membranes do not form a closed compartment, while the closure of this compartment appears clear from another study (Sagulenko et al., 2014) using analogous methods for cryoelectron tomography. Spatial localization of transcription and translation processes as recently reported (Gottshall et al., 2014) may be consistent with a closed compartment. Resolution of this question may require techniques, such as super-resolution microscopy as well as comparative study of the effects of growth conditions on the structure. The question has implications for resources in that the existence of a genuine nucleus-like structure may imply that such a compartment could also be applied to design of synthetic cells with simple nucleus-like compartmentation for sequestration of specialized functions. This of course would depend on our eventual knowledge of the genetic basis for such a compartment derived from study of *Gemmata* genomics and cell biology. In any case, the complex internal membranes of *Gemmata* (Acehan et al., 2014) may also be useful for increasing surface area for internal processes in designed cells.

Planctomycetes also possess a type of internal compartment occurring in other bacterial phyla—the protein shell-bounded bacterial microcompartment (BMC), and as in other bacteria this can harbor metabolic enzymes. In planctomycetes, such as *Planctomyces limnophilus* this BMC function is a specialized form of metabolism involving degradation of plant and algal cell wall sugars including L-fucose and L-rhamnose, and a conserved BMC gene cluster is found widespread in other planctomycetes also (Erbilgin et al., 2014). These may be of considerable value in industrial biotechnology context, for example, for synthetic biology involving use of microcompartments in cell design (Erbilgin et al., 2014). In *G. obscuriglobus*, a process of protein uptake (Fig. 1.4) analogous to the endocytosis process of large molecule uptake in eukaryote cells has been reported (Lonhienne et al., 2010). This process is energy dependent and appears to involve both cell surface receptors and vesicle formation from the cytoplasmic membrane.

Inside the anammoxosome, nitrite (NO_2^-) is reduced to nitric oxide (NO) catalyzed by the cytochrome c d1 nitrite::nitric oxide reductase NirS. NO is subsequently condensed with ammonium (NH_4^+) via hydrazine synthase to produce hydrazine N_2H_4, resulting in formation of a bond between two nitrogen atoms. Hydrazine is then oxidized to N_2 via a set of hydrazine dehydrogenases, and the resulting electrons are transferred via soluble cytochrome c carriers to membrane coenzyme Q, then to the membrane cytochrome bc 1 complex, and from there to more soluble cytochromes and finally to nitrite reductase and hydrazine synthase. As a result of this electron transfer, protons are transferred

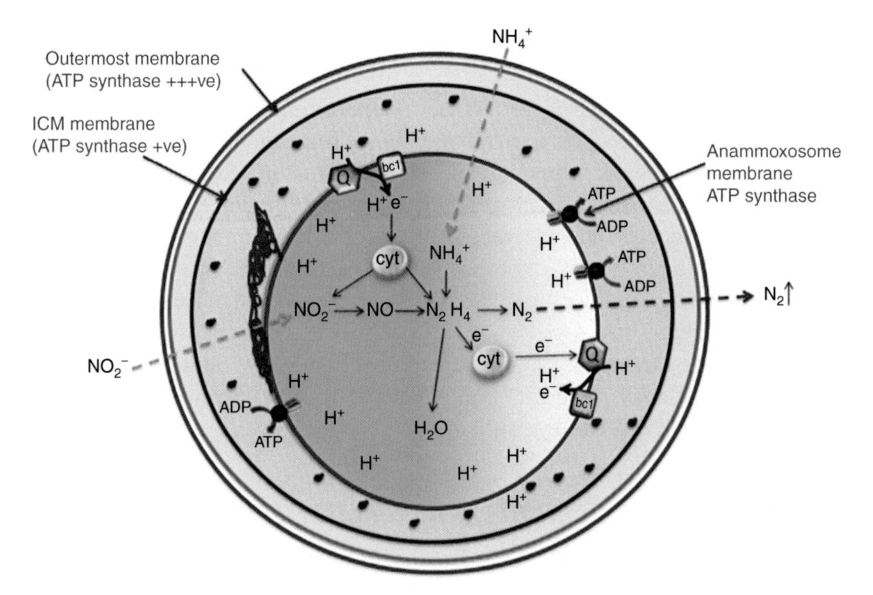

FIGURE 1.4 Model for biochemistry of an anammox planctomycete in relation to cell compartments and especially the role of the anammoxosome organelle, its enzymes, and its membrane in generation of an internal protonmotive force across the anammoxosome membrane. Ammonium and nitrite from the external medium are necessary substrates, and are transported across the cell wall *(dark blue)*, cytoplasmic membrane *(dark green)*, paryphoplasm *(light green)*, and ICM *(purple)*, as well as the riboplasm *(light orange)* and finally the anammoxosome membrane *(dark blue)*. The nucleoid DNA *(red)* within the riboplasm is bound to the anammoxosome membrane. The anammox reactions take place in the anammoxosome *(shaded blue)* or associated with its bounding membrane. *Modified from Fig. 4 of Fuerst, J.A., Sagulenko, E., 2013. Nested bacterial boxes: nuclear and other intracellular compartments in planctomycetes. J. Mol. Microbiol. Biotechnol. 23 (1–2), 95–103, author owns copyright.*

across the anammoxosome membrane, with the anammoxosome interior being positively charged relative to the riboplasm (light orange) external to the anammoxosome. This proton gradient is used to drive ATP synthesis via protons flowing through ATP synthases with their ATP-synthesizing domains in the riboplasm and their proton-translocating domain in the anammoxosome membrane. Synthesized ATP is released into the riboplasm, where it is used for biosynthesis (Fuerst and Sagulenko, 2013).

The vesicles contain a protein which is a homolog of the clathrin-like MC proteins important for membrane curvature in eukaryote endocytosis and may be other endocytosis-related proteins (Fuerst and Sagulenko, 2014; Lonhienne et al., 2010; Santarella-Mellwig et al., 2010). This macromolecule uptake process may conceivably be useful for developing genetic systems for modification of planctomycetes, including manipulation of any strains of industrial significance, or as a simple indicator system for screening compounds capable of inhibiting endocytosis in its most rudimentary form, for example, those capable of inhibiting MC-like proteins.

HOW MIGHT THE UNUSUAL CHARACTERISTICS OF PLANCTOMYCETES BE CONSIDERED AS RESOURCES?

The phylum *Planctomycetes* offers new opportunities for completely novel types of resources for application not previously found in other groups of bacteria. These include their compartmentalized cell structure, the correlated ability to take up proteins from the external milieu, unusual components of membranes depending on species, and enzymes with potential uses in industry. In addition and most significantly so far, there is the unique ability of anammox planctomycetes to oxidize ammonium anaerobically and thus remediate nitrogen-rich wastewater, which has already been extensively applied in wastewater treatment technology and is subject of a separate section below. Some of these properties may be ideally placed for collaborations with newly emerging areas of microbiology and cell biology, such as synthetic biology.

COMPARTMENTALIZED CELLS AS A POTENTIAL RESOURCE FOR SYNTHETIC BIOLOGY

Synthetic biology aims at using parts and modules from existing organisms to generate modified or new types of organisms with specific functions. In the past it is mostly genes and groups of genes and their associated regulatory elements which have been considered as part of the engineering "bricks" for synthetic organisms. However, recently organelles have been proposed as potential elements for such engineering, including such bacterial organelles as microcompartments (Giessen and Silver, 2015) and magnetosomes (Borg et al., 2015) as well as the pirellulosomes and anammoxomes unique to planctomycetes (Giessen and Silver, 2015).

The magnetosome compartment has for example been used in combination with a "nanotrap" system to redirect protein localization in a magnetotactic bacterium changing the taxis behavior of the bacterium. Microcompartments with protein shells harbor enzymes and such encapsulated enzymes can be conceivably used as modules to construct new catalytic functions for cells. Planctomycetes harbor such microcompartments containing enzymes degrading sugars from plant and algal cell walls, for example, L-fucose and L-rhamnose (Erbilgin et al., 2014) and they could be useful in such encapsulation designs. But further than use of such protein shell compartments, the unique membrane-bounded compartments in planctomycetes could be used as elements in synthetic cell designs, for example, the anammoxosome with its ammonium-oxidizing enzyme system, and the mechanistic basis for the pirellullosome and nuclear body compartments may reveal new ways to compartmentalize other bacteria for specific functions and transfer of materials within the cell. A limiting factor is our poor understanding of how such compartments are formed, but if such understanding can be achieved via focused cell biology research on planctomycetes then "engineered spatial control" of cellular metabolism could eventually generate increased yield of useful compounds (Giessen and Silver, 2015). The first step in this process might be development of genetic systems for planctomycetes which are only just emerging (Jogler and Jogler, 2013; Jogler et al., 2011), as well as further research on isolation of compartments and organelles for direct use biotechnologically (Neumann et al., 2014).

UNUSUAL COMPONENTS OF PLANCTOMYCETE CELLS

Sterols and Hopanoids

The complexly compartmented planctomycete *G. obscuriglobus* has been shown to synthesize the sterols lanosterol and parkeol (Pearson et al., 2003), compounds occurring relatively early in the pathway of sterols universal in eukaryotes. Although sterol synthesis also occurs sporadically but rarely in some other bacteria, such as the methylotrophs and myxobacteria, the pathway in *G. obscuriglobus* seems to be one of the simplest known in nature.

Several different hopanoids, which may have similar functions to sterols in modifying membrane fluidity, occur in planctomycetes, including bacteriohopanetetrol, diplopterol, diploptene, and a C27 hopanoid ketone (Sinninghe Damsté et al., 2004). The hopanoid-containing planctomycetes include strictly anaerobic bacteria capable of anaerobic ammonium oxidation. Hopanoids are potentially useful as biomarkers for geochemistry and palaeochemistry investigations of the first appearance of bacteria in earth history, and planctomycetes have been suggested as providing clues to potential origins for hopanoids in Proterozoic rocks and even in petroleum (Sinninghe Damsté et al., 2004).

Ladderane Lipids

Shortly after the discovery that planctomycetes were the active bacterium in bioreactors performing the anammox anaerobic oxidation of ammonium, a new type of lipid was found in the membranes of anammox planctomycetes, the ladderane lipids. These lipids contain a ladder-like hydrocarbon chain composed of linked cyclobutane rings bonded to a glycerol- or methanol-based backbone, and have been hypothesized to influence the permeability of the membrane surrounding the anammoxosome compartment (Sinninghe Damste et al., 2002). They were proposed to prevent toxic intermediate hydrazine from leaking into the compartment containing ribosomes and DNA, or to prevent proton leakage across the anammoxosome membrane, but ladderanes do not appear to form a perfect barrier since anammox intermediates are detectable outside the cells (Kartal et al., 2013).

Molecular dynamic modeling of ladderane function also indicates that hydrazine may not be prevented from diffusing across such anamoxosome membranes but that such membranes may be even more permeable to hydrazine than ones without (Chaban et al., 2014). However, the only experimental evidence on permeability indicates that fluorophores do not cross the anammoxosome, consistent with a barrier to low MW compounds (Sinninghe Damste et al., 2002). Their location and function may not be unique to the anammoxosome. They may be linked by either ester or ether links to the glycerol backbone. It has been suggested that ladderanes may be useful in optoelectronics applications—for example, by allowing linked chromophores to transfer electrons via π–π interactions along the ladder, and in materials science in general since ladderanes provide frameworks for long-range electron transfer through σ-bonds, and also function as molecular rods or stiff spacers between functional groups (Nouri and Tantillo, 2006). Although ladderanes can be synthesized, since massive quantities of biomass are produced in anammox bioreactors treating waste, planctomycete ladderanes might provide an economic source or starting compound of any industrially useful lipids of this unusual composition class.

Hydrocarbons

Planctomyces maris (now reclassified as *Gimesia maris*—see Scheuner et al., 2014) has been reported to produce the olefinic long chain 9 double-bond hydrocarbon 3,6,9,12,15,19,22,25,28-hentriacontanonaene, and also has a set of *ole* genes needed for its synthesis (Sukovich et al., 2010). It may prove worthwhile to investigate other planctomycetes for synthesis of such hydrocarbons, including anammox planctomycetes where large scale biomass is already produced that could provide an industrial scale source.

SULFATASES AND THEIR USES IN STEREOCHEMISTRY/ TRANSFORMATIONS IN ORGANIC CHEMISTRY

One of the remarkable characteristics of marine planctomycete species is their possession of a wide diversity of genes for sulfatases. This may relate to their habitat in association with algae synthesizing sulfated polysaccharides but much evidence remains to be acquired concerning confirmation of such a suggested function, and other functions remain possible. Studies of pure cultures of *Rhodopirellula baltica* (Glockner et al., 2003) and several other *Rhodopirellula* species (Wegner et al., 2013) confirmed the presence of genes for many sulfatases in these organisms. Marine metagenomes have also suggested correlation of marine planctomycete genomes with genes for sulfatases (Woebken et al., 2007). *R. baltica* (formerly known as *Pirellula* sp. strain 1) genome was originally annotated as possessing 110 genes encoding proteins with significant similarity to sulfatases in other organisms—82 of these with similarity to prokaryotic species and 28 with similarity to eukaryotic sulfatases (Glockner et al., 2003). A later study found more than 1000 open reading frames that were annotated as sulfatases in *R. baltica SH1*, and phylogenetic analysis indicated a high diversity with 22 separate similarity clusters, possibly correlated with high sulfated polysaccharide diversity in the natural habitat of the species

(Wegner et al., 2013). Other *Rhodopirellula* species, such as *R. sallentina* appear to also possess many genes for sulfatases (Wegner et al., 2014).

In the case of *R. baltica* SH1, sulfatase genes are correlated with ability of the bacteria harboring them to degrade sulfated polysaccharides, such as chondroitin sulfate, lambda-carrageenan, and fucoidan (Wegner et al., 2013).

Some of these sulfatases may prove useful in industrial chemical conversions involving stereochemical transformations. The sulfatases of *R. baltica* have been investigated for their promise in developing specialized protocols in industrial chemical transformations (Gadler and Faber, 2007; Gadler et al., 2006). Approaches for complete conversion of a racemic mixture of organic compounds into a single stereoisomer product—a process termed deracemization—have become a prime target in "white biotechnology" utilizing enzyme biocatalysis (Gadler and Faber, 2007). The alkyl sulfatases of *R. baltica* can be used in the enantioselective hydrolysis of *rac*-*sec*-alkyl sulfate esters, and the unique property of these planctomycete sulfatases is that configuration is retained—*sec*-alkyl sulfate esters can be hydrolyzed with high enantioselectivity and with strict retention of stereoisomer configuration (Gadler and Faber, 2007), making this reaction useful for deracemization. It has been proposed to use this sulfatase in combination with a stereocomplementary inverting enzyme from an archaeal *Sulfolobus* spp. in a deracemization strategy to achieve maximum yield of product from racemic mixtures of substrate (Wallner et al., 2005). With future development of genetic systems for planctomycetes, some of which are in the initial stages of being developed (Erbilgin et al., 2014; Jogler et al., 2011; Schreier et al., 2012), one can envisage such enzymes being optimized further for industrial purposes.

WHAT OTHER PROPERTIES DO PLANCTOMYCETES HARBOR RELEVANT TO NEW RESOURCES?

Polyketide Genes and Potential Natural Product Synthesis in Planctomycetes

Availability of many at least partially complete and some complete genome sequences for planctomycete species mean that the potential of these organisms to produce novel secondary metabolites can be assessed before chemical or biological assay of compounds (Jeske et al., 2013). From such an analysis it has been found that the potential for synthesis of bacteriocins and for polyketides and nonribosomal peptides is present in planctomycetes, including *Schelsneria paludicola* (13 metabolite synthesis gene clusters), *Blastopirellula marina* (12), *G. obscuriglobus* (12), *Singulisphaera acidiphila* (10) and *R. baltica* (10) with the highest number of such gene clusters, and an anammox planctomycete species and the planctomycete with the smallest genome, *Phycisphaera mikurensis*, the lowest (2 clusters each). From phenotypic microarray analysis, it was suggested that certain carbon sources associated with algae are likely to stimulate secondary metabolite production and provide the basis for optimal screening assays for natural products from new strain collections.

Planctomycetes have the genetic potential to synthesize long-chain fatty acids via a FAS/PKS mechanism, and the type produced by *P. limnophilus* (now reclassified as *Planctopirus limnophilus* see Scheuner et al. (2014) may be unique among the bacteria (Shulse and Allen, 2011). There is a possibility that these may provide precursors for olefinic hydrocarbon biosynthesis (Shulse and Allen, 2011).

Enzymes Other Than Sulfatases

A carboxyl esterase (TtEst2) has been reported to occur in the thermophilic planctomycete *Thermogutta* (Sayer et al., 2015). It is active on carboxylate esters, and is active up to 70°C, which may be useful in applied contexts. It has an unusual active site for the alpha-beta hydrolase family tree, lacking most of the "cap" region of the active site which is thus exposed to solvent. Due to its somewhat divergent nature, this enzyme may be useful for understanding relationships in esterases between substrate preference and the structure of the active site pocket. A gene for esterase synthesis has been retrieved from an activated sludge metagenome and found to have closest relationship (33% amino acid identity) to a *G. obscuriglobus* esterase/lipase protein; the properties of the cloned enzyme suggested biotechnology potential, implying that it may also be worthwhile investigating the *G. obscuriglobus* enzyme for such potential (Zhang and Han, 2009).

R. baltica is the source of a polysaccharide lyase which has been subject of an X-ray structural biology study after isolation (Dabin et al., 2008). Regarding polysaccharide-degrading enzymes, GH10 endo-beta-xylanases beta-xylanases have been detected widely in genomes of many planctomycetes including members of the *Isosphaera-Singulisphaera* cluster (Naumoff et al., 2014). Ecological stable-isotope probing studies indicate that uncultured planctomycetes may be the primary degraders of complex heteropolysaccharides in soils (especially those synthesized by other bacteria as exopolymers) and thus might be prime candidates for exploration of polysaccharide lyases and related enzymes for application after retrieval and expression of cloned genes from the metagenome (Wang et al., 2015).

Unusual methylated ornithine lipids have been found in planctomycetes from low pH habitats *Telmatocola sphagniphila*, a *Gemmata*-like strain SP5, *Singulisphaera acidiphila* and *Singulisphaera rosea* (Moore et al., 2015)—the lipids are mono-, di-, and trimethylated at the ε-nitrogen position of the ornithine head group, and it is postulated that this provides an alternative mechanism for membrane stability in acid but very low phosphate conditions. A direct molecular ecology/lipid analysis study of *Sphagnum* wetlands similar to those from which the acidophilic planctomycetes were isolated could detect trimethyornithine lipids at the same oxic/noxic interface where planctomycetes rRNA sequences were abundant (Moore et al., 2015). An enzyme OL *N*-methyltransferase OlsG (Sinac_1600) responsible for methylation of ornithine lipids has been characterized from the planctomycete *Singulisphaera acidiphila* (Escobedo-Hinojosa et al., 2015).

ANAMMOX PLANCTOMYCETES AS A MAJOR EXAMPLE OF PLANCTOMYCETES AS A MICROBIAL RESOURCE

Application of anammox planctomycetes in wastewater remediation has been a major example of the use of planctomycetes as a microbial resource. These species (all only in mixed cultures and thus of "Candidatus" nomenclatural status) are members of phylum Planctomycetes but form a relatively deep branch somewhat divergent from other planctomycetes. They all perform oxidation of ammonium anaerobically and chemoautotrophically, so the process performed was termed anammox for *an*aerobic *amm*onium *ox*idation (Jetten et al., 2001; Kuenen, 2008; van Niftrik and Jetten, 2012). Discovery and confirmation of this metabolism was a brilliant triumph for predictive microbial physiology and bioenergetics since it had been hypothesized to exist long before its experimental confirmation (Broda, 1977; Kuenen, 2008). The process was confirmed as a biochemical transformation in wastewater reactors some years before the role of planctomycetes was confirmed (Mulder et al., 1995). The process is unique to these planctomycetes and so far, unknown in other living organisms including bacteria and archaea. The overall reaction involves oxidation of ammonium to molecular dinitrogen with the use of nitrite as an electron acceptor, and much progress has been made regarding the nature of the reaction intermediates, such as hydrazine and nitric oxide, of enzymes involved in intermediate transformations and of the role of cytochromes and electron transfer (Kartal et al., 2012). Although anammox has effectively only been observed in bioreactors with a mixed microbial community though considerably enriched in anammox planctomycetes, the responsible planctomycetes can be purified via Percoll gradient centrifugation (Kartal et al., 2011a). Methods for their enrichment and cultivation have been described in excellent detail by the major Nijmegen anammox research group responsible for so many advances in this field (Jetten et al., 2005). Their pure culture, however has so far not been achieved. Many aspects of the biochemistry of the reaction have been elucidated including its mechanism (Kartal et al., 2011b; Dietl et al., 2015) and the application of metagenomics to enriched cultures such as "*Candidatus* Kuenenia stuttgartiensis" has been especially informative regarding the gene set involved (Strous et al., 2006). Enzyme isolation and characterization have been major achievements since anammox process proteins can be especially difficult to work with due to gel formation of cell extracts and extracellular polymers (Cirpus et al., 2006)—some of these problems have been effectively overcome (Cirpus et al., 2006). The reaction depends on the toxic compound hydrazine as an intermediate and a hydrazine synthase enzyme. Hydrazine synthase seems to act very slowly, and may be responsible for the slow growth rate of anammox planctomyetes which can have generation times in the order of 2 weeks (Kartal et al., 2012). It has been suggested that this enzyme may have to be expressed at high levels to allow substantial metabolic activity at all, and that accumulation of such levels within an organelle may be one function of the anammoxosome. The hydrazine synthase structure has been determined, and is revealed as a multiprotein complex—an elongated dimer of heterotrimers, with active sites connected via a system of tunnels, and implying a two-step mechanism for hydrazine synthesis involving a three-electron reduction of nitric oxide to hydroxylamine and its subsequent condensation with ammonia, yielding hydrazine in one of the active centres (Dietl et al., 2015). One of the remarkable features of all anammox plantomycetes examined so far is the occurrence of an internal membrane-bounded organelle, the anammoxosome, which has been determined via immunogold electron microscopy, at least in the model organism "*Candidatus* Kuenenia stuttgartiensis," to contain the major anammox process enzymes, including a hydroxylamine oxidoreductase-like protein with its redox partner kustc0457, hydrazine dehydrogenase, hydroxylamine oxidase, nitrite oxidoreductase, and hydrazine synthase (de Almeida et al., 2015). These enzymes are restricted to the anammoxosome, consistent with the catabolic role of this organelle, and in the case of NXR at least the enzyme is associated with ultrastructurally visible tubules (de Almeida et al., 2015), a feature of anammoxosomes previously noted from electron micrographs (Lindsay et al., 2001). The compartmentalization of these planctomycete cells plays a key role in recent models for anammox biochemistry—ATP synthase appears to be present in the bounding membrane of the anammoxosome (van Niftrik et al., 2010) and there is evidence also that cytochromes are contained within the organelle (van Niftrik et al., 2008), consistent with development of a protonmotive force across the anammoxosome membrane and consequent proton passage across the membrane via ATP synthase to generate ATP (Fig. 1.4). This is thus an energy-generating organelle, in some senses an analog in bacteria of the eukaryote mitochondrion, also an internal energy-generating organelle (Jogler, 2014). We have noted earlier that anammox planctomycetes possess ladderane lipids though the exact function of these lipids for the anammoxosome and anammox process is still unclear.

The anammox process forms the basis for several published or patented processes for nitrogen-rich wastewater remediation, such as CANON (Completely autotrophic nitrogen removal over nitrite) SHARON-Anammox, OLAND (oxygen-limited autotrophic nitrification-denitrification), DEMON (deammonification), and SNAD (simultaneous partial nitrification, ANAMMOX, and denitrification) (Chen et al., 2009; Kuai and Verstraete, 1998; Third et al., 2001; van Dongen et al., 2001; Wett, 2007) and several large-scale industrial plants in different regions of the globe use the anammox process in one of these combinations, for examples, see van der Star et al. (2007), Lackner et al. (2014). The two major reactions utilized in such processes are aerobic nitritation, aerobic

ammonia oxidation serving to supply nitrite electron acceptor to the second major reaction, anaerobic anammox oxidation of ammonium:
"nitritation"—partial nitrification:

$$2NH_4^+ + 1.5O_2 \rightarrow NH_4^+ + NO_2^- + H_2O + 2H^+$$

anammox:

$$NH_4^+ + NO_2^- \rightarrow N_2 + 2H_2O$$

Such a process can be organized as a two-reactor partial nitritation-anammox (PNA) process with each reaction (partial oxidation of ammonium to nitrite and anoxic anammox ammonium oxidation) largely confined to a separate reactor (e.g., in the SHARON-anammox process) or can be combined in one reactor where nitrite production and anammox occur in one aerated reactor (with anaerobic microniches for anammox) (van der Star et al., 2007). CANON, for example, is a one-reactor process where sludge granules combine aerobic and anaerobic zones via a granular sludge gas-lift reactor, having the advantage that N-rich wastewater can be fed directly to a single reactor and with no need for nitrite addition (Third et al., 2001). A one reactor denitrification-anammox process can also be possible combining anaerobic denitrification of nitrate for generation of nitrite electron acceptor for the anammox reaction. The anammox reaction also contributes to nitrogen removal in other types of nitrogen remediation systems, such as rotating disk contactors and the OLAND process (oxygen-limited autotrophic nitrification-denitrification) where biofilms develop with both aerobic layers with nitrifiers and anaerobic layers with anammox planctomycetes, a topology also seen in sludge granules. Two-reactor processes may return to fashion in proposed sewage treatment processes using side-stream partial nitritation combined with anammox (Perez et al., 2015). However, CANON one-reactor processes with granular sludge formation have very low utility consumption and have low volume requirements relative to nitrogen load (Bozkurta et al., 2016).

A full-scale plant successfully applying the anammox process at Rotterdam in The Netherlands, using the SHARON-anammox process and an anammox reactor with two compartments is used for treating digested sludge from the Dokhaven municipal waste treatment plant. Its maximum designed/practical loading is in the order of 500/700 kg of nitrogen per day, though it took more than 1000 days from startup to achieve stable operation at that level reflecting the slow biomass growth of anammox organisms (van der Star et al., 2007). Assay of anammox planctomycetes using quantitative PCR was critical to assessing active biomass growth and performance in this reactor. Other notable large scale industrial plants utilizing some version of anammox include plants for remediating potato processing effluent in the Netherlands, at Olburgen and Bergen op Zoom, and a plant for processing tannery waste at Lichtenvoorde, and one-reactor processes using DEMON at several plants in The Netherlands and Germany including one at Apeldoorn with a massive 2900 m^3 reactor volume (DeMooij et al., 2010; Lackner et al., 2014). Interestingly, the PNA- and granular sludge-based single reactor plant at Olburgen is used as a source of inoculum for PNA reactors for wastewater treatment in other parts of the world by the Paques b.v. company, involved with anammox development at Rotterdam and since the earliest commercial development of anammox (Speth et al., 2016). A trend since 2006 for industrial wastewater treatment has been for such one-reactor systems, a trend driven by lower investment costs (Lackner et al., 2014). There has been increasing anammox research and wastewater treatment application in China, with exceptionally large scale plants, such as the Tongliao Meihua one stage reactor for treating effluent from monosodium glutamate production with a design capacity of 11,000 kg of nitrogen per day, as well as plants designed to treat landfill leachate, and effluent from pharmaceutical waste, yeast production, and a distillery (Ali et al., 2013). One interesting and valuable feature of anammox is that it is autotrophic and thus there is no added carbon substrate like methanol needed and no carbon dioxide generated (at least during the anammox stage), and anammox itself does not generate emission of the other relevant greenhouse gas in aerobic nitrification, nitrous oxide. Anammox has much less energy consumption than conventional nitrification since it does not require aeration, unlike such aerobic nitrification. The cost of the SHARON-amammox process relative to nitrogen mass is at the least less than half the cost per kilogram of nitrogen for other processes used for N-removal from sludge digestion liquors (Kartal et al., 2004). It has been suggested that anammox could enable a sewage treatment process to be energy generating if combined with methane generation from the sludge produced (Kartal et al., 2010). The many different processes applying the anammox planctomycetes for varying types of wastewater and wastewater treatment processes are summarized regarding industrial applications or applications still being developed in some excellent recent surveys and assessments (Bozkurta et al., 2016; Lackner et al., 2014).

There has been an expansion of research activity on the factors governing the efficiency of anammox applications, and the potential for the process to be used where wastewater contains varying contaminating inhibitors of the process (Carvajal-Arroyo et al., 2013; Jin et al., 2013; Pereira et al., 2014; Vlaeminck et al., 2012), and claims of a process combining sulfate and ammonium removal (Ali et al., 2013). The wider application of the process to improve preexisting processes, such as sewage treatment and retrofitting of existing wastewater treatment processes to incorporate purposely designed anammox stages is now receiving greater attention (Bozkurta et al., 2016; Kartal et al., 2010). In one scheme suggested, sewage would be treated in a very high-load activated sludge system, where soluble organic matter is converted to biomass that can be separated in a settler, removing most organic matter, and enabling its use in the generation of biogas (methane) in a digestion process. Effluents from first and digester stage would then be combined and treated in a granular-sludge anammox reactor (Kartal et al., 2010). Effluent from this scheme would be expected to be below the required effluent standards for nitrogen discharge to the environment, and calculations show that the wastewater treatment process with anammox in the main line would *yield* 24 watt hours per person per day (W/p/d), compared to a *consumption* of 44 W/p/d in conventional treatment (Kartal et al., 2010). An engineering optimization approach comparing different solutions incorporating anammox into retrofitting existing wastewater treatment plants resulted in the selection of a high-rate oxic reactor coupled with anammox technology as an optimal solution, since there was low utility consumption and high biogas production, but one-reactor solutions were assessed as not very different in efficiency outcome depending on local considerations (Bozkurta et al., 2016).

The OLAND system approach has been already applied to landfill leachate and digestates from sewage sludge, specific industrial streams and concentrated black water, and is being more widely applied with technical-scale plants for treatment of source-separated domestic wastewater, pretreated manure and sewage, and liquors from organic waste bioenergy plants (Vlaeminck et al., 2012). Already it has been reported that using OLAND to treat sludge digestates in the side stream of a conventional wastewater treatment plant lowers the overall plant energy requirements by about 50% (Siegrist et al., 2008). It has even been claimed that bringing OLAND into the main sewage treatment stream could even allow electrical energy recovery and savings to exceed electrical energy input as mentioned earlier (Vlaeminck et al., 2012). Such engineering-enhanced extension of anammox applications will also benefit from greater understanding of the planctomycete species involved including the role of the anammoxosome organelle and of the anammox metabolism and enzymes. The goal of obtaining pure cultures of anammox planctomycetes has so far been out of reach but fulfilling that goal might open the benefits of genome manipulation for process improvement.

ANAMMOX PLANCTOMYCETES ARE RESOURCES PROVIDING GLOBAL ECOSYSTEM SERVICES IN THE NITROGEN CYCLE

Anammox planctomycetes play an important role on the global cycling of nitrogen in the environment, especially within marine ecosystems, where they are a major microbial component of the world's oceanic oxygen minimum zones (OMZs) and marine sediments (Francis et al., 2007; Kuypers et al., 2005; Lam et al., 2009; Op den Camp et al., 2006; Schmid et al., 2007). OMZs are expected to increase with global warming so understanding their implications for nutrient cycling in the oceans is important for modeling climate change effects (Lam et al., 2009). Marine anammox plancomycetes comprise a previously unsuspected important missing link in the global nitrogen cycle. It has been estimated that 50% of the molecular dinitrogen in the atmosphere originating from the oceans may have its origin in the ammonium-oxidizing activities of anammox plantomycetes (Dietl et al., 2015). This activity may be linked to other elements of the marine ecosystem, such as the sinking behavior of zooplankton during diurnal cycles (Bianchi et al., 2014). The anammox oxidation of ammonium may in addition have a role to play in nitrogen remediation of nitrogen-polluted groundwater (Smith et al., 2015). This is important for modeling groundwater contributions to eutrophication of freshwater bodies, and perhaps via manipulation of anammox processes in such groundwater to enable remediation.

CONCLUSIONS

Planctomycetes have proved to provide new models for our understanding of cell organization, and they have also provided insight into completely new ways of living previously unknown for bacteria or indeed for any other form of life. Planctomycetes have already proved to be useful microbial resources concerning the activity of anammox planctomycetes in remediation of N-rich wastewater, where they have been applied on an industrial scale. They promise to be useful in many other diverse ways dependent on their unique chemical and enzymatic components and their unique cell structure and biology. Their further exploration as resources for biopharmaceutical discovery programs and as a source of new mechanisms and models for synthetic biology, it is suggested, will continue to fulfil the considerable promise indicated successfully already by the anammox process.

ACKNOWLEDGMENTS

Research on planctomycetes in the author's group has been funded by the Australian Research Council. The author is thankful to Drs. Evgeny Sagulenko, Benjamin Yee, Margaret Butler, Hiroshi Izumi, Cheryl Jenkins, and Margaret Lindsay for their contributions to the planctomycete research in his laboratory.

REFERENCES

Acehan, D., Santarella-Mellwig, R., et al., 2014. A bacterial tubulovesicular network. J. Cell. Sci. 127 (Pt. 2), 277–280.

Ali, M., Chai, L.Y., et al., 2013. The increasing interest of ANAMMOX research in China: bacteria, process development, and application. Biomed. Res. Int. 2013, 134914.

Bauer, M., Lombardot, T., et al., 2004. Archaea-like genes for C1-transfer enzymes in Planctomycetes: phylogenetic implications of their unexpected presence in this phylum. J. Mol. Evol. 59 (5), 571–586.

Bauld, J., Staley, J.T., 1976. *Planctomyces maris* sp.nov.: a marine isolate of the Planctomyces-Blastocaulis group of the budding bacteria. J. Gen. Microbiol. 97, 45–55.

Bengtsson, M., Øvreås, L., 2010. Planctomycetes dominate biofilms on surfaces of the kelp *Laminaria hyperborea*. BMC Microbiol. 10, 261.

Bianchi, D., Babbin, A.R., et al., 2014. Enhancement of anammox by the excretion of diel vertical migrators. Proc. Natl. Acad. Sci. USA 111 (44), 15653–15658.

Borg, S., Popp, F., et al., 2015. An intracellular nanotrap redirects proteins and organelles in live bacteria. MBio 6 (1).

Bozkurta, H., van Loosdrechtb, M.C.M., et al., 2016. Optimal WWTP process selection for treatment of domestic wastewater—a realistic full-scale retrofitting study. Chem. Eng. J. 286, 447–458.

Brochier, C., Philippe, H., 2002. Phylogeny: a non-hyperthermophilic ancestor for bacteria. Nature 417 (6886), 244.

Broda, E., 1977. Two kinds of lithotrophs missing in nature. Z. Allg. Mikrobiol. 17, 491–493.

Carvajal-Arroyo, J.M., Sun, W., et al., 2013. Inhibition of anaerobic ammonium oxidizing (anammox) enrichment cultures by substrates, metabolites and common wastewater constituents. Chemosphere 91 (1), 22–27.

Chaban, V.V., Nielsen, M.B., et al., 2014. Insights into the role of cyclic ladderane lipids in bacteria from computer simulations. Chem. Phys. Lipids 181, 76–82.

Chen, H., Liu, S., et al., 2009. The development of simultaneous partial nitrification, ANAMMOX and denitrification (SNAD) process in a single reactor for nitrogen removal. Bioresour. Technol. 100 (4), 1548–1554.

Chistoserdova, L., 2013. The distribution and evolution of C1 transfer enzymes and evolution of the planctomycetes. In: Fuerst, J.A. (Ed.), Planctomycetes: Cell Structure, Origins and Biology. Humana Press, New York, pp. 195–209.

Chistoserdova, L., Jenkins, C., et al., 2004. The enigmatic planctomycetes may hold a key to the origins of methanogenesis and methylotrophy. Mol. Biol. Evol. 21 (7), 1234–1241.

Cirpus, I.E., Geerts, W., et al., 2006. Challenging protein purification from anammox bacteria. Int. J. Biol. Macromol. 39 (1–3), 88–94.

Dabin, J., Jam, M., et al., 2008. Expression, purification, crystallization and preliminary X-ray analysis of the polysaccharide lyase RB5312 from the marine planctomycete *Rhodopirellula baltica*. Acta Crystallogr. Sect. F Struct. Biol. Cryst. Commun. 64 (Pt. 3), 224–227.

Davis, K.E., Joseph, S.J., et al., 2005. Effects of growth medium, inoculum size, and incubation time on culturability and isolation of soil bacteria. Appl. Environ. Microbiol. 71 (2), 826–834.

de Almeida, N.M., Neumann, S., et al., 2015. Immunogold localization of key metabolic enzymes in the anammoxosome and on the tubule-like structures of Kuenenia stuttgartiensis. J. Bacteriol. 197 (14), 2432–2441.

Dedysh, S.N., Kulichevskaya, I.S., 2013. Acidophilic planctomycetes: expanding the horizons of new planctomycete diversity. In: Fuerst, J.A. (Ed.), Planctomycetes: Cell Structure, Origins and Biology. Humana Press, New York, pp. 141–164.

DeMooij, H.W., Thomas, G., et al., 2010. Ammoniacal nitrogen removal from sludge liquors—operational experience with the DEMON process. In: Fifteenth European Biosolids and Organic Resources Conference.

Devos, D.P., Ward, N.L., 2014. Mind the PVCs. Environ. Microbiol. 16 (5), 1217–1221.

Devos, D.P., Jogler, C., et al., 2013. The 1st EMBO workshop on PVC bacteria-Planctomycetes-Verrucomicrobia-Chlamydiae superphylum: exceptions to the bacterial definition? Antonie Van Leeuwenhoek 104 (4), 443–449.

Dietl, A., Ferousi, C., et al., 2015. The inner workings of the hydrazine synthase multiprotein complex. Nature 527 (7578), 394–397.

Elshahed, M.S., Youssef, N.H., et al., 2007. Phylogenetic and metabolic diversity of Planctomycetes from anaerobic, sulfide- and sulfur-rich Zodletone Spring, Oklahoma. Appl. Environ. Microbiol. 73 (15), 4707–4716.

Erbilgin, O., McDonald, K.L., et al., 2014. Characterization of a planctomycetal organelle: a novel bacterial microcompartment for the aerobic degradation of plant saccharides. Appl. Environ. Microbiol. 80 (7), 2193–2205.

Escobedo-Hinojosa, W.I., Vences-Guzman, M.A., et al., 2015. OlsG (Sinac_1600) is an ornithine lipid *N*-methyltransferase from the planctomycete *Singulisphaera acidiphila*. J. Biol. Chem. 290 (24), 15102–15111.

Francis, C.A., Beman, J.M., et al., 2007. New processes and players in the nitrogen cycle: the microbial ecology of anaerobic and archaeal ammonia oxidation. ISME J. 1 (1), 19–27.

Franzmann, P.D., Skerman, V.B., 1984. *Gemmata obscuriglobus*, a new genus and species of the budding bacteria. Antonie Van Leeuwenhoek 50 (3), 261–268.

Fuerst, J.A., 1995. The planctomycetes: emerging models for microbial ecology, evolution and cell biology. Microbiology 141 (Pt. 7), 1493–1506.

Fuerst, J.A., 2005. Intracellular compartmentation in planctomycetes. Annu. Rev. Microbiol. 59, 299–328.

Fuerst, J.A., 2013. The PVC superphylum: exceptions to the bacterial definition? Antonie Van Leeuwenhoek 104 (4), 451–466.

Fuerst, J.A., Sagulenko, E., 2011. Beyond the bacterium: planctomycetes challenge our concepts of microbial structure and function. Nat. Rev. Microbiol. 9 (6), 403–413.

Fuerst, J.A., Sagulenko, E., 2013. Nested bacterial boxes: nuclear and other intracellular compartments in planctomycetes. J. Mol. Microbiol. Biotechnol. 23 (1–2), 95–103.

Fuerst, J.A., Sagulenko, E., 2014. Towards understanding the molecular mechanism of the endocytosis-like process in the bacterium *Gemmata obscuriglobus*. Biochim. Biophys. Acta 1843 (8), 1732–1738.

Fuerst, J.A., Webb, R.I., 1991. Membrane-bounded nucleoid in the eubacterium *Gemmata obscuriglobus*. Proc. Natl. Acad. Sci. USA 88 (18), 8184–8188.

Fukunaga, Y., Kurahashi, M., et al., 2009. *Phycisphaera mikurensis* gen. nov., sp. nov., isolated from a marine alga, and proposal of *Phycisphaeraceae* fam. nov., *Phycisphaerales* ord. nov. and *Phycisphaerae* classis nov. in the phylum Planctomycetes. J. Gen. Appl. Microbiol. 55 (4), 267–275.

Gadler, P., Faber, K., 2007. New enzymes for biotransformations: microbial alkyl sulfatases displaying stereo- and enantioselectivity. Trends Biotechnol. 25 (2), 83–88.

Gadler, P., Glueck, S.M., et al., 2006. Biocatalytic approaches for the quantitative production of single stereoisomers from racemates. Biochem. Soc. Trans. 34 (Pt 2), 296–300.

Giessen, T.W., Silver, P.A., 2015. Encapsulation as a strategy for the design of biological compartmentalization. J. Mol. Biol.

Gimesi, N., 1924. Hydrobiologiai Tanulmanyok (Hydrobiologische Studien). *Planctomyces Bekefii* Gim.nov. et sp,. Budapest, Kiadja a Magyar Ciszterci Rend.

Giovannoni, S.J., Schabtach, E., et al., 1987. *Isosphaera pallida*, gen and comb-nov, a gliding, budding eubacterium from hot springs. Arch. Microbiol. 147 (3), 276–284.

Glockner, F.O., Kube, M., et al., 2003. Complete genome sequence of the marine planctomycete *Pirellula* sp. strain 1. Proc. Natl. Acad. Sci. USA 100 (14), 8298–8303.

Gottshall, E.Y., Seebart, C., et al., 2014. Spatially segregated transcription and translation in cells of the endomembrane-containing bacterium *Gemmata obscuriglobus*. Proc. Natl. Acad. Sci. USA 111 (30), 11067–11072.

Jenkins, C., Staley, J.T., 2013. History, classification and cultivation of the planctomycetes. In: Fuerst, J.A. (Ed.), Planctomycetes: Cell Structure, Origins and Biology. Humana Press, New York, pp. 1–38.

Jeske, O., Jogler, M., et al., 2013. From genome mining to phenotypic microarrays: planctomycetes as source for novel bioactive molecules. Antonie Van Leeuwenhoek 104 (4), 551–567.

Jeske, O., Schuler, M., et al., 2015. Planctomycetes do possess a peptidoglycan cell wall. Nat. Commun. 6, 7116.

Jetten, M.S., Wagner, M., et al., 2001. Microbiology and application of the anaerobic ammonium oxidation ('anammox') process. Curr. Opin. Biotechnol. 12 (3), 283–288.

Jetten, M., Schmid, M., et al., 2005. Anammox organisms:enrichment, cultivation and environmental analysis. Methods Enzymol. 397, 34–57.

Jetten, M.S., den Camp, H.J., O., et al., 2015. "*Candidatus Brocadiales*" ord. nov. Bergey's Manual of Systematics of Archaea and Bacteria., John Wiley in assoc with Bergey's Manual Trust. 1.

Jin, R.C., Yang, G.F., et al., 2013. The effect of sulfide inhibition on the ANAMMOX process. Water Res. 47 (3), 1459–1469.

Jogler, C., 2014. The bacterial 'mitochondrium'. Mol. Microbiol. 94 (4), 751–755.

Jogler, M., Jogler, C., 2013. Toward the development of genetic tools for planctomycetes. In: Fuerst, J. A. (Ed.), Planctomycetes: Cell Structure, Origins and Biology. Humana Press, New York, pp. 141–164.

Jogler, C., Glockner, F.O., et al., 2011. Characterization of *Planctomyces limnophilus* and development of genetic tools for its manipulation establish it as a model species for the phylum Planctomycetes. Appl. Environ. Microbiol. 77 (16), 5826–5829.

Jun, S.R., Sims, G.E., et al., 2010. Whole-proteome phylogeny of prokaryotes by feature frequency profiles: an alignment-free method with optimal feature resolution. Proc. Natl. Acad. Sci. USA 107 (1), 133–138.

Kalyuzhnaya, M.G., Korotkova, N., et al., 2005. Analysis of gene islands involved in methanopterin-linked C1 transfer reactions reveals new functions and provides evolutionary insights. J. Bacteriol. 187 (13), 4607–4614.

Kartal, B., van Niftrik, L., et al., 2004. Application,eco-physiology and biodiversity of anaerobic ammonium-oxidizing bacteria. Rev. Environ. Sci. Biotechnol. 3, 255–264.

Kartal, B., Kuenen, J.G., et al., 2010. Engineering. Sewage treatment with anammox. Science 328 (5979), 702–703.

Kartal, B., Geerts, W., et al., 2011a. Cultivation, detection and ecophysiology of anaerobic ammonium-oxidizing bacteria. Methods Enzymol. 486, 89–108.

Kartal, B., Maalcke, W.J., et al., 2011b. Molecular mechanism of anaerobic ammonium oxidation. Nature 479 (7371), 127–130.

Kartal, B., van Niftrik, L., et al., 2012. Anammox—growth physiology, cell biology, and metabolism. Adv. Microb. Physiol. 60, 211–262.

Kartal, B., de Almeida, N.M., et al., 2013. How to make a living from anaerobic ammonium oxidation. FEMS Microbiol. Rev. 37 (3), 428–461.

Kovaleva, O.L., Merkel, A.Y., et al., 2015. *Tepidisphaera mucosa* gen. nov., sp. nov., a moderately thermophilic member of the class *Phycisphaerae* in the phylum *Planctomycetes*, and proposal of a new family, *Tepidisphaeraceae* fam. nov., and a new order, *Tepidisphaerales* ord. nov. Int. J. Syst. Evol. Microbiol. 65 (Pt 2), 549–555.

Kuai, L., Verstraete, W., 1998. Ammonium removal by the oxygen-limited autotrophic nitrification-denitrification system. Appl. Environ. Microbiol. 64 (11), 4500–4506.

Kuenen, J.G., 2008. Anammox bacteria: from discovery to application. Nat. Rev. Microbiol. 6 (4), 320–326.

Kulichevskaya, I.S., Ivanova, A.O., et al., 2007. *Schlesneria paludicola* gen. nov., sp. nov., the first acidophilic member of the order *Planctomycetales*, from Sphagnum-dominated boreal wetlands. Int. J. Syst. Evol. Microbiol. 57 (Pt 11), 2680–2687.

Kulichevskaya, I.S., Ivanova, A.O., et al., 2008. *Singulisphaera acidiphila* gen. nov., sp. nov., a non-filamentous, Isosphaera-like planctomycete from acidic northern wetlands. Int. J. Syst. Evol. Microbiol. 58 (Pt. 5), 1186–1193.

Kulichevskaya, I.S., Baulina, O.I., et al., 2009. *Zavarzinella formosa* gen. nov., sp. nov., a novel stalked, *Gemmata*-like planctomycete from a Siberian peat bog. Int. J. Syst. Evol. Microbiol. 59 (Pt 2), 357–364.

Kulichevskaya, I.S., Serkebaeva, Y.M., et al., 2012. *Telmatocola sphagniphila* gen. nov., sp. nov., a novel dendriform planctomycete from northern wetlands. Front Microbiol. 3, 146.

Kuypers, M.M., Lavik, G., et al., 2005. Massive nitrogen loss from the Benguela upwelling system through anaerobic ammonium oxidation. Proc. Natl. Acad. Sci. USA 102 (18), 6478–6483.

Lackner, S., Gilbert, E.M., et al., 2014. Full-scale partial nitritation/anammox experiences—an application survey. Water Res. 55, 292–303.

Lage, O.M., Bondoso, J., 2014. Planctomycetes and macroalgae, a striking association. Front Microbiol. 5, 267.

Lagkouvardos, I., Jehl, M.A., et al., 2014. Signature protein of the PVC superphylum. Appl. Environ. Microbiol. 80 (2), 440–445.

Lam, P., Lavik, G., et al., 2009. Revising the nitrogen cycle in the Peruvian oxygen minimum zone. Proc. Natl. Acad. Sci. USA 106 (12), 4752–4757.

Liesack, W., König, H., et al., 1986. Chemical composition of the peptidoglycan-free cell envelopes of budding bacteria of the *Pirella/Planctomyces* group. Arch. Microbiol. 145 (4), 361–366.

Lindsay, M.R., Webb, R.I., et al., 2001. Cell compartmentalisation in planctomycetes: novel types of structural organisation for the bacterial cell. Arch. Microbiol. 175 (6), 413–429.

Lonhienne, T.G., Sagulenko, E., et al., 2010. Endocytosis-like protein uptake in the bacterium *Gemmata obscuriglobus*. Proc. Natl. Acad. Sci. USA 107 (29), 12883–12888.

Moore, E.K., Villanueva, L., et al., 2015. Abundant trimethylornithine lipids and specific gene sequences are indicative of planctomycete importance at the oxic/anoxic interface in sphagnum-dominated northern wetlands. Appl. Environ. Microbiol. 81 (18), 6333–6344.

Mulder, A.A., van de Graaf, A.A., et al., 1995. Anaerobic ammonium oxidation discovered in a denitrifying fluidized bed reactor. FEMS Microbiol. Ecol. 16, 177–183.

Naumoff, D.G., Ivanova, A.A., et al., 2014. Phylogeny of beta-xylanases from Planctomycetes. Mol. Biol. (Mosk.) 48 (3), 508–517.

Neumann, S., Jetten, M.S., et al., 2011. The ultrastructure of the compartmentalized anaerobic ammonium-oxidizing bacteria is linked to their energy metabolism. Biochem. Soc. Trans. 39 (6), 1805–1810.

Neumann, S., Wessels, H.J., et al., 2014. Isolation and characterization of a prokaryotic cell organelle from the anammox bacterium *Kuenenia stuttgartiensis*. Mol. Microbiol. 94 (4), 794–802.

Nouri, D.H., Tantillo, D.J., 2006. They came from the deep: syntheses, applications, and biology of ladderanes. Curr. Org. Chem. 10, 2055–2074.

Op den Camp, H.J., Kartal, B., et al., 2006. Global impact and application of the anaerobic ammonium-oxidizing (anammox) bacteria. Biochem. Soc. Trans. 34 (Pt 1), 174–178.

Pearson, A., Budin, M., et al., 2003. Phylogenetic and biochemical evidence for sterol synthesis in the bacterium *Gemmata obscuriglobus*. Proc. Natl. Acad. Sci. USA 100 (26), 15352–15357.

Pereira, A.D., Leal, C.D., et al., 2014. Effect of phenol on the nitrogen removal performance and microbial community structure and composition of an anammox reactor. Bioresour. Technol. 166, 103–111.

Perez, J., Isanta, E., et al., 2015. Would a two-stage N-removal be a suitable technology to implement at full scale the use of anammox for sewage treatment? Water Sci. Technol. 72 (6), 858–864.

Pilhofer, M., Aistleitner, K., et al., 2013. Discovery of chlamydial peptidoglycan reveals bacteria with murein sacculi but without FtsZ. Nat. Commun. 4, 2856.

Sagulenko, E., Morgan, G.P., et al., 2014. Structural studies of planctomycete *Gemmata obscuriglobus* support cell compartmentalisation in a bacterium. PLoS One 9 (3), e91344.

Santarella-Mellwig, R., Franke, J., et al., 2010. The compartmentalized bacteria of the planctomycetes-verrucomicrobia-chlamydiae superphylum have membrane coat-like proteins. PLoS Biol. 8 (1), e1000281.

Santarella-Mellwig, R., Pruggnaller, S., et al., 2013. Three-dimensional reconstruction of bacteria with a complex endomembrane system. PLoS Biol. 11 (5), e1001565.

Sayer, C., Isupov, M.N., et al., 2015. Structural studies of a thermophilic esterase from a new Planctomycetes species, *Thermogutta terrifontis*. FEBS J. 282 (15), 2846–2857.

Scheuner, C., Tindall, B.J., et al., 2014. Complete genome sequence of *Planctomyces brasiliensis* type strain (DSM 5305(T)), phylogenomic analysis and reclassification of Planctomycetes including the descriptions of *Gimesia* gen. nov., *Planctopirus* gen. nov. and *Rubinisphaera* gen. nov. and emended descriptions of the order *Planctomycetales* and the family *Planctomycetaceae*. Stand. Genomic. Sci. 9, 10.

Schlesner, H., 1989. *Planctomyces brasiliensis* sp.nov., halotolerant bacterium from a salt pit. Syst. Appl. Microbiol. 12, 159–161.

Schlesner, H., 1994. The development of media suitable for the microorganisms morphologically resembling *Planctomyces* spp., *Pirellula* spp., and other *Planctomycetales* from various aquatic habitats using dilute media. Syst. Appl. Microbiol. 17, 135–145.

Schlesner, H., Hirsch, P., 1986. Assigment of the genera *Planctomyces* and *Pirella* to a bew family *Planctomycetaceae* fam.nov. and description of the order *Planctomycetales* ord.nov. Syst. Appl. Microbiol. 8, 174–176.

Schmid, M.C., Risgaard-Petersen, N., et al., 2007. Anaerobic ammonium-oxidizing bacteria in marine environments: widespread occurrence but low diversity. Environ. Microbiol. 9 (6), 1476–1484.

Schreier, H.J., Dejtisakdi, W., et al., 2012. Transposon mutagenesis of *Planctomyces limnophilus* and analysis of a pckA mutant. Appl. Environ. Microbiol. 78 (19), 7120–7123.

Shulse, C.N., Allen, E.E., 2011. Widespread occurrence of secondary lipid biosynthesis potential in microbial lineages. PLoS One 6 (5), e20146.

Siegrist, H., Salzgeber, D., et al., 2008. Anammox brings WWTP closer to energy autarky due to increased biogas production and reduced aeration energy for N-removal. Water Sci. Technol. 57 (3), 383–388.

Sinninghe Damste, J.S., Strous, M., et al., 2002. Linearly concatenated cyclobutane lipids form a dense bacterial membrane. Nature 419 (6908), 708–712.

Sinninghe Damsté, J.S., Rijpstra, W.I.C., et al., 2004. The occurrence of hopanoids in planctomycetes: implications for the sedimentary biomarker. Org. Geochem. 35, 561–566.

Slobodkina, G.B., Panteleeva, A.N., et al., 2016. *Thermostilla marina* gen. nov., sp. nov., a thermophilic, facultatively anaerobic planctomycete isolated from a shallow submarine hydrothermal vent. Int. J. Syst. Evol. Microbiol. 66 (2), 633–638.

Smith, R.L., Bohlke, J.K., et al., 2015. Role of anaerobic ammonium oxidation (anammox) in nitrogen removal from a freshwater aquifer. Environ. Sci. Technol. 49 (20), 12169–12177.

Speth, D.R., In 't Zandt, M.H., et al., 2016. Genome-based microbial ecology of anammox granules in a full-scale wastewater treatment system. Nat. Commun. 7, 11172.

Stackebrandt, E., Wehmeyer, U., et al., 1986. 16S ribosomal RNA- and cell wall analysis of *Gemmata obscuriglobus*, a new member of the order Planctomycetales. FEMS Microbiol. Lett. 37 (3), 289–292.

Storesund, J.E., Ovreas, L., 2013. Diversity of Planctomycetes in iron-hydroxide deposits from the Arctic Mid Ocean Ridge (AMOR) and description of *Bythopirellula goksoyri* gen. nov., sp. nov., a novel Planctomycete from deep sea iron-hydroxide deposits. Antonie Van Leeuwenhoek 104 (4), 569–584.

Strous, M., Pelletier, E., et al., 2006. Deciphering the evolution and metabolism of an anammox bacterium from a community genome. Nature 440 (7085), 790–794.

Sukovich, D.J., Seffernick, J.L., et al., 2010. Widespread head-to-head hydrocarbon biosynthesis in bacteria and role of OleA. Appl. Environ. Microbiol. 76 (12), 3850–3862.

Third, K.A., Sliekers, A.O., et al., 2001. The CANON system (completely autotrophic nitrogen-removal over nitrite) under ammonium limitation: interaction and competition between three groups of bacteria. Syst. Appl. Microbiol. 24 (4), 588–596.

van der Star, W.R., Abma, W.R., et al., 2007. Startup of reactors for anoxic ammonium oxidation: experiences from the first full-scale anammox reactor in Rotterdam. Water Res. 41 (18), 4149–4163.

van Dongen, U., Jetten, M.S., et al., 2001. The SHARON-Anammox process for treatment of ammonium rich wastewater. Water Sci. Technol. 44 (1), 153–160.

van Niftrik, L., Jetten, M.S., 2012. Anaerobic ammonium-oxidizing bacteria: unique microorganisms with exceptional properties. Microbiol. Mol. Biol. Rev. 76 (3), 585–596.

van Niftrik, L., Geerts, W.J., et al., 2008. Linking ultrastructure and function in four genera of anaerobic ammonium-oxidizing bacteria: cell plan, glycogen storage, and localization of cytochrome C proteins. J. Bacteriol. 190 (2), 708–717.

van Niftrik, L., Geerts, W.J., et al., 2009. Cell division ring, a new cell division protein and vertical inheritance of a bacterial organelle in anammox planctomycetes. Mol. Microbiol. 73 (6), 1009–1019.

van Niftrik, L., van Helden, M., et al., 2010. Intracellular localization of membrane-bound ATPases in the compartmentalized anammox bacterium '*Candidatus* Kuenenia stuttgartiensis'. Mol. Microbiol. 77 (3), 701–715.

van Teeseling, M.C., Mesman, R.J., et al., 2015. Anammox Planctomycetes have a peptidoglycan cell wall. Nat. Commun. 6, 6878.

Vlaeminck, S.E., De Clippeleir, H., et al., 2012. Microbial resource management of one-stage partial nitritation/anammox. Microb. Biotechnol. 5 (3), 433–448.

Wallner, S.R., Bauer, M., et al., 2005. Highly enantioselective sec-alkyl sulfatase activity of the marine planctomycete *Rhodopirellula baltica* shows retention of configuration. Angew. Chem. Int. Ed. Engl. 44 (39), 6381–6384.

Wang, J., Jenkins, C., et al., 2002. Isolation of *Gemmata*-like and *Isosphaera*-like planctomycete bacteria from soil and freshwater. Appl. Environ. Microbiol. 68 (1), 417–422.

Wang, X., Sharp, C.E., et al., 2015. Stable-Isotope probing identifies uncultured planctomycetes as primary degraders of a complex heteropolysaccharide in soil. Appl. Environ. Microbiol. 81 (14), 4607–4615.

Ward, N.L., 2015. *Planctomycetia* class. nov. Bergey's Manual of Systematics of Archaea and Bacteria. John Wiley & Sons, Ltd, in association with Bergey's Manual Trust.

Wegner, C.E., Richter-Heitmann, T., et al., 2013. Expression of sulfatases in *Rhodopirellula baltica* and the diversity of sulfatases in the genus *Rhodopirellula*. Mar. Genomics 9, 51–61.

Wegner, C.E., Richter, M., et al., 2014. Permanent draft genome of *Rhodopirellula sallentina* SM41. Mar. Genomics 13, 17–18.

Wett, B., 2007. Development and implementation of a robust deammonification process. Water Sci. Technol. 56 (7), 81–88.

Woebken, D., Teeling, H., et al., 2007. Fosmids of novel marine Planctomycetes from the Namibian and Oregon coast upwelling systems and their cross-comparison with planctomycete genomes. ISME J. 1 (5), 419–435.

Yarza, P., Yilmaz, P., et al., 2014. Uniting the classification of cultured and uncultured bacteria and archaea using 16S rRNA gene sequences. Nat. Rev. Microbiol. 12 (9), 635–645.

Yilmaz, P., Yarza, P., et al., 2015. Expanding the world of marine bacterial and archaeal clades. Front. Microbiol. 6, 1524.

Yoon, J., Jang, J.H., et al., 2014. *Algisphaera agarilytica* gen. nov., sp. nov., a novel representative of the class *Phycisphaerae* within the phylum *Planctomycetes* isolated from a marine alga. Antonie Van Leeuwenhoek 105 (2), 317–324.

Zhang, T., Han, W.J., 2009. Gene cloning and characterization of a novel esterase from activated sludge metagenome. Microb. Cell. Fact. 8, 67.

FURTHER READING

Slobodkina, G.B., Kovaleva, O.L., et al., 2015. *Thermogutta terrifontis* gen. nov., sp. nov. and *Thermogutta hypogea* sp. nov., thermophilic anaerobic representatives of the phylum *Planctomycetes*. Int. J. Syst. Evol. Microbiol. 65 (Pt 3), 760–765.

Chapter 2

A Flavor of Prokaryotic Taxonomy: Systematics Revisited

Paul De Vos*, Fabiano Thompson†, Cristiane Thompson,† and Jean Swings*,†**

**Ghent University, Gent, Belgium; **Federal University of Rio de Janeiro (UFRJ), Rio de Janeiro, Brazil; †SAGE-COPPE-UFRJ, Rio de Janeiro, Brazil*

INTRODUCTION

Prokaryotic (microbial/bacterial) taxonomy (systematics) covers three main subtopics: (1) classification, that is, ordering of organisms into delineated groups based on similarities, (2) nomenclature, that is, naming these groups according to the bacterial code (Lapage et al., 1992; Parker et al., 2015), and (3) identification, that is, verifying whether an unknown belongs to one of the groups defined in (1) and named in (2) for which a combination of methods is usually applied. Taxonomy and systematics are synonymous and used as such. In prokaryotic taxonomy, the overall classification system is a hierarchic grouping. It concerns from highest to the lowest: domain, phylum, class, order, family, genus, species, and subspecies.

Prokaryotic systematics is a scientific discipline that evolves with the technological tools that become available. This has an effect on the species definition which forms the cornerstone in prokaryotic systematics.

Since the first determination of the first whole genome sequence of *Haemophilus influenzae* (Fleischmann et al., 1995) became available, high throughput sequencing (HTS) has been introduced and is more cost effective. At the same time, developments of user friendly bioinformatic analytical tools led to new insights in content and functioning of the sequenced prokaryotic genomes. It is therefore not surprising that whole genome sequencing (WGS) data tend to underpin more and more prokaryotic systematics because it is undoubtedly the best resource to reveal natural relationships of prokaryotes. However, the value of WGS for taxonomic purposes is not fully explored yet. The conservative nature of systematics per se and conservative reflexes of (some) taxonomists, although valuable to some extent, slow down the introduction of WGS in prokaryotic taxonomy.

Microbial Resources. http://dx.doi.org/10.1016/B978-0-12-804765-1.00002-3

Further, PCR and subsequent sequencing methodologies and analyses allow to detect sequences of an unknown in samples unraveling their biological composition and demonstrating the nearly endless prokaryotic diversity on earth. This raises the question why so few species have been described formally. Moreover, only a fraction of the prokaryotic diversity is accessible in culture. The actually detected genetic diversity remains still a small part of the natural diversity, the greatest part being the microbial "dark matter."

THE PROKARYOTIC SPECIES CONCEPT

Most microbiologists will agree that (1) a microbial species is an independently evolving population, (2) the events that sculpture species may vary in the different groups, and (3) gaps exist between species that allow their differentiation.

Unlike in the classification of higher organisms, the biological species concept cannot be applied for prokaryotes because of their replication system(s) and the tremendous plasticity of their genomes frequently affected by internal and external drivers allowing instant and versatile adaptations to changing environments and exploration of multiple niches.

In the early days of microbiology, prokaryotic (bacterial) diversity was based on morphology, analogous to Eukaryotic species. Introduction of growing and isolation facilities resulted in more complex characterization of the bacterial phenotype as deduced by physiological and biochemical tests. In the 1960s and 1970s of the last century, numerical interpretations were introduced on phenotypic data sets (Sneath and Sokal, 1973) and user friendly miniaturized phenotypic methodologies, such as API and other commercial systems became accessible. Computerized groupings, mostly dendrograms, based on a variety of algorithms visualized groups for taxonomic interpretation. A recent approach in this context is the BIOLOG phenotype micro array (Bochner et al., 2001) that allows generating a phenotypic array of gene functioning. Although very valuable, this phenotypic micro array is relatively expensive and has a limited flexibility of incubation conditions. The strength of the numerical algorithms is that they allow roughly the same principles to be applied for interpretation of visualized digital profiles obtained via gel electrophoresis of cell proteins, gas chromatographic analysis of methylated ethyl esters of fatty acids (FAME) (e.g., MIDI Sasser, 1990), and matrix assisted desorption/ionization time of flight (MALDI TOF Clark et al., 2013) of rRNA proteins.

The exploration of DNA, RNA, and proteins starting from the 1960s of last century and gradually lead to a better insight in prokaryotic phylogenetic diversity and evolution. The idea of a prokaryotic classification underpinned by natural (phylogenetic) relationships of the taxa became introduced for the first time in the species concept integrating genomic data in combination with the existing traditionally phenotypic elements (Wayne et al., 1987). This phylogenetic concept introduced the 97% rule of 16S rRNA gene sequence similarity for species discrimination—accompanied with laborious and not standardized overall physicochemical DNA:DNA hybridization (DDH) value of 70% (with a maximum thermostability difference of 5°C). Later the rRNA sequence similarity changed to 98.7% (Stackebrandt and Elbers, 2006). Despite this improvement, the taxonomic resolution of 16S rRNA gene sequences is not discriminatory at the species level (or below), defined at 70% DNA: DNA hybridization.

Description of species must not only comply with the phylogenetic species concept as formulated by Wayne et al. (1987), but also be distinguishable by phenotypic differences from the most closely related species.

The requirements of laborious DDH together with phenotypic differentiation limited the number of prokaryotic species descriptions. With the tree of live reconstruction (Woese et al., 1990) as a starting point, a consensus classification was generated with 16S rRNA sequence similarity as the cornerstone (Garrity et al., 2004; Ludwig and Klenk, 2005). Until now it remains the pragmatic basis for prokaryotic systematics. Recently, Hug et al. (2016) have presented a new view on the tree of life obtained after a comprehensive analysis of 16 ribosomal protein sequences. Striking features are the large number of lineages without isolated representatives together with the candidate phyla radiation (CPR), which comprise the majority of life's current biodiversity. Only cultivation-independent genome approaches reveal this biodiversity. The concept of Candidatus as provisional status was first introduced by Murray and Stackebrandt (1995) and recently taken up by Konstantinidis and Rossellò Mòra (2015) in an attempt to accommodate uncultured prokaryotes of which only genomic data are available. It might be argued that this proposal is of little use, if not blatantly unworkable because the formal description of taxa demands cultured representatives as references. We propose that taxonomists accept the principle that the researchers who discover a new genome sequence have the right to name it following a simple legitimate taxonomic procedure when they fulfil a number of technical quality standards. Massive generation of WGS stimulated exploration to replace DDH by user-friendly alternative (Goris et al., 2007). Today DDH of 70% is replaced by the ANI (Average Nucleotide Identity) boundary value of 95% (Kim et al., 2014) but also by calculation of de DDH value from WGS data (Auch et al., 2010).

The Wayne et al. (1987) species concept has been evaluated by Stackebrandt et al. (2002) and by Gevers et al. (2005). In the light of new methodologies that became available several recommendations were made. Promising genomic methodologies, such as sequencing of housekeeping genes (MLST—Multilocus Sequence Typing and MLSA—Multilocus Sequence Analysis), DNA profiling (e.g., AFLP), and DNA arrays were put forward. DDH was kept for the definition of a species. For the MLST and MLSA approach, selection of the best suited housekeeping genes is best based on (1) single copy genes, (2) genes spread over the genome, (3) the sequence stretch must be long enough to allow sufficient differentiation, (4) coding genes, and (5) a minimal number of genes must be taken into account for either separate or concatenated tree construction. It is clear that WGS information will facilitate selection of these genes to build datasets that can be used for differentiation and identification both at the species (in combinations with set boundaries) and infra species levels. It is important to pay attention to eventual bias of the implementation of methods, such as MLST or MLSA. Due to the vast exchange (lateral gene transfer—LGT) and rearrangement of genome, prokaryotic evolution and phylogeny is seen by some as a phylogenetic network (Popa and Dagan, 2011). The frequency of LGT is dependent on the phylogenetic distance between partners, niche, and stress factors. LGT is undoubtedly a powerful evolutionary force that reshaped genomes throughout evolution, allowing the combination of metabolic functions beyond species boundaries (Daubin and Szöllösi, 2016; Koonin, 2016). These elements are important to understand speciation in prokaryotes and have to be taken into account in constructing the prokaryotic phylogenomic frame.

Whether MLST or MLSA of housekeeping genes will remain as a parallel approach to underpin specific and sub specific groupings is uncertain once complete genomic data reach full exploitation in taxonomy.

Further massive generation of WGS stimulated by various initiatives (e.g., GEBA project of the Joint Genome Insitute—http://jgi.doe.gov/geba-phase-iii) will support a genomic species concept (Konstantinidis et al., 2006; Rosselló-Móra and Amann, 2001; Varghese et al., 2015). Culture collections should continue to play a central role in these international initiatives. It is obvious that the WGS based taxonomy has potentialities to reach a much finer taxonomic resolution, that is, down to strain level. The evolution toward a genomic prokaryote taxonomy may leave the microbiologist unsatisfied because of the reduced attention for microbial function. Indeed, an in-depth analysis of the prokaryotic genomes will be needed to give an answer to the questions (1) which organisms are out there, (2) doing what, and (3) together with which organisms. The prokaryotic species are then considered as a group of different populations or strains that share enough characteristics to be united in a taxonomic unit in accordance with the accepted (species) definition of the moment.

Prokaryotic systematics is a communication tool to report and understand the presence and role/function of prokaryotic life forms in a variety of ecosystems and to support diagnostics. Genomic analyses undoubtedly will evolve in revealing these functionalities from the sequence not only in their potentialities (gene content) but in the gene expression as influenced by niche conditions.

REMARKS ON THE GENERALIZED USE OF 16S RRNA SEQUENCES

The 16S RNA backbone for the taxonomic outline is in practice not applicable to discriminate between species as the gene is extremely conserved. Furthermore, it is part of a more extended rrp operon together with the 23S and 5S rRNA genes of which a variable number (2–15) of multiple copies are present in the prokaryotic genome (Pei et al., 2010). Although not all 16S rRNA gene copies show intra genomic sequence variations that may interfere with the cut off value of 1.3% dissimilarity (Stackebrandt and Elbers, 2006). Millions of

16S rRNA gene sequences have been deposited in international databases; for the vast majority of them there is no culture available in an international public culture collection. This prevents comparisons and checks. It is also troublesome that for taxonomic references (type strains, pathotypes, etc.) multiple sequences of the 16S rRNA gene have been deposited, either from what should be the same or equivalent material that show variations and uncertainties in base composition due to technical mistakes as well as a bias in linking of the sequence and the authentic microbial material. The all living species tree project and updates thereof has the merit to provide the user with technically correct generated 16S sequences for the type strains (Chun et al., 2007; Yarza et al., 2010). Via StrainInfo.net (http://www.straininfo.net/), the user can find out if anomalous matches occur between the sequences of equivalent type material that one is using (De Smet et al., 2013). These problems with confusing sequence depositions are well known, nevertheless curation of the content in the international repositories is lacking. Because of these quality problems the direct use of sequences from international databases in general and of 16S rRNA gene sequences in particular must be done with care and for taxonomic studies and identifications, the Silva (https://www.arb-silva.de/) and EzTaxon cloud (http://www.ezbiocloud.net/eztaxon) are highly recommended.

CHEMOTAXONOMIC AND PHENOTYPIC CHARACTERIZATION

Phenotypic or chemotaxonomic differentiation from neighboring species is obligatory for any new species description. The outcomes are greatly affected by the testing methodology. Unfortunately, a variety of methods for testing is often combined, which results in conflicting data. Cultivation dependent factors (growth medium, temperature, pH, etc.) require maximal standardization for comparative analysis. Unfortunately by doing so, the ability of the natural phenotypic flexibility of the biological material is not addressed. Furthermore, new species descriptions are often based on one or at the most a few strains that are in many cases derived from the same ecological niche. It is well known that niche adaptations are strain and not species dependent and hence measured phenotypic characteristics of, for example, the type strain may be strain linked. The phenotypic differentiation tests are unreliable in many cases and therefore phenotypic characterization is of limited use to differentiate the prokaryote species.

Chemotaxonomic approaches, in particular MALDI-TOF patterns and to a lesser extend FAME patterns are for reasons of built—in stringent standardization less prone to external factors. However, both techniques do not directly reflect phenotypic functionalities. MALDI TOF is very well suited for dereplication and provides stable groupings. Databases covering large numbers of strains per species offer an important extension and increase the taxonomic reliability. In taxonomic studies linked with isolation campaigns or extensive cumulative research on isolates, MALDI TOF is the preferred technique for dereplication.

Mining the genome sequences to predict the phenotype of the sequenced strain (in silico phenotyping) avoids costly experimental phenotypic screenings. Identifying or predicting the genes involved in phenotypes is known as gene-trait matching. Recently, a complete in silico pipeline was outlined for the consistent annotation of bacterial genomes followed by automated gene-trait matching (Dutilh et al., 2013). It illustrates the versatility of gene-trait matching and its power for identifying genes associated to specific bacterial phenotypes.

In conclusion, whatever approach to prokaryotic systematics is followed it will always be a pragmatic one that will be preferred.

PROKARYOTE NOMENCLATURE

An important feature of prokaryote nomenclature was that it was internationally agreed to start again from 1980 (Skerman et al., 1980) with the publication of the "Approved list of bacterial names" as amended in 1989 (Skerman et al., 1989). The naming of the species is subjected to the rules of the Prokaryotic Code (formerly Bacterial Code) (Lapage et al., 1992; Parker et al., 2015). Validation of the name of a new prokaryotic species or new combination has to be published in the *International Journal of Systematic and Evolutionary Microbiology* (IJSEM) either as full paper/note or has to appear on the validation list (VL) if published elsewhere. The whole process demands that a type strain is designated and that the material is deposited with authenticated certification in at least two independent public culture collections (De Vos and Trüper, 2000). The process of naming and validation follows a number of strict rules and recommendations. Although it is conservative in nature, it has the important merit of reducing chaos in systematics of prokaryotes. The type strain (indicated by superscript "T" behind the strain number) is not a "typical strain" reflecting necessarily all characteristics of the species, but is only the name bearer of the taxon. It is the reference for new isolates to be compared with in order to find out if a new isolate belongs to the same species or not. Conflicts are discussed by the Judicial Commission (JC) as part of the International Committee on Systematics of Prokaryotes (ICSP) that is member of IUMS (International Union of Microbial Societies) which is itself a member of ICSU (International Council for Science). Taxonomic rearrangements often concern validly named species or higher taxa. In general, the Prokaryotic Code covers the overall process of renaming taxa or part of taxa by a number of priority rules and recommendations. It does not entirely eliminate ambiguity. One of the confusions that may occur with renaming is that a valid name is kept. Although it is recommended to use the most recent name (after renaming), the original name can still be legitimately used, herewith, two (or more) different names are created for the same organisms that are classified in different taxa (e.g., *Pseudomonas acidovorans*, *Comamonas acidovorans*, and *Delftia acidovorans*). It is important to notice that the, name status can be traced relatively easy from the Euzéby website—http://www.bacterio.net/. Another anomaly concerns the partial reclassification of a species when a number of strains is transferred into a new or already existing species on the basis of new data. Although the biological content of the donating and the accepting species changes, their names are kept (or a new name is created if the transferred strains are regarded as belonging to a new species). The confusion here is created by the fact that the name of the taxon does no longer covers its former biological and/or phylogenetic content. These two examples show also how difficult it will be to realize the computerized integration of various elements/taxa and their changes.

IDENTIFICATION OF PROKARYOTES

Identification of an unknown isolate or sequence relies on the robustness of the former two parts of prokaryotic systematics (classification and naming). The identification result is limited to the species and subspecies level, the latter being the lowest level covered by the Prokaryotic Code (Parker et al., 2015).

A first step in the identification of an unknown is usually to find the most similar 16S rRNA gene sequence through a comparative BLAST (Altschul et al., 1990, 1997) search in international sequence data bases EMBL (http://www.ebi.ac.uk/), NCBI (http://www.ncbi.nlm.nih.gov/), EzTaxon (http://www.ezbiocloud.net/), and Silva (https://www.arb-silva.de/). The international genomic data bases from EMBL and NCBI are not filtered while the SILVA and EzTaxon make use of high quality sequences and the outcome will be easier to interpret and more reliable. The result is then a phylogenetic tree that gives an insight in the phylogenetic position of the unknown. Once this is done, a more in depth allocation is possible. The final identification may be obtained by comparative analysis of, for example, a set of housekeeping genes or by a polyphasic characterization for which the methods to be applied depend on the availability of databases comprising commercialized systems, for example API, BIOLOG, FAME (Fatty Acid Methyl Ester) analysis, VITEK, MALDI-TOF, or other in-house facilities and related databases. When the phylogenetic position of the unknown turns out to be part of a taxon for which at present no culture representative is available, a formal identification cannot be achieved. Formal description of taxa demands cultured representatives of taxonomic references. It is important to stimulate development of new culturing techniques and exploring extended panels of media and incubations conditions to obtain pure cultures or enriched cultures (Lagier et al., 2015a,b; Stewart, 2012). An important shortcoming in the quality of the data of the international databases originates from the informal and unchecked distribution of reference material (including type strains) between microbiologists and its use in taxonomic research. Also the material transfers (without authenticity quality checks) between public culture collections of "equivalent reference material," is subject of human handling mistakes of various nature (e.g., simple mistakes in numbering, mixing of cultures, purifying, and distributing contaminants as authentic material). This has been demonstrated by the Ghent group and its results are reflected in the bioportal StrainInfo.net. In particular, the Seqrank tool (De Smet et al., 2013) allows to verify whether 16S rRNA gene sequences of equivalent subcultures show unacceptable variations despite their high technical quality. Knowledge of the strain exchange history (Verslyppe et al., 2011) may allow to spot the origin of discrepancies of the material. In a next curation, data from this wrong material should be disconnected from the real reference material.

Identification of bacterial cultures at the species level is often not sufficient in the medical, agricultural, or biotechnological fields. A step further in the identification may be reached via genomic in house or public databases of housekeeping sequences linked to a limited number of related taxa. These specialized databases are generally well supported by extensive genomic studies on vast amounts of isolates plus taxonomic references (type strains, pathovar, or biovar etc.) and represent the known variety. As result, a sequenced based barcoding may be obtained accompanied by a stepwise standardized procedure leading to fine identification as in the case of Qbank (http://www.q-bank.eu/Bacteria/), plant quarantine bacteria listed by EPPO (European and Mediterranean Plant Protection Organization (https://www.eppo.int/). However, for maintenance and further expansion of this and similar initiatives that started via public funding (e.g., EU funding in this case Q-BOL see http://www.qbol.org/en/qbol.htm) need sufficient financial support in order to keep up with new emerging wild types worldwide.

Taxonomic changes have an impact on the identification, diagnosis, and treatment of plant, animal, or human pathogens. As long as no fully automated systems exist, regular taxonomic updates have to be announced, for example, for the medical bacteria (Janda, 2016).

THE GAP BETWEEN PROKARYOTIC DIVERSITY AND TAXONOMY OF PROKARYOTES

At present, well over 10,000 prokaryotic species have been described using the species definition and principles of polyphasic taxonomy (PT). Konstantinidis and Rossellò Mòra (2015) calculated that the total number of prokaryotic species would not exceed 10 million. Indeed, a wide gap!

Whereas huge scientific efforts are devoted to different fields of microbiology in order to determine its diversity, its overwhelming metabolic and genomic diversity, its functional role, in most cases prokaryote taxonomy plays a minor role. A quick look to the genome announcements of ASM reveal that the far majority of the cultures for which a WGS is announced are not present in public culture collections or in public databases (DB). This illustrates the second gap, growing each day between formal PT of prokaryotes and the forest of acronyms, numbers, and codes of isolates, genomes, genes, and features in the different DB. In depth, ecological studies reveal that not the artificially delineated prokaryotic species but that genomic subclusters (clades, OTUs, CTU's) characterize the real microbial biodiversity. This leads to an informal nomenclature in DB creating vast "terrae incognitae." Indeed, in private and public DB nameless OTU's, strains, isolates, sequences, and WGS (Beiko, 2015) form the great majority. The international initiative of GEBA is active on determining the WGS of the type strains of all known species (more than 11,000 genomes), covering a small but significant part of this gap. Prokaryote taxonomy is coming to a stage of profound reflection on its principles and its relevance and how to bring about its second Renaissance pushed for by the technological progress that we are witnessing.

A RENAISSANCE IN PROKARYOTE TAXONOMY?

Some microbial taxonomists (including the authors) claim that a Renaissance lies ahead of us whereas others point to the growing gap with the unnamed items in the databases. In European cultural history, the Renaissance (French for "rebirth") period, from the 14th and the 17th century, is characterized by an unprecedented flowering of arts, sciences, philosophy, and literature. This period changed life in Europe drastically. The Renaissance is expected to come from the introduction of HTS technology and the development of a new bioinformatics toolbox. The realization of the old dream of our predecessors to build a genome based prokaryote taxonomy (GPT) is coming true. Although the coming of age of GPT is welcomed by many (Hugenholtz et al., 2016; Thompson et al. 2013, 2015), others prefer to stick close to the practice of orthodox PT. A few prokaryotic taxonomists are even considering our discipline as their own hunting field.

HURDLES UNDERWAY

The development of GPT has to cope with three main hurdles. The *first hurdle* is that in current practice of PT, the species descriptions is based on a cutoff of 70% in DDH (DNA-DNA hybridization), which leads to an underspeciation of prokaryotes. The species often comprise assemblages of ecologically and genomically distinct populations. The generally used 16S rRNA gene on the other hand lacks resolution at the species level, even at the 98.7% level. Universal cutoff levels to delineate species do not make much ecological sense since speciation is considered as a dynamic process leading to sister taxa as separate gene flow units (Shapiro and Polz 2014). Moreover, in orthodox PT, a species description requires differentiating phenotypic features from the neighboring species. It becomes also clear that the cornerstone of taxonomy, the prokaryote species, needs updated revision as the last reevaluation of the species definition already dates back to 2005 (Coenye et al., 2005; Gevers et al., 2005). The *second hurdle is that* the description of microbial diversity and giving names comprise an ever-growing canon, translated in increasing numbers of rules (Konstantinidis and Rossellò Mòra, 2015; Rosselló-Móra and Amann, 2015; Tindall et al., 2010). This resulted in a growing gap between taxonomy and microbial ecology, between formal and informal taxonomies. Hugenholtz et al. (2016) refer in this respect to the increasing amounts of recognized microbial dark matter.

The question arises whether we really do need names for taxa? Most microbial ecologists mainly use numbers and codes that replace names, and many do not care at all. Numbers/Codes and conventional binomial names can perfectly coexist and operate together (Casiraghi et al., 2016) and that is exactly the ongoing development. A reasonable number of names are useful and needed. But the rules for giving names should be simplified instead of being complexified. The *third hurdle* is the lack of any open discussion. We are no longer convinced of the scientific founding of prokaryote PT as it is still practiced today, in an environment where metagenomics and WGS becomes standard practice. An open discussion will contribute to the construction of a more performing universal GPT in the coming years.

The current canon of orthodox PT as it is practiced today is impeding progress both in the description of new biodiversity and in the development of taxonomy as a scientific discipline. A recent illustration is found in the Gaget et al. (2015a,b) case: after a comprehensive taxonomic study of the cyanobacterium *Planktothrix*, the authors proposed three new species and a practical solution to an old problem of cyanobacterial taxonomy where botanical and bacterial codes are in conflict (Gaget et al., 2015b). This publication is followed by a prompt comment (Oren, 2015) stating that the proposal does little to solve the problems but instead it further adds to the confusing state of cyanobacterial nomenclature. The answer of the authors (Gaget et al., 2015a,b) was very clear and deserves all our respect and attention: "…we believe that it is of utmost importance in this modern age of science to single out as soon as possible, well described cyanobacterial taxa, represented by axenic type strains under the ICNP, for ready recognition of reliable material for genome sequencing projects, metagenomic studies, or any other research relying on such strains as reference material…" (Gaget et al., 2015a).

TOWARD A NEW "GOLD STANDARD"

The "gold standard" for prokaryote species delineation is still DDH, in spite of the demonstration that techniques, such as MLSA, Average Amino Acid Identity (AAI), Average Nucleotide identity (ANI), and Genome-to-Genome Distance (GGD) are portable and have greater discriminatory power. Vandamme and Peeters (2014) stated that "DDH had been historically introduced to approach WGS derived information as closely as possible and now that we have direct access to WGS information, we want it to mimic the results obtained through (physicochemical) DDH experiments". Their statement points to *the paradox of keeping DDH as a standard, attempting to translate the old 70% DDH species threshold into modern WGS based thresholds, even though the information derived from the latter techniques is superior to DDH.* Recently, Kim et al. (2014) confirmed the 95%–96% ANI and 98.65% 16S rRNA values for a conventional bacterial species. But the real paradox is that keeping this circular reasoning (i.e., finding back what we already defined) will not allow great progress on the genomic definition of the bacterial species.

WGS IS THE BASIC UNIT FOR GENOMIC PROKARYOTE TAXONOMY (GPT)

These hurdles refrain many microbiologists from formally describing new microbial species. The challenge is to keep up with progress in environmental and evolutionary microbiology and with the needs of clinical microbiologists and epidemiologists. PT is just too slow and needs to be revisited and rethought. GPT will take into account the tsunami of novel insights of genomics and other "omic" sciences (Kämpfer and Glaeser, 2012). The Renaissance comes from the acceptance of the WGS as the basic unit of GPT, thus transforming prokaryote taxonomy from a 16S rRNA based into a genome based one.

The framework of WGS allows identification of sequence clusters at high genotypic resolution based on variation in protein coding genes distributed across the genomes. Analyses of environmental metagenomes and the microbiomes have shown that microbial communities consist of genotypic clusters of closely related organisms. These clusters seem to display cohesive environmental associations and dynamics that differentiate them from other such clusters coexisting in the same samples (Shapiro and Polz, 2014). In face of such new concepts, there is a mounting uneasiness with the definition of the microbial species itself. Evolving microbial genomes show a rapid and highly variable flux of genes, suggesting that extensive gene loss and horizontal gene transfer are the dominant evolutionary processes (Puigbò et al., 2014). GPT will solve the problem of the frequent observation that even closely related genomes can have high gene content variation that gives rise to phenotypic variation. GPT is adjusting to HTS, addressing the needs of its users in microbial ecology and clinical microbiology, in a new paradigm of open access (Beiko, 2015). GPT builds further on the already established genomic tools and calculators, for example, GGD, AAI, ANI, Karlin genomic signature, supertree analysis, codon usage bias, metabolic pathway content, core-genome analysis, pan genome family trees, in silico proteome analysis, genotype-to-phenotype derived metabolic features. GPT is expected to provide a predictive operational framework for reliable automated and openly available identification and classification (Thompson et al., 2015). The Integrated Microbial Genome System (IMG) (Chen et al., 2016) is a great leap forward.

High-throughput genome sequence technologies (HTS, e.g. Illumina) constituted a revolution in microbial diversity studies. Through the process of binning of contigs or scaffolds derived from the same strain, complete genomes can be reconstructed. Recent studies have obtained dozens of new metagenome-assembled genomes from complex environmental samples (Almstrand et al., 2016; Haroon et al., 2016; Hugerth et al., 2015; Pinto et al., 2016). The abundance of these genomes across different environments and their metabolic and ecofunctional potential can now be inferred from metagenomics. A post genomic development is constituted by the Unipept application (Mesuere et al., 2012, 2015), which offers a fast and easy to use pipeline by indexing tryptic peptides obtained from WGS data that are turned into taxonomic groupings. GPT parallels these developments.

GENOMIC PROKARYOTE TAXONOMY (GPT)

It becomes clear that GPT will be founded on: (1) *WGS as basic unit*. From the WGS of an isolate, the phylogenomic position is inferred and its genomic similarity (ANI/AAI, GGD and phenotype/metabolism) determined. The type can be a culture, DNA or a WGS. The principle that the one who discovers a new genome has the right to name it according to a simplified procedure would be a great progress. (2) *A new prokaryote species concept, to be agreed upon*, taking into account recent genomic concepts, for example, the gene flow unit OTU and CTU. The adaptive power of the prokaryotes in changing ecological conditions reveals ecological variants, for example, ecotypes or populations that occupy the same inches and that can be distinguished by sequencing based techniques (Cohan, 2002; Koeppel et al., 2003). At the same time, there is a need to reconcile phylogeny and taxonomy and to tackle the polyphyletic taxa and the unevenness of rank assignments (Hugenholtz et al., 2016), and (3) *an efficient and simplified procedures for GPT operating in a bottom-up approach from WGS > clade > species > genus>* ….GPT should englobe and maintain the existing information, integrating it with new data on DNA, genomes, isolates/strains (cultured and uncultured), also Candidatus as well as reconstructed genomes from metagenomes (Hugerth et al., 2015). But we are not there yet (Garrity, 2016).

CONCLUSIONS

The aim of prokaryote taxonomy is to put order in the world of microorganisms. It became a full-fledged scientific discipline but is at present unable to cope with the estimated prokaryotic diversity. The Renaissance of GPT started based on WGS but has still to address some hurdles. GPT is underway and real progress may be expected from collaborations with ecologists, bioinformaticians, and clinical microbiologists.

International dimension: the established networks of Biological Resource Centers and Culture collections and their individual members will be part of the Renaissance if they are willing and able to reengineer in order to adapt to the ongoing technological changes and to the future needs of its public, medical, and industrial users.

Ethical dimension: public funders, publishers, editors, and authors should adhere to and subscribe the principle that sequence data and biological cultures and/or material should be publicly available.

ACKNOWLEDGMENTS

We thank Research Foundation Flanders (FWO), BELSPO (Belgian Science Policy Office) CAPES, CNPq, and FAPERJ for funding.

REFERENCES

Almstrand, R., Pinto, A.J., Figueroa, L.A., Sharp, J.O., 2016. Draft genome sequence of a novel *Desulfobacteraceae* member from a sulfate-reducing bioreactor metagenome. Genome Announc. 4 (1), e01540-15.

Altschul, S.F., Gish, W., Miller, W., Myers, E.W., Lipman, D.J., 1990. Basic local alignment search tool. J. Mol. Biol. 215, 403–410.

Altschul, S.F., Madden, T.L., Schäffer, A.A., Zhang, J., Zhang, Z., Miller, W., Lipman, D.J., 1997. Gapped BLAST and PSI-BLAST: a new generation of protein database search programs. Nucleic Acids Res. 25 (17), 3389–3402.

Auch, A.F., von Jan, M., Klenk, H.-P., Göker, M., 2010. Digital DNA-DNA hybridization for microbial species delineation by means of genome-to-genome sequence comparison. Stand. Genom. Sci. 28, 117–134.

Beiko, R.G., 2015. Microbial malaise: how can we classify the microbiome? Trends Microbiol. 23, 671–679.

Bochner, B.R., Gadzinsky, P., Panomitros, E., 2001. Phenotype microarrays for high throughput phenotypic testing and assay gene function. Genome Res. 11, 1241–1255.

Casiraghi, M., Galimberti, A., Sandionigi, A., Bruno, A., Labra, M., 2016. Life without names. Evol. Biol. 43, 582.

Chen, I.M.A., Markowitz, V.M., Palaniappan, K., Szeto, E., Chu, K., Huang, J., Ratner, A., Pillay, M., Hadjithomas, M., Huntemann, M., Mikhailova, N., 2016. Supporting community annotation and user collaboration in the integrated microbial genomes (IMG) system. BMC Genomics 17 (1), 307–324.

Chun, J., Lee, J.H., Jung, Y., Kim, M., Kim, S., Kim, B.K., Lim, Y.W., 2007. EzTaxon: a web-based tool for the identification of prokaryotes based on 16S ribosomal RNA gene sequences. Int. J. Syst. Evol. Microbiol. 57, 2259–2261.

Clark, A.E., Kaleta, E.J., Arora, A., 2013. WolkDM. Matrix-assisted laser desorption/ionization time-of-flight mass spectrometry: a fundamental shift in the routine practice of clinical microbiology. Clin. Microbiol. Rev. 26, 547–603.

Coenye, T., Gevers, D., Van de Peer, Y., Vandamme, P., Swings, J., 2005. Towards a prokaryotic taxonomy. FEMS Microbiol. Rev. 29, 147–167.

Cohan, F.M., 2002. What are bacterial species? Ann. Rev. Microbiol. 56, 457–487.

Daubin, V., Szöllösi, G.J., 2016. Horizontal gene transfer and the history of life. Cold Spring Harb. Perspect. Biol. 8 (4), a018036.

De Smet, W., De Loof, K., De Vos, P., Dawyndt, P., De Baets, B., 2013. Filtering and ranking techniques for automated selection of high-quality 16S rRNA gene sequences. Syst. Appl. Microbiol. 36, 549–559.

De Vos, P., Trüper, H.G., 2000. Judicial commission of the International Committee on Systematic Bacteriology. Int. J. Syst. Evol. Microbiol. 50, 2239–2244.

Dutilh, B.E., Backus, L., Edwards, R.A., Wels, M., Bayjanov, J.R., Van Hijum, S.A., 2013. Explaining microbial phenotypes on a genomic scale: GWAS for microbes. Brief Funct. Genomics 12, 366–380.

Fleischmann, R., Adams, M., White, O., Cleyton, R.A., Kirkness, E.F., Kerlavage, A.R., et al., 1995. Whole-genome random sequencing and assembly of *Haemophilus influenzae* Rd. Science 269, 496–512.

Gaget, V., Welker, M., Rippka, R., Tandeau de Marsac, N., 2015a. Response to: "Comments on: "A polyphasic approach leading to the revision of the genus Planktothrix (Cyanobacteria) and its type species, *P. agardhii*, and proposal for integrating the emended valid botanical taxa, as well as three new species, *Planktothrix paucivesiculata* sp. nov.ICNP, *Planktothrix tepida* sp. nov.ICNP, and *Planktothrix serta* sp. nov. ICNP, as genus and species names with nomenclature standing under the ICNP," by V. Gaget, M. Welker, R. Rippka, and N. Tandeau de Marsac, Syst Appl Microbiol (2015), http://dx.doi.org/10.1016/j.syapm.2015.02.004", by A. Oren [Syst. Appl. Microbiol. (2015), doi:10.1016/j.syapm.2015.03.002]. Syst. Appl. Microbiol. 38(5), 368–370.

Gaget, V., Welker, M., Rippka, R., Tandeau de Marsac, N., 2015b. A polyphasic approach leading to the revision of the genus Planktothrix (Cyanobacteria) and its type species, P. agardhii, and proposal for integrating the emended valid botanical taxa, as well as three new species, Planktothrix paucivesiculata sp. nov.ICNP, Planktothrix tepida sp. nov.ICNP, and Planktothrix serta sp. nov.ICNP, as genus and species names with nomenclature standing under the ICNP. Syst. Appl. Microbiol. 38(3), 141–158.

Garrity, G.M., 2016. A genomics driven taxonomy of Bacteria and Archaea: are we there yet? J. Clin. Microbiol. 54, 1956–1963.

Garrity, G. M., Bell, J. A., Lilburn, T. G., 2004. Taxonomic outline of the procaryotes. In: Bergey's Manual Of Systematic Bacteriology, second ed. release 5.0.Springer, New York, pp. 1–399.

Gevers, D., Cohan, F.M., Lawrence, J.G., Lawrence, J.G., Spratt, B.G., Coenye, T., Feil, E.J., et al., 2005. Re-evaluating prokaryotic species. Nat. Rev. Microbiol. 3, 733–739.

Goris, J., Konstantinidis, K.T., Klappenbach, J.A., Coeneye, T., Vandamme, P.H., Tiedje, J.M., 2007. DNA-DNA hybridization values and their relationship to whole-genome sequence similarities. Int. J. Syst. Evol. Microbiol. 57, 81–91.

Haroon, M.F., Thompson, L.R., Stingl, U., 2016. Draft genome sequence of uncultured SAR324 bacterium lautmerah10, binned from a red sea metagenome. Genome Announc. 4, e01711–e01715.

Hug, L.A., Baker, B.J., Anantharman, K., Brown, C.T., Probst, J., Castelle, J.C., et al., 2016. A new tree of life. Nat. Microbiol. 1, 16048.

Hugenholtz, P., Skarshewski, A., Parks, D.H., 2016. Genome-based microbial taxonomy coming of age. Cold Spring Harb. Perspect. Biol. 8 (6), a018085.

Hugerth, L.W., Larsson, J., Alneberg, J., Lindh, M.V., Legrand, C., Pinhassi, J., et al., 2015. Metagenome-assembled genomes uncover a global brackish microbiome. Genome Biol. 16, 279.

Janda, J.M., 2016. Taxonomic update on proposed nomenclature and classification changes for bacteria of medical importance, 2015. Diagn. Microbiol. Infect. Dis. 83 (1), 82–88.

Kämpfer, P., Glaeser, S.P., 2012. Prokaryotic taxonomy in the sequencing era—the polyphasic approach revisited. Environ. Microbiol. 14, 291–317.

Kim, M., Oh, S.-C., Park, S.-C., Chun, J., 2014. towards a taxonomic coherence between average nucleotide identity and 16S rRNA gene sequence similarity for species demarcation of prokaryotes. Int. J. Syst. Evol. Microbiol. 64, 346–351.

Koeppel, A.F., Wertheim, J.O., Barone, L., Gentile, N., Krizanc, D., Cohan, F.M., 2003. Speedy speciation in a bacterial microcosm: a new species can arise as adaptations within a species. ISME J. 7, 1080–1091.

Konstantinidis, K.T., Rossellò Mòra, R., 2015. Classifying the uncultivated microbial majority: a place for metagenomic data in the Candidatus proposal. Syst. Appl. Microbiol. 18, 223–230.

Konstantinidis, K.T., Ramette, A., Tiedje, J.M., 2006. The bacterial species definition in the genomic era. Philos. Trans. R. Soc. Lond. B Biol. Sci. 361, 1929–1940.

Koonin, E.V., 2016. Horizontal gene transfer: essentiality and evolvability in prokaryotes, and roles in evolutionary transitions. F1000 Res. 5doi: 10.12688/f1000research.8737.1.

Lagier, J.-C., Edouard, S., Pagnier, I., Mediannikov, O., Dancourt, M., Raoult, D., 2015a. Current and past strategies for bacterial culture in clinical microbiology. Clin. Microbiol. Rev. 28, 208–236.

Lagier, J.-C., Hugon, P., Kehlaifia, S., Fournier, P.-E., La Scola, B., Raoult, D., 2015b. The rebirth of culture in microbiology through the example of culturomics to study human gut microbiota. Clin. Microbiol. Rev. 28, 237–264.

Lapage, S.P., Sneath, P.H.A., Lessel, E.F., Skerman, V.B.D., Seeliger, H.P., Clark, W.A., 1992. International Code of Nomenclature of Bacteria. ASM Press, Washington DC.

Ludwig, W., Klenk, H.-P., 2005. Overview: a phylogenetic backbone and taxonomic frame for Procaryotic Systematics. Brenner, D.J., Krieg, N.R., Staley, T.S., Garrity, G.M. (Eds.), Bergeys Manual of Systematic Bacteriology, vol. 2, second ed Springer, New York, pp. 49–69.

Mesuere, B., Devreese, B., Debyser, G., Aerts, M., 2012. Unipept: tryptic peptide-based biodiversity analysis of metaproteome samples. J. Proteome Res. 11, 5773–5780.

Mesuere, B., Debyser, G., Aerts, M., Devreese, B., Vandamme, P., Dawyndt, P., 2015. The unipept metaproteomics analysis pipeline. Proteomics 15 (8), 1437–1442.

Murray, R.G.E., Stackebrandt, E., 1995. Taxonomic note: implementation of the provisional status Candidatus for incompletely describe procaryotes. Int. J. Syst. Evol. Microbiol. 45, 186–187.

Oren, A., 2015. Response to: "Comments on: "A polyphasic approach leading to the revision of the genus Planktothrix (Cyanobacteria) and its type species, P. agardhii, and proposal for integrating the emended valid botanical taxa, as well as three new species, Planktothrix paucivesiculata sp. nov.ICNP, Planktothrix tepida sp. nov.ICNP, and Planktothrix serta sp. nov.ICNP, as genus and species names with nomenclature standing under the ICNP," by V. Gaget, M. Welker, R. Rippka, and N. Tandeau de Marsac, Syst. Appl. Microbiol. (2015), http://dx.doi.org/10.1016/j.syapm.2015.02.004". Syst. Appl. Microbiol. 38,159–160.

Parker, C.T., Tindall, B.J., Garrity, G.M., 2015. International code of nomenclature of Prokaryotes. Int. J. Syst. Evol. Microbiol., Published on line.

Pei, A.Y., Oberdorf, W.E., Nossa, C.W., Argawal, A., Chokshi, P., Gerz, E.A., et al., 2010. Diversity in 16S rRNA genes within individual prokaryotic genomes. Appl. Environ. Microbiol. 76 (15), 5333.

Pinto, A.J., Sharp, J.O., Yoder, M.J., Almstrand, R., 2016. Draft genome sequences of two novel Acidimicrobiaceae members from an acid mine drainage biofilm metagenome. Genome Announc. 4 (1), e01563-15.

Popa, O., Dagan, T., 2011. Trends and barriers to lateral gene transfer in prokaryotes. Curr. Opin. Microbiol. 14, 615–623.

Puigbò, P., Lobkovsky, A.E., Kristensen, D.M., Wolf, Y.I., Koonin, E.V., 2014. Genomes in turmoil:quantification of genome dynamics in prokaryote supergenomes. BMC Biol. 12, 66.

Rosselló-Móra, R., Amann, R., 2001. The species concept for prokaryotes. FEMS Microbiol. Rev. 25 (1), 39–67.

Rosselló-Móra, R., Amann, R., 2015. Past and future species definitions for Bacteria and Archaea. Syst. Appl. Microbiol. 38, 209–216.

Sasser, M., 1990. Identification of bacteria by gas chromatography of cellular fatty acids. Technical note 101. 1990 revised in 2001. Available from: http://www.microbialid.com/PDF/TechNote_101.pdf

Shapiro, B.J., Polz, M.F., 2014. Ordering microbial diversity into ecologically and genetically cohesive units. Trends Microbiol. 22, 235–247.

Skerman, V.D.B., McGowan, V., Sneath, P.H.A., 1980. Approved lists of bacterial names. Int. J. Syst. Bacteriol. 30, 225–420.

Skerman, V.D.B., McGowan, V., Sneath, P.H.A., 1989. Approved List of Bacterial Names (Amended). ASM press, Washington DC.

Sneath, P.H.A., Sokal, R.R., 1973. Principles and Practice of Numerical Taxonomy. WH Freeman and Company, San Francisco.

Stackebrandt, E, Elbers, J., 2006. Taxonomic parameter revisited: tarnished gold standards. Microbiol. Today. 33, 152–1555.

Stackebrandt, E., Frederiksen, W., Garrity, G.M., Grimont, P.A.D., Kämpfer, P., Maiden, M.C.J., et al., 2002. Report of the ad hoc committee for the re-evaluation of the species definition in bacteriology. Int. J. Syst. Evol. Microbiol. 52, 1043–1047.

Stewart, E.J., 2012. Growing unculturable bacteria. J. Bact. 194, 4151–4160.

Thompson, C.C., Chimetto, L., Edwards, R.A., Swings, J., Stackebrandt, E., Thompson, F., 2013. Microbial genomic taxonomy. BMC Genomics 14, 913–920.

Thompson, C.C., Amaral, G.R., Campeão, M., Edwards, R.A., Polz, M.F., Dutilh, B.E., et al., 2015. Microbial taxonomy in the post-genomic era: rebuilding from scratch? Arch. Microbiol. 197, 359–370.

Tindall, B.J., Rosselló-Móra, R., Busse, H.-J., Ludwig, J., Kämpfer, P., 2010. Notes on thecharacterization of prokaryote strains for taxonomic purposes. Int. J. Syst. Evol. Microbiol. 60, 249–266.

Vandamme, P., Peeters, C., 2014. Time to revisit polyphasic taxonomy. Antonie Van Leeuwenhoek 106, 57–65.

Varghese, N.J., Mukherjee, S., Ivanova, N., Konstantinidis, K.T., Mavrommatis, K., et al., 2015. Microbial species delineation using whole genome sequences. Nucleic Acids Res. 43 (14), 6761–6771.

Verslyppe, B., De Smet, W., De Baets, B., De Vos, P., Dawyndt, P., 2011. Make Histri: reconstructing the exchange history of bacterial and archael type strains. Syst. Appl. Microbiol. 37, 328–336.

Wayne, L.G., Brenner, D.J., Colwell, R.R., Grimont, P.A.D., Kandler, O., Krichevsky, M.I., Moore, L.H., et al., 1987. Report of the ad hoc committee on reconciliation of approaches to bacterial systematics. Int. J. Syst. Evol. Microbiol. 37, 463–464.

Woese, C.R., Kandler, O., Wheelis, M.L., 1990. Towards a natural system of organisms: proposal for the domains Archaea, Bacteria and Eucarya. PNAS 87, 4576–4579.

Yarza, P., Ludwig, W., Euzéby, J., Amann, R., Schleifer, K.-H., Glöckner, F.O., et al., 2010. Update of the all-species tree project based on 16S and 23S rRNA sequence analysis. Syst. Appl. Microbiol. 33, 291–299.

Chapter 3

Bioactive Actinomycetes: Reaching Rarity Through Sound Understanding of Selective Culture and Molecular Diversity

İpek Kurtböke
GeneCology Research Centre and Faculty of Science, Health, Education and Engineering, University of the Sunshine Coast, Maroochydore DC, QLD 4558, Australia

INTRODUCTION

Actinomycetes have been an inexhaustible source for the production of bioactive compounds to be used in medicine and agriculture ever since their "majesty" and "versatility" was revealed in the 1940s (Bérdy, 2005, 2012; Demain, 1988, 2014; Kurtböke, 2012). The discovery, development, and exploitation of antibiotics was one of the most significant advances in medicine in the 20th century and in a "golden era" lasting from the 1940s to the late 1960s, the members of the order have provided mankind with a wide range of structurally diverse and effective agents for the treatment of microbial infections (Demain, 1988; Hopwood, 2007). Even after the "golden era" was over and rediscovery of the known compounds dominated the scene (Baltz, 2006), the fascinating diversity of bioactive compounds from actinomycetes, ranging from immunosuppressors, immunostimulants to enzyme inhibitors, ensured the continuation of biodiscovery research from this group of bacteria. Even for the most in-depth studied genus *Streptomyces,* mathematical modeling studies have predicted that more than 100,000 different and previously unknown antibiotics could be discovered from this genus (Watve et al., 2001). As a result, isolation strategies targeting new bioactive isolates for therapeutic and industrial applications adapted the use of objective approaches to detect and isolate previously unknown or rare actinomycete diversity in different ecosystems (Goodfellow, 2010; Goodfellow and Fiedler, 2010; Kurtböke, 2003, 2012, 2016; Tiwari and Gupta, 2012, 2013). Streptomycetes with different properties, for example, halophilic, osmophilic, acidophilic, and alkaliphilic were also included

Microbial Resources. http://dx.doi.org/10.1016/B978-0-12-804765-1.00003-5

in the search as a previously untapped source for discovery of new bioactive compounds (Goodfellow and Williams, 1986). Actinomycetes inhabiting marine and extreme environments have also proved to be a novel source of bioactive compounds due to their unique metabolic and physiological capabilities that not only ensure their survival in extreme habitats, but also offer the potential for the production of metabolites that are not observed in their terrestrial counterparts (Bull et al., 2005; Fenical, 1993; Fenical and Jensen, 2006; Kurtböke et al., 2015; Zotchev et al., 2016; Romano et al., 2016). In the light of the earlier presented information, this chapter will review past and current approaches employed in the search for bioactive members of actinomycetes and will emphasize the value of objective approaches stemming from sound understanding of the morphology, physiology, taxonomy, and metabolism of these microorganisms.

Order Actinomycetales: Growth Morphology, Differentiation, and Current Classification

Intraclass relatedness of the class actinobacteria currently reveals the presence of nine orders based on 16s rRNA gene sequence comparison (refer to LPSN Bacterio.net, accessed 30-01-2017 at http://www.bacterio.net/-classifphyla.html). Each order has specific traits and the order *Actinomycetales* is known for industrially important compound production (e.g., antibiotics and enzymes) (Demain, 1988; Hopwood, 2007). In this chapter the generic name "actinomycetes" will be used as the chapter content is specifically related to the members of the order *Actinomycetales*. The term "actinomycetes" will also include the newly created orders *Frankiales, Geodermatophiales, Kineosporiales and Micrococcales* (refer to LPSN Bacterio.net, accessed 30-01-2017 at http://www.bacterio.net/-classifphyla.html) as they were part of the order *Actinomycetales* until recently.

Members of the actinomycetes constitute a diverse group of Gram positive bacteria, with most of its genera holding the ability to form branching hyphae at some stage of their development (Cross, 1982; Lechevalier and Lechevalier, 1985). However, advances in the molecular studies in the 1980s revealed that the possession of branched hyphae should not automatically place bacterial genera within this class (e.g., *Thermoactinomyces*), nor should the inability of a species to form branching filaments (e.g., *Arthrobacter*, *Cellulomonas*, and *Rothia*) necessarily exclude it from this class (Stackebrandt and Woese, 1981).

In earlier studies members of actinomycetes were placed into two broad groups of unequal size (Lechevalier and Lechevalier, 1985). The first group composed of mainly fermentative organisms that were most often associated with the natural cavities of man and animals. On the other hand, the second much larger group composed of oxidative organisms for which soil is mostly the basic reservoir and from where they were disseminated into nonterrigenous environments (Cross, 1982; Kurtböke et al., 2017; Gerber, 1979). Species within the first group do not produce distinctive morphological features other than a branching mycelium. Mycelial development may be transitory, occurring only during active growth and fragmenting immediately when the growth rate slows; in extreme cases appearing as branched single cells; or it may occur under special growth conditions. Their cell walls do not contain diaminopimelic acid (DAP). Whereas, in contrast, the organisms in the second group are morphologically more intricate than the first group, and contain DAP in their peptidoglycan. They mainly develop in the mycelial state and reproduce through formation of unicellular spores, differentiated either singly or in chains at the tips of the hyphae, such as *Streptomyces* and *Actinomadura*. Some of them are sporangia-forming members with flagellated sporangiospores, for example, *Actinoplanes*, while micromonosporae form single conidia on their substrate mycelium. *Dermatophilus* and *Geodermatophilus* form a substrate mycelium which divides both transversely and longitudinally to give a multilocular sporangium. This large and complex second group contain many genera distinguished by a combination of structural and chemical properties (Lechevalier and Lechevalier, 1985). The most industrially-important genera, such as *Streptomyces*, *Micromonospora*, and *Actinoplanes* originate from this second group (Bérdy, 2005, 2012; Kurtböke, 2012).

After the discovery of the first antibiotics from the members of the order, an intensive search for new and bioactive species started throughout the world. Thousands of species were isolated using variety of media, which were relatively nonselective and the resulting isolates were screened for bioactivity. Mainly large and obvious streptomycete colonies were selected resulting in the need of an international effort to redescribe the 458 strains of streptomycetes representing the "golden era" of biodiscovery. Between 1962 and 1970, 42 research groups from 18 different nations joined the *International Streptomyces Project* (ISP) and the strains were called the ISP-strains and media used to characterize them were called ISP-media. Project participants used the following key criteria to define the clusters within this genus (1) development of vegetative and aerial mycelia, (2) mass color of mature sporulation on aerial mycelium, (3) morphology of spore chains, (4) color of vegetative mycelia, and (5) diffusible pigment into the growth media (Higashide, 1988).

Starting from the 1980s the application of new and reliable biochemical, chemical, and molecular biological techniques revolutionized actinomycete systematics (Goodfellow and Minnikin, 1985; Kroppenstedt, 1985; Stackebrandt and Schleifer, 1984; Stackebrandt and Woese, 1981). At the same time numerical databases containing extensive information on the biochemical, nutritional, and physiological requirements of the actinomycete taxa were developed and statistically-validated probability matrices provided sound platforms to define streptomycete clusters (Williams et al., 1981, 1983a,b, 1985). All these emerging identification schemes later resulted in the adaptation of polyphasic approaches that combine various traits of streptomycetes deriving from numerical taxonomy, chemotaxonomy, and molecular methods. Most recently

TABLE 3.1 Current Classification of the Order Actinomycetales[a]

Phylum Actinobacteria
- Class Actinobacteria
  - Subclass Actinobacteridae
    - Order Actinomycetales
      1. Suborder Actinomycineae
         - Family Actinomycetaceae
      2. Suborder Actinopolysporineae
         - Family Actinopolysporaceae
      3. Suborder Catenulisporineae
         - Family Actinospicaceae
         - Family Catenulisporaceae
      4. Suborder Corynebacterineae
         - Family Corynebacteriaceae
         - Family Dietziaceae
         - Family Mycobacteriaceae
         - Family Nocardiaceae
         - Family Segniliparaceae
         - Family Tsukamurellaceae
      5. Suborder Glycomycineae
         - Family Glycomycetaceae
      6. Suborder Jiangellineae
         - Family Jiangellaceae
      7. Suborder Micromonosporineae
         - Family Micromonosporaceae
      8. Suborder Propionibacterineae
         - Family Nocardioidaceae
         - Family Propionibacteriaceae
      9. Suborder Pseudonocardineae
         - Family Pseudonocardiaceae
      10. Suborder Streptomycineae
          - Family Streptomycetaceae
      11. Suborder Streptosporangineae
          - Family Nocardiopsaceae
          - Family Streptosporangiaceae
          - Family Thermomonosporaceae

[a]*Taken from http://www.bacterio.net/-classifphyla.html#Actinobacteria*

advancing genomic methods were also included in the polyphasic approaches as well (Kämpfer, 2014; Kämpfer and Glaeser, 2012; Stackebrandt, 2011). Current holistic understanding of sound actinomycete classification grew on the "foundations of the findings of early workers supported by the scaffolding provided by modern technology" (Cross and Alderson, 1988).

Interestingly, although 11 different suborders and 22 families currently exist (Table 3.1), the growth morphology of the members of order *Actinomycetales* still persistently fall into four distinctive structures (Table 3.2) (Williams and

TABLE 3.2 Distinctive Growth Morphologies of Actinomycetes

Group 1: No aerial mycelium; substrate mycelium soon fragmenting into various-sized rod-coccoid elements; colony texture is soft, bacterial like	*Example genera: Agromyces, Oerskovia, Rhodococcus*
Group 2: No aerial mycelium; substrate mycelium not fragmenting; colony texture is tough	*Example genera: Micromonospora; Actinoplanes, Dactylosporangium*
Group 3: Dry, powdery-cottony aerial mycelium formed; substrate mycelium fragmenting into various sized rod-coccoid elements; colony texture is moderately soft	*Example genera: Nocardia, Nocardioides*
Group 4: Dry, powdery-cottony aerial mycelium formed; substrate mycelium not fragmenting; colony texture is tough	*Example genera: Streptomyces, Actinomadura, Thermomonospora, Microbispora, Microtetraspora, Planomonospora*

Source: From Williams, S.T., Wellington, E.M.H., 1982a. Actinomycetes. In: Page, A.L., Miller, R.H., Keeney, D.R. (Eds.), Methods of Soil Analysis, Part 2, Chemical, Microbiological Properties, second ed., pp. 969–987.

Wellington, 1982a). Descriptions of these distinctive structures are illustrated in Table 3.2 and they still facilitate rapid grouping of the isolates at the genus/family level. In the early years, morphological examinations using traditional bacteriological stained smear methods resulted in the generation of misleading or incomplete information, however, with the use of the "inclined cover-slip" technique, undisturbed or fragmenting substrate mycelium, aerial mycelial, and spore chain types became easily distinguishable (Cross, 1989).

Color changes during colony growth, maturation, and spore mass formation have also provided some guidance since the late 1950s. The Tresner-Backus color series, adopted by the participants in the 1962 Montreal Workshop (Tresner and Backus, 1963) was used extensively for color determinations in the early days of grouping of the isolates. However, as more new actinomycete genera and species were described, use of only the color criterion became unfeasible. Nevertheless, for some genera, even in the presence of current advance molecular identification techniques, distinctive colony colors still facilitate their rapid recognition on the isolation plates. Examples include the members of genus *Saccharomonospora* (Fig. 3.1) that initially form a white aerial mycelium, which turns gray-green to dark green on continued sporulation, a distinguishable characteristic of this genus (Goodfellow and Cross, 1983). Such a distinguishable characteristic can be used to select them on isolation plates and differentiate them from other white *Thermoactinomyces* colonies if coinhabited substrates are studied. However, molecular sequencing

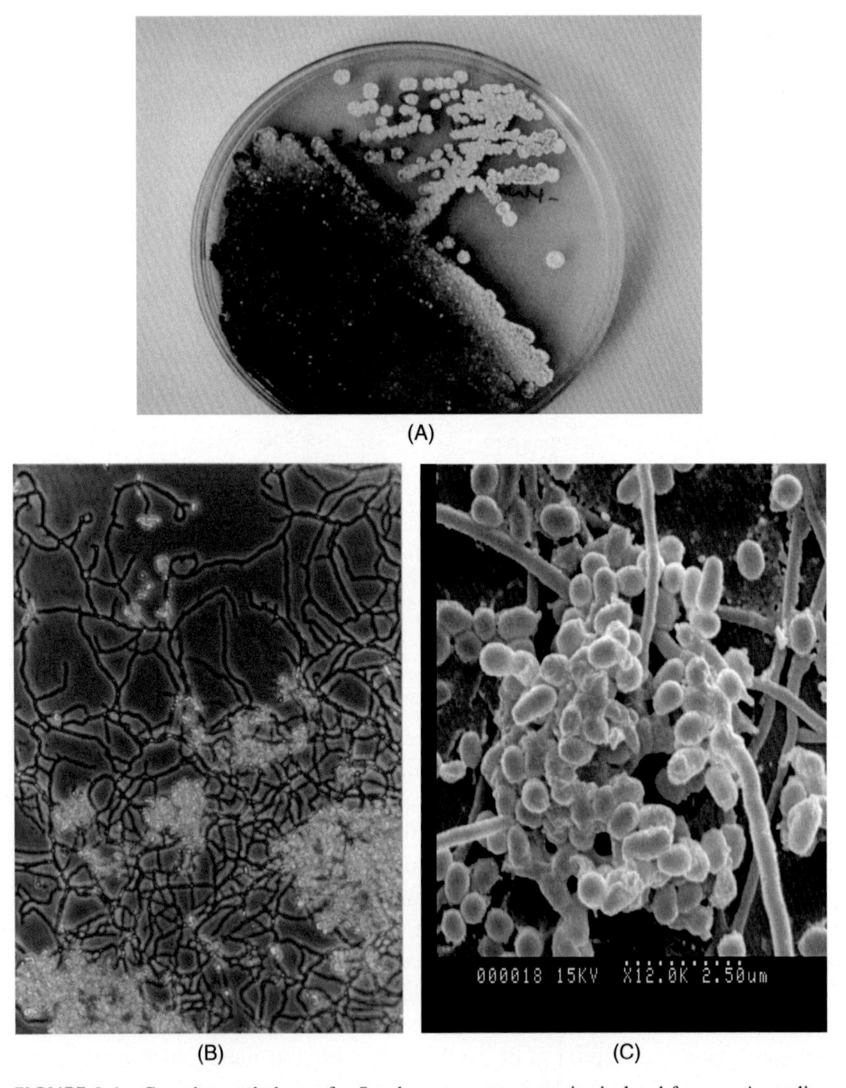

FIGURE 3.1 Growth morphology of a *Saccharomonospora* species isolated from an Australian bagasse sample (Kurtböke, 1996; Kurtböke et al., 1999). (A) on Tryptone soy agar (TSA) plate, (B) under phase contrast microscope (40× objective) using the "inclined cover-slip" technique, and (C) under the scanning electron microscopy (SEM).

complementary data is also required when *Micropolyspora* species exhibiting similar green colony pigmentation might also be present in the substrate under study (Goodfellow and Cross, 1983). These observations again highlight the importance of polyphasic approaches for the reliable identification of genera and species (Stackebrandt, 2011; Kämpfer, 2014).

Employment of Selective Pressures to Isolate Rare or Previously Nonencountered Members of the Known Taxa

Selectivity Based on Taxon-Specific Search

In the early years of biodiscovery genera reported to be abundant in terrestrial environments, such as *Streptomyces* and *Micromonospora* were targeted for isolations as they proved to be the prolific sources of antibiotics, enzymes, and enzyme inhibitors. The number of antibiotic compounds reported were 3541 from streptomycetes and 269 from micromonosporae by the mid-1980s (Bérdy, 1974, 1984; Goodfellow and Williams, 1986; Nisbet, 1982) (see Chapter 7 for the key antibiotics used in medicine originating from these compounds). Isolations then were mostly conducted without introducing selectivity, such as the use of conventional soil dilution techniques where aliquots were inoculated onto nonselective media to provide conditions which would allow the growth of a wider range of genera. But such moves were rarely successful as nonselective media often favored streptomycetes while grossly underestimating slower growing actinomycete taxa. None of the media were selective for actinomycetes [e.g., malt-extract-yeast-extract (Pridham et al., 1957), glycerol-arginine (Porter et al., 1960)], and growth of other bacteria on isolation plates was always a problem. This was exaggerated by the spread of motile bacteria over the medium surface. Moreover, the surface of an isolation plate was a highly competitive environment and the slow-growing actinomycete genera present in substrates in relatively low numbers were at a considerable disadvantage (Cross, 1982; Goodfellow and Williams, 1986; Wellington and Cross, 1983; Williams and Wellington, 1982a,b). The use of complex sources based on the degradative abilities of most actinomycetes, such as starch and chitin were the first examples of media to favor actinomycete growth while reducing the numbers of swarming bacteria on the surface of the media used. Examples include Starch-casein (Küster and Williams, 1964) and chitin media (Lingappa and Lockwood, 1962).

In later years, the search and discovery efforts between the mid-1970s to mid-80s revealed the antibiotic production capabilities of nonstreptomycete and nonmicromonosporae species [e.g., *Actinosynnema* (25 antibiotics), *Streptoalloteichus* (14 antibiotics), *Saccharopolyspora* (33 antibiotics), *Streptosporangium* (26 antibiotics), *Actinomadura* (51 antibiotics)] and selective isolation efforts were then directed to recover these taxa with low isolation frequency (Bérdy, 1974, 1984; Goodfellow and Williams, 1986; Nisbet, 1982). Consequently, a target-directed approach efforts concentrating on a particular genus or closely related family members was used so that specific isolation methods and culture media could be developed for the targeted genera. Research laboratories around the world developed specific expertize on the isolation of certain groups of bioactive actinomycete genera or families. Examples include *Micromonospora* (Luedemann and Brodsky, 1964, 1965), *Actinoplanes* and related genera (Couch and Bland, 1974), and *Actinomadura* (Lavrova et al., 1972;

Preobrazhenskaya et al., 1975). An alternative approach was to make isolation media more selective by adding compounds to inhibit competing bacteria including nontargeted members of actinomycetes. A variety of antimicrobials proved useful in reducing the number of competitors and revealed previously masked new and rare species and genera. Examples include isolation of new species of *Actinomadura* (Athalye et al., 1981; Preobrazhenskaya et al., 1975) and *Micromonospora* (Ivantiskaya et al., 1978) through exploitation of their resistance characteristics to a variety of antibiotics. During these years, most of the selective agents were discovered accidentally or by incorporation of different inhibitory agents or their combinations which provided a strong starting point for subsequent selective isolation protocol designs used by researchers all around the world (Table 3.3).

Preinoculation techniques were implemented to increase the number of actinomycete propagules in the sample (enrichment) or to reduce the nonactinomycete content of the sample by physical and chemical treatments without substantially reducing the number of actinomycete propagules. Examples include nutrient enrichment methods used for *Nocardia* and *Rhodococcus* species (Gavrilenko et al., 1979; Nesterenko et al., 1977, as well as the baiting (Kane, 1966; Palleroni, 1980) and rehydration (Makkar and Cross, 1982) techniques to isolate actinoplanetes, which rarely appear on conventional spread plates inoculated with soil and litter samples (Table 3.4). Such approaches proved to be useful and by mid-1980s isolation of bioactive *Actinoplanes* species resulted in discovery of 95 new antibiotic compounds (Bérdy, 1974, 1984; Goodfellow and Williams, 1986; Nisbet, 1982), including Lipiarmycin (Parenti et al., 1975), Teichomycins (Parenti et al., 1978), and Gardimycin (Parenti et al., 1976) by the Research Laboratories of the Gruppo Lepetit S.p.A., in Milan, ItaIy. Their concentrated efforts in the family *Micromonosporaceae* resulted in the discovery of a new bioactive genus *Dactylosporangium* (Thiemann et al., 1967) of this family as well as discovery of a new bioactive species of *Actinoplanes* producing distinct red pigment (Beretta, 1973).

As the aerial spores of most actinomycete genera resist desiccation and show a slightly higher resistance to wet or dry heat than the corresponding vegetative hyphae (Cross, 1982), various treatments of samples before plating were also applied, including heating of soil at 40°C for 4–16 h to isolate streptomycetes (Williams et al., 1972), and 100°C for 15 min to isolate *Actinomadura* species (Athalye et al., 1981). Nonomura and Ohara in Japan looked for new genera or rare species by relying on their experience to recognize the unusual while employing pretreatments including dry heating, specialized growth media, and long incubation times to isolate new species of *Actinomadura*, *Microbispora*, *Microtetraspora*, *Streptosporangium*, and *Thermomonospora* (Nonomura, 1969, Nonomura and Ohara 1971a,b,c,d,). Although such treatments did not completely reduce the number of associated bacteria, the numbers of Gram-negative bacteria were significantly reduced. These bacteria were known to mask growth of targeted

TABLE 3.3 Selective Agents Used for the Target-Directed Isolation of Particular Actinomycete Taxa

Selective agent	*Actinomycetales* genera selected	References
Nitrofurazone[a]	*Streptomyces*	Yoshioka (1952)
Rose Bengal[b]	*Streptomyces*	Ottow and Glathe (1968)
Streptomycin[b]	*Actinomadura*	Lavrova (1971)
Tellurite[a]	*Actinoplanes*	Willoughby (1971)
Rubomycin[b]	*Actinomadura*	Lavrova et al. (1972)
Tetracyclines[b]	*Nocardia asteroids*	Orchard and Goodfellow (1974)
Bruneomycin[a]	*Actinomadura*	Preobrazhenskaya et al. (1975)
Benzoate[b]	*Micromonospora*	Sandrak (1977)
Penicillin +NaCl	*Streptomyces*	Mackay (1977)
Gentamicin[b]	*Micromonospora*	Ivantiskaya et al. (1978)
Dihydroxymethl-furatriazine[b]	*Microtetraspora*	Tomita et al. (1980)
Nalidixic acid + penicillin + tellurite	*Rhodococcus*	Barton and Hughes (1981)
Rifampicin[c]	*Actinomadura*	Athalye et al. (1981, 1985)
Kanamycin[c]	*Thermomonospora chromogena*	McCarthy and Cross (1981)
Tunicamycin[a]	*Micromonospora*	Wakisaka et al. (1982)
Raffinose-histidine	*Streptomyces chromofuscus, S. cyaneus, S. rochei*	Vickers et al. (1984)
Rifampicin[c]	*Streptomyces atroolivaceus, S. diastaticus*	Vickers et al. (1984)
Rifampicin[c]	*Streptomyces anulatus, S. exfoliates, S. griseoruber*	Goodfellow (1988)
Oxytetracycline[c]	*Streptoverticillium*	Hanka et al. (1985)
Vancomycin[c]	*Amycolatopsis*	Lechevalier et al. (1986)
Novobiocin, streptomycin	*Glycomyces*	Labeda (1987)
Benzylpenicillin + nalidixic acid[c]	*Saccharothrix*	Labeda (1987)

[a]*discovered accidentally;*
[b]*discovered by examination of different compounds;*
[c]*discovered by design; either by visual scanning of databases or using DIACHAR program (Sneath, 1980).*
Source: Adapted from Kurtböke, D.I., 1990. New approaches to the isolation of non-streptomycete actinomycetes from soil. PhD Thesis. University of Liverpool, UK.

TABLE 3.4 Early Pretreatment Techniques Used for the Target-Directed Isolation of Particular Actinomycete Taxa

Pretreatment	*Actinomycetales* genera selected	References
Enrichment		
Low molecular weight alchols	*Nocardia, Rhodococcus*	Gavrilenko et al. (1979)
n-Alkanes	*Nocardia, Rhodococcus*	Nesterenko et al. (1977)
Paraffin wax	*Nocardia asteriodes*	Gordon and Hagan (1936)
Chemotaxis	*Actinoplanes*	Palleroni (1980)
Baiting (pollen, keratin)	*Actinoplanes*	Couch (1939), Kane (1966)
Physical		
Ageing (300–1000years)	*Micromonospora*	Cross and Attwell (1974)
Dry heat (soil) 120°C, 1 h	*Microbispora, Streptosporangium*	Nonomura and Ohara (1969)
Wet heat (isolation plate) 110°C, 10 min	*Streptomyces*	Agate and Bhat (1963)
Wet heat (soil suspension)	*Rhodococcus, Streptomyces, Micromonospora*	Rowbotham and Cross (1977)
Air drying (soil)	*Streptomyces*	Meiklejohn (1957)
Soil successively dried and moistened then dry heat at 120°C	*Streptomyces*	Okamura et al. (1979)
Suspension in an air followed by Andersen sampler entrapment	*Micropolyspora, Streptomyces, Thermomonospora*	Gregory and Lacey (1963); Lacey and Dutkiewicz (1976)
Differential centrifugation	*Streptomyces, Frankia*	Řeháček (1956); Baker et al. (1979); Reibach et al. (1981)
Filtration (water)	*Streptomyces*	Burman et al. (1969)
Flotation (water)	*Streptomyces, Rhodococcus*	Al-Diwany et al. (1978)
Flotation (litter), drying and rehydration	*Actinoplanes*	Makkar and Cross (1982)
Chemical		
Phenol	*Streptomyces*	Lawrence (1956); Speer and Lynch (1969); Panthier et al. (1979)
Chlorine	*Streptomyces*	Burman et al. (1969)
Quaternary ammonium compounds	*Mycobacterium, Rhodococcus*	Phillips and Kaplan (1976); du Moulin and Stottmeier (1978)

Source: Adapted from Kurtböke, D.I., 1990. New approaches to the isolation of non-streptomycete actinomycetes from soil. PhD Thesis. University of Liverpool, UK.

bacteria by swarming over the surface of isolation plates. Sequential drying of the agar plates also represented the best inoculation method using the surface spread technique resulting in the control of such undesired bacterial swarming (Vickers and Williams, 1987).

All these early experiences generated in-depth understanding that target-directed isolation of rare actinomycetes should involve (1) information on the nature of the selected substrate, (2) pretreatment of the material to bait the targeted taxa and remove unwanted background bacteria, (3) selection of appropriate growth media, (4) incubation conditions, (5) colony selection and purification (Goodfellow and Williams, 1986). Even today in the era of advance molecular techniques, colony selection and purification relies on expertize regarding colony morphology when attempting to detect and isolate actinomycete colonies from isolation media.

Starting from the 1980s advancements in the use of computer-assisted or numerical taxonomical approaches resulted in the generation of in-depth information on the nutritional, physiological, and antibiotic sensitivity profiles of different actinomycete clusters. Such data were then exploited for the design of selective isolation media including for the purposes of reexamining previously studied habitats to discover new relatives of the previously detected genera (Williams et al., 1984). The growing links between sound identification systems and design of selective isolation protocols then became increasingly clearer (Goodfellow and Williams, 1986) and information obtained from these cluster analyses was extensively used for the isolation of *Streptomyces* (Williams et al., 1981, 1983a,b), *Actinomadura* and *Nocardiopsis* (Athalye et al., 1985), *Mycobacterium*, *Rhodococcus* and *Nocardia* (Tsukamura et al., 1979), and *Thermomonospora* and other monosporic and oligosporic actinomycetes (McCarthy and Cross, 1984a,b).

Combinations of benzyl penicillin (5–10 μg/mL) with nalidixic acid (15 μg/mL) were used to select *Saccharothrix* species, whereas novobiocin (25 μg/mL) and streptomycin (15 μg/mL) was used for *Glycomyces* species. Addition of vancomycin allowed the selection of new strains of *Amycolatopsis* (Labeda, 1987; Lechevalier et al., 1986), based on the observation that actinomycetes are resistant to their own antibiotics (Cundliffe, 1989; Demain, 1974; Glöckner and Wolf, 1984; Hotta and Okami, 1996; Yamamoto et al., 1981). Hanka et al. (1985) by using a medium containing oxytetracycline and Attwell et al. (1985) by using a medium containing lysozyme were able to increase the proportion of *Streptoverticillium* colonies on isolation plates. Lysozyme resistance in *Streptoverticillium* was reported to be due to the binding of the enzyme to the mycelium of the streptoverticillia, in contrast to the sensitivity exhibited by the majority of streptomycetes, in which the inactivation of the enzyme results from the actions of a soluble inhibitor or protease (Attwell et al., 1985).

Most notably, Williams et al. (1983a,b) showed that there were clear qualitative differences between commonly detected streptomycetes as compared with the to-be-detected ones in the same soil samples when objectively formulated media were used. When the results were compared using reference data from the data matrixes, the observed increase or decrease in numbers of a major cluster on the new media compared with that on the controls were in correlation with the selection of media constituents or the effects of their combined composition. Remarkable differences in the utilization of raffinose and histidine by *Streptomyces cyaneus* and *S. albidoflavus* was utilized as a selective pressure to reduce the numbers of the latter species on the isolation plates since this species has been one of the most commonly detected species in soil. Selective media design using raffinose-histidine combination to supply the major carbon and nitrogen sources thus became possible (Vickers et al., 1984). Similarly, incorporation of rifampicin into starch-casein agar resulted in an increase of *S. diastaticus* numbers indicating that competition had previously excluded them from the medium. Information derived from data bases, such as the rifampicin resistance displayed by *S. diastaticus* enabled the design of species-specific and highly selective media resulting in an increased detection of this species on isolation plates (Vickers et al., 1984). *Streptomyces violaceoniger* species-group were also recovered on selective media designed utilizing numerical taxonomic data on the *Streptomyces*-cluster #32 of Williams et al. (1983a,b). Kanamycin resistance resulted in the selective isolation of *S. tenjimariensis*, the producer of a then new antibiotic named istamycin (Hotta et al., 1980). With the antibiotic resistance data obtained from the numerical classification of different taxa, selective media containing tetracyclines were also designed for the selective isolation of nocardiae and these approaches revealed the abundant presence of this group of actinomycetes in soil (Orchard and Goodfellow, 1974, 1980). Orchard et al. (1977) pointed out that the infrequent occurrence of nocardiae on certain media was not due to their absence in soil, but was due to the unsuitability of the media chosen for their isolation. Since they do not degrade chitin and rarely use egg albumin (Cross et al., 1976), casein or starch (Goodfellow, 1971), they have not been frequently isolated on chitin agar (Lingappa and Lockwood, 1962), starch-casein (Küster and Williams, 1964) or egg albumin (Waksman, 1961). Salt tolerance of streptomycetes was also used to select against common bacteria and common streptomycetes while revealing the presence of tolerant ones in selected substrates by the use of starch-casein agar supplemented with 4.6% w/v sodium chloride (Mackay, 1977).

From the late 1980s onward efforts increasingly utilized target-directed approaches utilizing the information generated from the past experiences mentioned in the earlier sections. Examples include the cross gradient plate method for differential isolation of actinomycetes with varied physiological properties by Hotta et al. (1992), or the use of polyvalent phages to decrease unwanted taxa on isolation plates (Kurtböke et al., 1992). Increasing the species numbers for monotypic genera has also been targeted, such as for the genus *Sporichthya*

(Suzuki et al., 1999) by incorporating gellan gum into the agar. By using baiting techniques with different chemo-attractants, the species numbers were also increased for genera represented with only few species, such as *Planomonospora* (Mertz, 1994). Physical characteristics of actinomycete taxa, such as resistance to heat and desiccation were also exploited and the microwave irradiation and electric pulses were used with significant success for the selective isolation of rare actinomycetes (Bulina et al., 1997, 1998; Terekhova, 2003). Design of Humic-acid-vitamin agar for isolation of soil actinomycetes by Hayakawa and Nonomura (1987) was another significant step toward the isolation of rare actinomycete species. This medium successfully reflected the soil environment and supplementation with vitamins encouraged the growth of many different and rare actinomycete taxa (Hayakawa, 2003).

Selectivity Through Exploitation of Bacteriophage–Host Interactions

Actinophages through their advantageous specificity to species, genera, and families, were also used for selective isolation purposes (Kurtböke, 2003, 2009, 2011; Kurtböke et al., 1992, 1993). Since the phage susceptibility relies on adsorption to receptors that are usually present in only a very limited number of closely related bacterial strains, phages were also effectively used as indicators of the presence of targeted taxa before selective isolations started (Kurtböke, 2003, 2005; Williams et al., 1993). Such specificity also facilitated their use as inhibitory agents to reduce the numbers of unwanted taxa on isolation plates (Kurtböke, 2003, 2009, 2011; Kurtböke et al., 1992, 1993) and select the ones which could not be cultivated in crowded plates. Examples include the use of a phage battery to selectively isolate termite gut-associated actinomycetes (Kurtböke and French, 2007) or detection of cellulolytic micromonosporae Fig. 3.2 (El-Tarabily et al., 1996) active against the cell wall of

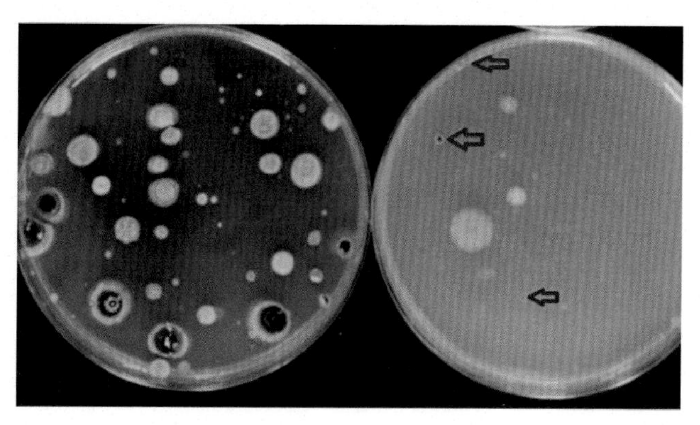

FIGURE 3.2 Removal of streptomycete colonies via the use of phage battery (El-Tarabily et al., 1996).

Phytophthora cinnamomi that infects eucalyptus trees and causes the jarrah-die back disease in Australia. After the removal of larger streptomycete colonies via phage battery on the isolation plates visual detection and isolation of such micromonosporae became feasible (plate 2, red arrows).

Since 2001, use of target-directed highly selective isolation techniques resulted in the generation of an actinomycete library at the University of the Sunshine Coast (Kurtböke, 2004). Examples of the isolates are shown in the Figs. 3.3–3.5. Fig. 3.3 examples relate to the Group 2 morphology (Table 3.2) with no aerial mycelium; substrate mycelium not fragmenting; colony texture is tough. Stable substrate mycelium can easily be seen in Fig. 3.3B–D. Fig. 3.4 relates to Group 1 and 3 members (Table 3.2). Fig. 3.4B illustrates a *Rhodococcus* isolate from near-shore foaming sea-water samples, with a typical Group 1 morphology (no aerial mycelium; substrate mycelium soon fragmenting into various-sized rod-coccoid elements; colony texture is soft, bacterial like). A Group 3 member *Nocardia* isolate (Fig. 3.4D) displays soft colony texture, with fragmenting substrate mycelium and dry, powdery-cottony aerial mycelium.

FIGURE 3.3 Growth morphology of the members of the family Micromonosporaceae (A) on Oatmeal (OAM) agar plates, (B) *Micromonospora* species under light microscope using the "inclined cover-slip" technique (Eccleston et al., 2008), (C) and (D) under the scanning electron microscopy (SEM).

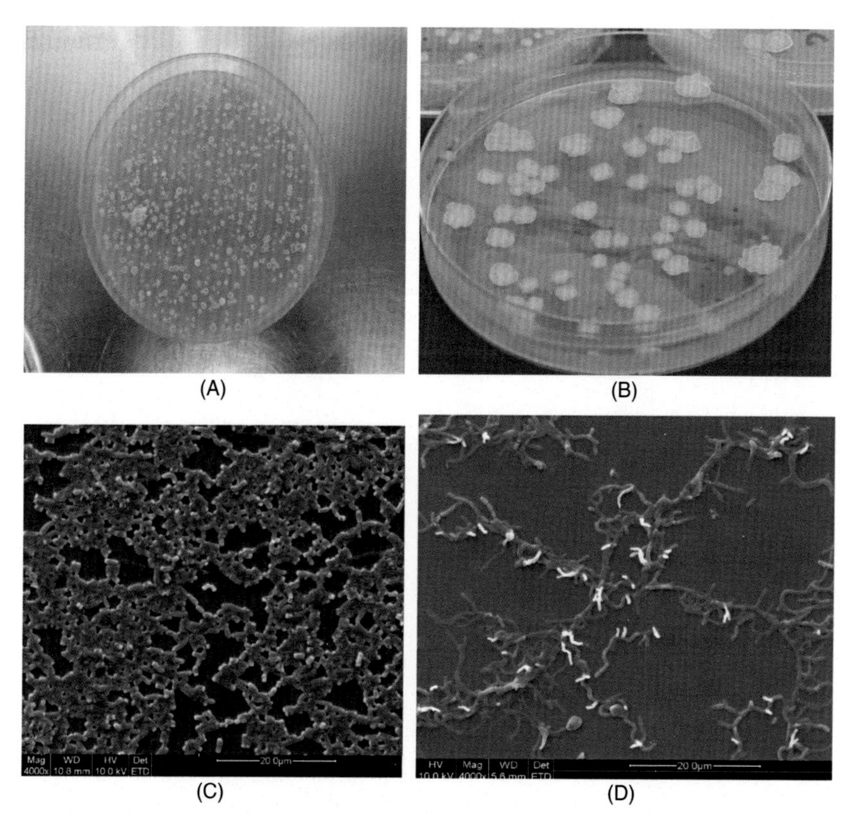

FIGURE 3.4 Colony morphologies of the species belonging to the *Nocardiaceae* and *Pseudonocardiaceae* families on (A) GYM, (B) Nocardia-Histidans agar, (C) SEM micrograph of a local *Rhodococcus* isolate, (D) SEM micrograph of a local *Nocardia* isolate. Fig. 3.3A shows a plate of previously isolated *Nocardiaceae* and *Pseudonocardiaceae* (Kurtböke, 2016) seeded into sterile seawater and re-streaked on Glucose-Yeast extract-Malt extract agar (GYM) agar (DSMZ Medium #65, www.dsmz.de) to illustrate the colony formation morphology. Fig. 3.3B shows the selective isolation of nocardiae on Nocardia-Histidans agar (Atlas, 2010) isolated directly from the near-shore foaming sea water samples (Kurtböke, 2016)

Selectivity Based on the Emerging Molecular Ecological Advances and Functional Diversity in the Environment

Premolecular era information on the numbers, types, and distribution of actinomycetes in natural habitats was limited. Hayakawa et al. (1988) noted the occurrence of rare actinomycetes, which were markedly higher than those previously reported and their distribution was interrelated with ecological factors, such as the water content and porosity of soil, nutrients, and climate. Climatic factors, such as temperature and rainfall that influence development of soil types were also reported to influence the microflora as well as the surrounding vegetation of the terrestrial environments (Hayakawa et al., 1988).

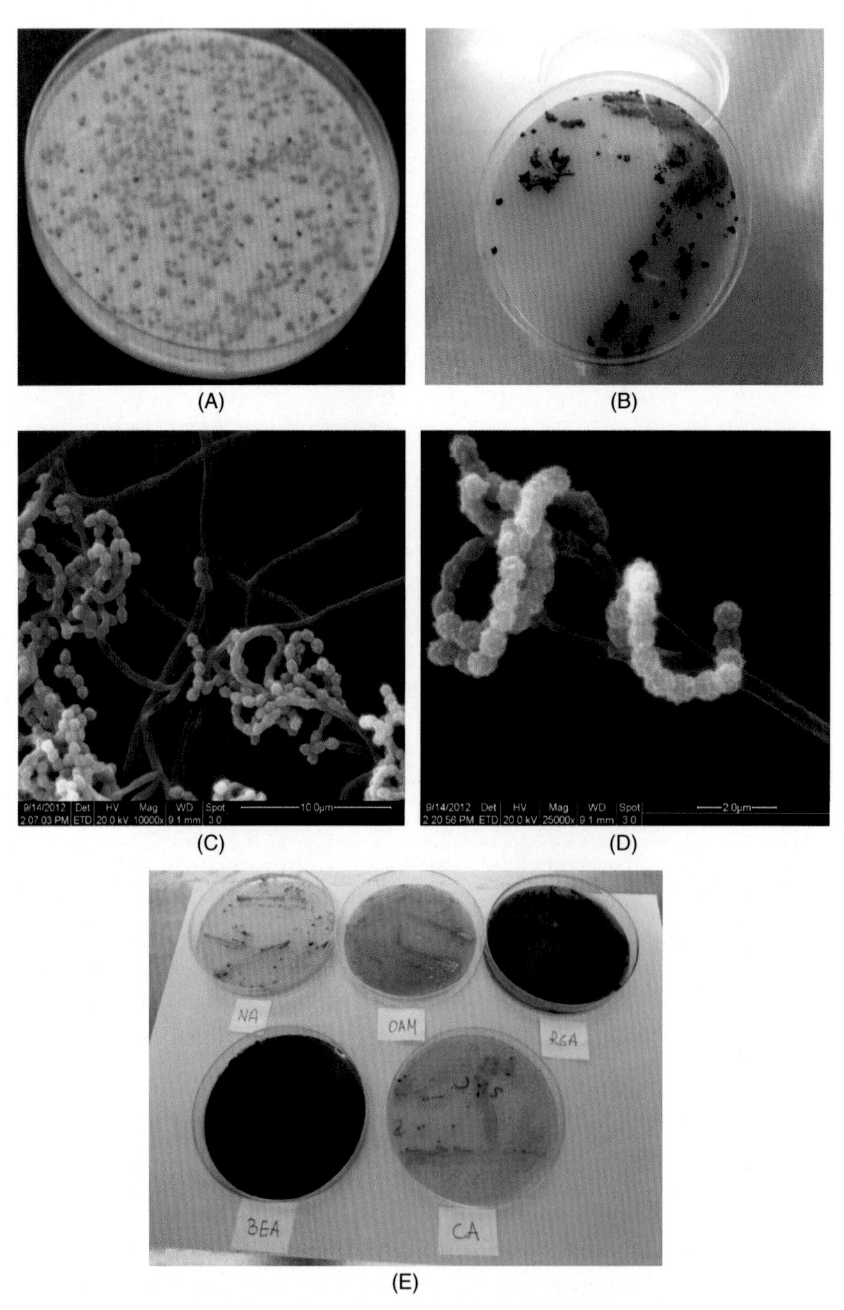

FIGURE 3.5 Growth morphologies of (A) enteric bacteria usually detected on Uriselect 4 selective isolation medium, (B) *Streptomyces* species (USC-633) on GYM agar, (C) and (D) sporulation on OAM medium observed under SEM microscopy, (E) growth characteristics of *Streptomyces* isolate (USC-633) on different media. *NA*, nutrient agar; *OAM*, oatmeal agar; *RGA*, ryegrass agar; *BEA*, Beef-extract agar, *CA*, cellulose agar.

Starting from the mid-1980s, members of the *Actinomycetales* were increasingly investigated for the breakdown and recycling of complex organic materials, such as proteins, nucleic acids, polysaccharides, and lignocellulosic material found in soils (McCarthy et al., 1986). Detection of these hydrolytic activities contributed toward the design of selective isolation media prepared with incorporation of specific substrates through which such degradative activities could be detected. Examples include the recovery of cellulase- and xylanase-producing streptomycetes and thermonomosporae (McCarthy, 1987) as well as lignocellulolytic actinomycetes from soil and plant litter (Crawford, 1978).

Conventional approaches using in vitro determination of the biological activities of isolates obtained at random from environment although had no ecological rationale, has been widely adopted to obtain industrially important species, such as the antibiotic producers. The assessment of overall or specific activities of actinomycetes in the environment was then made by direct means, such as (1) direct measurement of activity in situ; (2) determination of specific activities of isolates obtained from selected, well characterized environments, and (3) determination of activities of isolates obtained at random from environment (Williams, 1986). For the direct measurement of activity in situ examples include determination of growth rates of streptomycetes in soil by fluorescence microscopy (Mayfield et al., 1972) and the measurement of acetylene reduction to assess nitrogenase activity of *Frankia* in root nodules (Schwintzer and Tjepkema, 1983).

From 1960s to late 1980s knowledge on the degree of specificity of interactions among actinomycetes with other microorganisms, plants, and animals has been extremely variable, as is the extent and accuracy of our ecological knowledge. For example, in the 1980s, ecological and epidemiological information on the genera *Actinomyces*, *Actinomadura*, and *Nocardia* was in evidence (Schall and Beaman, 1984) but in contrast, potential interaction of sapropyhic actinoflora with humans or animals was missing (Williams, 1982). Similarly, biological control interactions between actinomycetes and root infecting fungi was circumstantial (Lechevalier, 1988; Williams, 1978, 1982). *Frankia* endophytic associations were well documented (Baker, 1987) but conversely the role of free-living saprophytes in the rhizoplane and rhizosphere was incomplete. As a result, our understanding remained on the surface due to the limitations of the past methods.

Starting from the late 1990s, powerful molecular tools have enabled the assessment of the gene content of a given environment (Chandler, 2008; Janssen, 2006) resulting in the definition of gene profiles. Definition of entire genome sequences and definition of the whole metagenome via accurate sequencing technologies as well as analysis of these gene pools via new statistical approaches has also become possible (Chandler, 2008). Metagenomics thus has provided an integrated view of our ecosystems and their response to environmental pressures. Such information has importance in terms of the analysis of microbial communities in specific niches, such as marine, extreme, and polluted environments (Chandler, 2008). Metagenomics has also enabled scientists to relate potential functions to specific microorganisms within multispecies of soil communities (Schmeisser et al., 2007). In addition, analyses of DNA sequences from cultivated microorganisms have revealed genome-wide, taxa-specific nucleotide compositional characteristics, referred to as genome signatures (Dick et al., 2009). These signatures will contribute toward advancing understanding of genome evolution and in the classification of metagenomic sequence fragments (Dick et al., 2009). Searching for these specific signatures in natural microbial communities will also generate further understanding on the impact of environmental factors that shape specific microbial communities (Dick et al., 2009). Moreover, development of prototype databases which can generate microbial diversity maps in relation to their geochemical and geological environment as well as to their geographic location (Stoner et al., 2001), such as the GenGIS database (Parks et al., 2013) will also facilitate detection of certain functional groups of microorganisms.

With the emerging molecular advances metagenomics or microbial environmental genomics is currently providing information that centres on the habitat rather than the species. Cumulative information is thus being generated to complement "organismal genomics" (Rodríguez-Valera, 2004). Metaproteomics approaches are also allowing detection of soil microorganisms by the analysis of collective proteins isolated directly from the targeted environments (Rajendhran and Gunasekaran, 2008). However, the importance of improved DNA and protein extraction efficiencies to be able to detect rare phylotypes and the need to develop target-directed strategies aimed at culturing frequently observed but yet-to-be-cultured bacteria have been noted by many researchers (Jensen and Lauro, 2008; Rajendhran and Gunasekaran, 2008). Another powerful molecular advancement has also been in the analysis of soil mobile metagenomics that will inevitably provide valuable knowledge on occurrence of horizontal gene transfer events and their implications for soil microbial ecology, evolution, and species differentiation (Rajendhran and Gunasekaran, 2008).

Functional metagenomics also can complement information concerning the specific functions of microorganisms at different sites. Examples include definition of the roles of effectors required to support host colonization (Rovenich et al., 2014), which can in turn facilitate design of rhizosphere-specific selective isolation techniques to culture microbial "coinhabitants" and "cooperators" based on the true understanding of interactions happening in these highly specialized zones. Information on the "functional microbiomes" in relation to plant fitness and disease protection can also contribute toward the recognition of host-specific factors that play roles in beneficial host-microbe interactions (Lakshmanan et al., 2014). Correlations between disease suppression and presence of rhizosphere-specific microorganisms can thus be generated and hence contribute toward design of target-directed selective isolation techniques, such as the one illustrated in Fig. 3.2 (El-Tarabily et al., 1996, 1997).

Molecular approaches can also reveal more detailed knowledge concerning uncharacterized microsymbionts that were resistant to isolation in pure culture, as exemplified by the identification of a greater diversity among *Alnus*-infective *Frankia* strains compared to other strains of this genus that infect other host plants (Normand et al., 1996). Normand et al. (2007) using molecular advances were able to retrieve information on the genome characteristics of facultatively symbiotic *Frankia* sp. strains reflecting host range and host plant biogeography. Taxon-specific gene information combined with functional gene information can provide an X-ray of potentially selective substrates and culturability can be thus be improved for hard to isolate bacteria (Davis et al., 2005; Janssen et al., 2002; Joseph et al., 2003; Sait et al., 2002). Recent design of traps for in situ cultivation of filamentous actinomycetes is one of the examples resulting in the successful isolation of previously nonencountered actinomycete taxa (Gavrish et al., 2008).

To be able to understand the impact of single species and complex communities on their associated host or natural environment, tracking of molecular interactions among microorganisms is necessary (Abreu and Taga, 2016). Functional imaging techniques are also increasing our understanding at deeper levels within complex microbial communities (Abreu and Taga, 2016). Research findings also are revealing the importance of helper bacteria, which stimulate the growth of previously uncultured bacteria through the production of siderophores or signaling compounds (D'Onofrio et al., 2010; Lewis et al., 2010; Audrain et al., 2015; Effmert et al., 2012).

Surveys of phenotypic characteristics of environmental bacteria are still limited to revealing in situ microbial strategies for growth and microbial diversity in the ecosystems (Ernebjerg and Kishony, 2012). Using an array of flat-bed scanners coupled with automated image analysis, Ernebjerg and Kishony (2012) examined the growth properties of the members of a microbial community from untreated soil. They observed that cells were most likely to form colonies when nearby colonies were present but not too dense, possibly indicating an offset by the induction of colony formation caused by the interactions between microbes. Their findings suggested new types of growth classification of soil bacteria and potential effects of species interactions on colony growth (Ernebjerg and Kishony, 2012). All this information thus facilitates the creation of simulated natural environments in which previously unculturable bacteria can form colonies in the presence of other microorganisms and subsequently be cultured (Kaeberlein et al., 2002). Study on the growth and metabolite secretion by *Streptomyces* species within permeable chambers surrounded by multispecies communities is also another example of such approaches (D'Onofrio et al., 2010).

Charlop-Powers et al. (2014) recently studied soil type-specific secondary metabolite gene cluster enrichment patterns and noted the presence of functionally similar meta-secondary metabolomes in similar soil types. They also observed that soil type also correlates with species diversity trends observed in these microbiomes. Generated new information is in line with the past observations that found correlations between the soil types and species diversity, when conventional techniques were used (Hayakawa et al., 1988). Adding onto these findings, Charlop-Powers et al. (2014) identified conserved soil type-specific secondary metabolome enrichment patterns despite significant sample-to-sample sequence variations. With the data generated they created chemical biogeographic distribution maps for bioactive families of microorganisms. Another important development is generation of sound understanding related to microbial dormancy in natural environments (Buerger et al., 2012; Keep et al., 2006; Dworkin and Shah, 2010). Examples include the use of diluted complete media for the restoration of the colony-forming ability of the dormant cystlike resting cells of *Arthrobacter globiformis* (Mulyukin et al., 2009). Understanding molecular level basis of environmental adaptations is also adding additional strength toward determining the behavior of microbial cells under extreme conditions, for example, linking the large mechanosensitive channel gene *mscl* of the obligate marine bacterium *Salinispora tropica* to its resistance to osmotic downshocks (Bucarey et al., 2012).

All the earlier listed molecular advances apply to the members of Domain Bacteria. However, subsets of data and related information for each order including the order *Actinomycetales* (e.g., selection of habitats for highest diversity or distinct diversity patterns, functional diversity, utilization of site-specific substrates, environmental adaptation, and ecophysiology) (Roberts et al., 2012; Vizcaino et al., 2014) will now bring additional strength enabling design of highly selective and target-directed isolation techniques to detect and culture previously undetected or uncultured members of the order (Hirsch et al., 2010). Using bioinformatics tools, information can be generated on a specific substrate and metagenome analysis can be conducted to define the diversity of a habitat before the isolation process starts, enabling target-directed and highly selective isolation methods to be used.

CONCLUSIONS

The pioneering discoveries from actinomycetes not only ensured that they were prolific sources for biodiscoveries but also sparked global interest in these organisms. Labor-intensive years involving culture resulted in the detection and isolation of many species of actinomycetes but also created the libraries of these organisms which subsequently generated currently present ecotaxonomic data. Molecular advances are now changing the scene but culturability is still an important aspect of biodiscovery, satisfying the needs of industry (Joint et al., 2010; Kurtböke, 2011, 2012). Current advances together with highly specific isolation methods generated via in-depth analysis of emerging molecular and metagenomic data will provide a sound basis for culturability. It might be timely to revisit the concept of the ISP project, and perhaps create a new one to be able to deduce clues on the whereabouts and functions of actinomycetes in the environment by effectively analyzing the sound information generated via the use of modern molecular methods. Consideration of computer-based cluster analysis utilizing newly generated data and correlating it to the past data might further facilitate creation of effective selective isolation methods.

Following effective metagenomic analysis of the environments of interest, such as the ones described earlier, target-directed selective isolation techniques can now be applied. However, adequate morphological and physiological descriptions at species level for each described genus should be available to assist target-directed selective isolation techniques to be utilized during examination of diverse substrates. Constant and satisfactory results again will be best obtained by an experienced actinomycetologist and gaining such expertize requires patience, perseverance, and careful observations over many years via visual and microscopic examinations (Cross and Alderson, 1988; Goodfellow and Cross, 1983; Goodfellow and Williams, 1983). One recent example has been the recognition of a near shore marine intertidal environment-associated *Streptomyces* isolate (USC-633) on an isolation plate among nonactinomycete taxa displaying similar colony colours (Fig. 3.5A and B) (Kurtböke et al., 2015). Following the screening of the area using indicator phages isolated to marine—derived streptomycetes as phage hosts combined with the use of diverse selective media such as the Uriselect 4 (Biorad), the detection and isolation of this species was possible (Kurtböke et al., 2015). The turquoise blue color colonies usually typical of *Enteroccoccus* species grow on this agar, but Group 4 morphology (Table 3.2) of streptomycete species is very different from colony morphology of enterococci species, and this medium resulted in the detection and isolation of this organism despite similar colors being displayed during their colony formations. Group 4 morphology of streptomycetes is recognized by the dry, powdery-cottony colony, and aerial mycelium formation (Fig. 3.5B and C), with a stable substrate mycelium and a tough colony texture which is very different to the growth morphology of *Enterococcus* species (Fig. 3.5B).

Once isolated, the isolate displayed impressive growth patterns and pigment productions in a wide range of media used to grow both terrestrial and marine actinomycetes (English et al., 2016) (Fig. 3.5E) perhaps indicating its ecological success in surviving the continually changing conditions of a near shore intertidal marine environment. Uriselect 4 selects for Gram-positive *Enterococcus* species and also contains silica which might be a reflection of the near shore marine environment and perhaps contributed toward the isolation of this unusual *Streptomyces* species on this highly selective medium.

As stated before the concept of selective isolation is not an application of a single method but it is a combination of adequate knowledge in ecotaxonomic disciplines (Williams et al., 1984). Current molecular advances will reveal the identities and functions of many uncultured microorganisms and facilitate the objective selection of target-directed isolation techniques to study diverse habitats (Bomar et al., 2011).

I would like to finish with a quote from the prominent actinomycetologists late Prof. Tom Cross and Prof. Grace Alderson who in their presentation in the *7th International Symposium on the Biology of Actinomycetes* (1988, Tokyo,

Japan) stated that "one should not underestimate the considerable influence exerted or enjoyed by powerful scientists such as Waksman and Krassilnikov during the middle years of this (20th) century. They simultaneously stimulated and blinkered a generation of actinomycetologists, and their effects are still evident to-day." I will proudly add that I am one of the actinomycetologists inspired by those who followed Waksman and Krassilnikov, such as the late Professors Tom Cross, Romano Locci, and Stanley Williams. By bringing early career researchers and pioneers together in this book I am hoping that the inspiration will continue for many more generations to come and the new species of actinomycetes will continue to provide fascinating biological and chemical diversities for the service of mankind.

ACKNOWLEDGMENTS

I would like to thank Prof. Manfred Rohde, GBF, Germany (Fig. 3.2C and D), Mrs Rachel Hancock, *Central Analytical Research Facility* operated by the *Institute for Future Environments*, Queensland, *University of Technology* (QUT), Brisbane, Australia (Fig. 3.4C and D and Fig. 3.5 C and D), and Mrs Gunta Jaudzems, *Electron Microscope Unit, Department of Ecology and Evolutionary Biology, Monash University*, Melbourne, Australia (Fig. 3.1C), for the electron micrographs of the actinomycetes. I also would like to thank Emeritus Professor John Fuerst for the peer-review he has kindly provided on my chapter.

REFERENCES

Abreu, N.A., Taga, M.E., 2016. Decoding molecular interactions in microbial communities. FEMS Microbiol. Rev. 40 (5), 648–663.

Agate, A.D., Bhat, J.V., 1963. A method for the preferential isolation of actinomycetes from soils. Antonie van Leeuwenhoek 29 (1), 297–304.

Al-Diwany, L.J., Unsworth, B.A., Cross, T., 1978. A comparison of membrane filters for counting *Thermoactinomyces* endospores in spore suspensions and river water. J. Appl. Bacteriol. 45 (2), 249–258.

Athalye, M., Lacey, J., Goodfellow, M., 1981. Selective isolation and enumeration of actinomycetes using rifampicin. J. Appl. Bacteriol. 51 (2), 289–297.

Athalye, M., Goodfellow, M., Lacey, J., White, R.P., 1985. Numerical classification of *Actinomadura* and *Nocardiopsis*. Int. J. Syst. Bacteriol. 35 (1), 86–98.

Atlas, R.M., 2010. Handbook of Microbiological Media. CRC press, Boca Raton, Florida.

Attwell, R.W., Surrey, A., Cross, T., 1985. Lysozyme sensitivity in *Streptoverticillium* and *Streptomyces* species. Zentralblatt für Bakteriologie, Mikrobiologie und Hygiene C3, 239–242.

Audrain, B., Farag, M.A., Ryu, C.M., Ghigo, J.M., 2015. Role of bacterial volatile compounds in bacterial biology. FEMS Microbiol. Rev. 39 (2), 222–233.

Baker, D.D., 1987. Relationships among pure cultured strains of *Frankia* based on host specificity. Physiol. Plantarum 70 (2), 245–248.

Baker, D., Torrey, J.G., Kidd, G.H., 1979. Isolation by sucrose-density fractionation and cultivation *in vitro* of actinomycetes from nitrogen-fixing root nodules. Nature 281, 76–78.

Baltz, R.H., 2006. Marcel Faber Roundtable: is our antibiotic pipeline unproductive because of starvation, constipation or lack of inspiration? J. Ind. Microbiol. Biotechnol. 33, 507–513.

Barton, M.D., Hughes, K.L., 1981. Comparison of three techniques for isolation of *Rhodococcus (Corynebacterium) equi* from contaminated sources. J. Clin. Microbiol. 13 (1), 219–221.

Beretta, G., 1973. *Actinoplanes italicus*, a new red-pigmented species. Int. J. Syst. Bacteriol. 23 (1), 37–42.

Bérdy, J., 1974. Recent developments of antibiotic research and classification of antibiotics according to chemical structure. Adv. Appl. Microbiol. 18, 309–406.

Bérdy, J., 1984. New ways to obtain new antibiotics. J. Antibiot. 7, 348–360.

Bérdy, J., 2005. Bioactive microbial metabolites. J. Antibiot. 58 (1), 1–26.

Bérdy, J., 2012. Thoughts and facts about antibiotics: where we are now and where we are heading. J. Antibiot. 65, 385–395.

Bomar, L., Maltz, M., Colston, S., Graf, J., 2011. Directed culturing of microorganisms using metatranscriptomics. MBio 2 (2), e00012–11.

Bucarey, S.A., Penn, K., Paul, L., Fenical, W., Jensen, P.R., 2012. Genetic complementation of the obligate marine actinobacterium *Salinispora tropica* with the large mechanosensitive channel gene *mscL* rescues cells from osmotic downshock. Appl. Environ. Microbiol. 78 (12), 4175–4182.

Buerger, S., Spoering, A., Gavrish, E., Leslin, C., Ling, L., Epstein, S.S., 2012. Microbial scout hypothesis, stochastic exit from dormancy, and the nature of slow growers. Appl. Environ. Microbiol. 78 (9), 3221–3228.

Bulina, T.I., Alferova, I.V., Terekhova, L.P., 1997. A novel approach to isolation of actinomycetes involving irradiation of soil samples with microwaves. Mikrobiologiya 66, 231–234.

Bulina, T.I., Terekhova, L.P., Tyurin, M.V., 1998. The use of electric pulses for selective isolation of actinomycetes from soil. Mikrobiologiya 64, 231–234.

Bull, A.T., Stach, J.E., Ward, A.C., Goodfellow, M., 2005. Marine actinobacteria: perspectives, challenges, future directions. Antonie Van Leeuwenhoek 87 (1), 65–79.

Burman, N.P., Oliver, C.W., Stevens, J.K., 1969. Membrane filtration techniques for the isolation from water, of *coli-aerogenes*, *Escherichia coli*, faecal streptococci, *Clostridium perfringens,* actinomycetes and microfungi. In: Shapton, D.A., Gould, G.W. (Eds.), Isolation Methods for Microbiologists. Academic Press, London, pp. 127–134.

Chandler, M., 2008. Microbiology: what now? Res. Microbiol. 159(1), 51–58.

Couch, J.N., 1939. Technique for collection, isolation and culture of chytrids. J. Elisha Mitchell Sci. Soc. 55, 208–214.

Charlop-Powers, Z., Owen, J.G., Reddy, B.V.B., Ternei, M.A., Brady, S.F., 2014. Chemical-biogeographic survey of secondary metabolism in soil. Proc. Natl. Acad. Sci. USA 111 (10), 3757–3762.

Couch, J.N., Bland, C.E., 1974. Family Actinoplanaceae Couch. In: Buchanan, R.E., Gibbons, N.E. (Eds.), Bergey's Manual of Determinative Bacteriology. eighth ed. The Williams and Wilkins Co, Baltimore, pp. 706–708.

Crawford, D.L., 1978. Lignocellulose decomposition by selected streptomyces strains. Appl. Environ. Microbiol. 35 (6), 1041–1045.

Cross, T., 1982. Actinomycetes: a continuing source of new metabolites. Dev. Ind. Microbiol. 23, 1–18.

Cross, T., 1989. Growth and examination of actinomycetes-some guidelines. Williams, S.T., Sharpe, M.E., Holt, J.G. (Eds.), Bergey's Manual of Systematic Bacteriology, vol. 4, Williams and Wilkins, Baltimore, pp. 2340–2343.

Cross, T., Alderson, G., 1988. What weight morphology in current actinomycete taxonomy? In: Okami, Y., Beppu, T., Ogawara, H. (Eds.), Biology of Actinomycetes'88. Japan Scientific Societies Press, Tokyo, Japan, pp. 216–220.

Cross, T., Attwell, R.W., 1974. Recovery of viable thermoactinomycete endospores from deep mud cores. In: Spore Research, Academic Press, London, pp. 11–20.

Cross, T., Rowbotham, T.J., Mishustin, E.N., Tepper, E.Z., Antoineportaels, F., Schall, K.P., Bickenbach, H.,1976. The ecology of nocardioform actinomycetes. In: Goodfellow, M., Brownell, G.H., Serrano, J.A., (Eds.), The Biology of the Nocardiae. Academic Press, London, pp. 337–371.

Cundliffe, E., 1989. How antibiotic-producing organisms avoid, suicide. Annu. Rev. Microbiol. 43(1), 207–233.

Davis, K.E., Joseph, S.J., Janssen, P.H., 2005. Effects of growth medium, inoculum size, and incubation time on culturability and isolation of soil bacteria. Appl. Environ. Microbiol. 71 (2), 826–834.

Demain, A.L., 1974. How do antibiotic-producing microorganisms avoid suicide? Ann. NY Acad. Sci. 235 (1), 601–612.

Demain, A.L., 1988. Actinomycetes: what have you done for us lately? In: Okami, Y., Beppu, T., Ogawara, H. (Eds.), Biology of Actinomycetes'88. Japan Scientific Societies, Tokyo, Japan, pp. 288–293.

Demain, A.L., 2014. Importance of microbial natural products and the need to revitalize their discovery. J. Ind. Microbiol. Biotechnol. 41 (2), 185–201.

Dick, G.J., Andersson, A.F., Baker, B.J., Simmons, S.L., Thomas, B.C., Yelton, A.P., Banfield, J.F., 2009. Community-wide analysis of microbial genome sequence signatures. Genome Biol. 10 (8), R85.

D'Onofrio, A., Crawford, J.M., Stewart, E.J., Witt, K., Gavrish, E., Epstein, S., Clardy, J., Lewis, K., 2010. Siderophores from neighboring organisms promote the growth of uncultured bacteria. Chem. Biol. 17 (3), 254–264.

du Moulin, G.C., Stottmeier, K.D., 1978. Use of cetylpyridinium chloride in the decontamination of water for culture of mycobacteria. Appl. Environ. Microbiol. 36 (5), 771–773.

Dworkin, J., Shah, I.M., 2010. Exit from dormancy in microbial organisms. Nat. Rev. Microbiol. 8 (12), 890–896.

Eccleston, G.P., Brooks, P.R., Kurtböke, D.I., 2008. The occurrence of bioactive micromonosporae in aquatic habitats of the Sunshine Coast, Australia. Marine Drugs 6 (2), 243–261.

Effmert, U., Kalderás, J., Warnke, R., Piechulla, B., 2012. Volatile mediated interactions between bacteria and fungi in the soil. J. Chem. Ecol. 38 (6), 665–703.

El-Tarabily, H.A., Sykes, M.L., Kurtböke, D.I., Hardy, G.E.St.J., Barbosa, A.M., Dekker, R.F.H., 1996. Synergistic effects of a cellulase-producing *Micromonospora carbonacea* and an antibiotic producing *Streptomyces violascens* on the suppression of *Phytophthora cinnanmomi* root rot of *Banksia grandis*. Can. J. Bot. 74, 618–624.

El-Tarabily, H.A., Hardy, G.E.St.J., Sivasithamparam, K., Hussein, A.M., Kurtböke, D.I., 1997. The potential for the biological control of cavity spot disease of carrots caused by *Pythium coloratum* by streptomycete and non-streptomycete actinomycetes. New Phytol. 137 (3), 495–507.

English, A.L., Boufridi, A., Quinn, R.J., Kurtböke, D.I., 2016. Evaluation of fermentation conditions triggering increased antibacterial activity from a nearshore marine intertidal environment-associated Streptomyces species. Synthetic and Systems Biotechnology, http://dx.doi.org/10.1016/j.synbio.2016.09.005.

Ernebjerg, M., Kishony, R., 2012. Distinct growth strategies of soil bacteria as revealed by large-scale colony tracking. Appl. Environ. Microbiol. 78 (5), 1345–1352.

Fenical, W., 1993. Chemical studies of marine bacteria: developing a new resource. Chem. Rev. 93, 1673–1683.

Fenical, W., Jensen, P.R., 2006. Developing a new resource for drug discovery: marine actinomycete bacteria. Nat. Chem. Biol. 2, 666–673.

Gavrilenko, M.N., Nesterenko, O.A., Nogina, T.M., 1979. Growth of nocardio-and coryneform bacteria on media with low molecular weight alcohols. Mikrobiologicheskii zhurnal 41 (5), 451–455.

Gavrish, E., Bollmann, A., Epstein, S., Lewis, K., 2008. A trap for in situ cultivation of filamentous actinobacteria. J. Microbiol. Methods 72 (3), 257–262.

Gerber, N., 1979. Odorous substances from actinomycetes. Dev. Ind. Microbiol. 20, 225–238.
Glöckner, C., Wolf, H., 1984. Mechanism of natural resistance to kirromycin-type antibiotics in actinomycetes. FEMS Microbiol. Lett. 25 (1), 121–124.
Goodfellow, M., 1971. Numerical taxonomy of some nocardioform bacteria. Microbiology 69 (1), 33–80.
Goodfellow, M., 1988. Numerical taxonomy and selective isolation of industrially important actinomycetes. Actinomycetologica 2 (1), 13–29.
Goodfellow, M., 2010. Selective isolation of Actinobacteria. In: Bull, A.T., Davies, J.E. (section Eds.), Isolation and Screening of Secondary Metabolites and Enzymes. Manual of Industrial Microbiology and Biotechnology. Baltz, R.H., Davies, J., Demain, A.L. (Eds.), pp. 13–27. ASM Press, Washington DC.
Goodfellow, M., Cross, T., 1983. Classification. In: Goodfellow, M., Mordarski, M., Williams, S.T. (Eds.), The Biology of Actinomycetes. Academic Press, London, pp. 7–164.
Goodfellow, M., Fiedler, H.-P., 2010. A guide to successful bioprospecting: informed by actinobacterial systematics. Antonie Van Leeuwenhoek 98, 119–142.
Goodfellow, M., Minnikin, D.E., 1985. Chemical Methods in Bacterial Systematics. Academic Press, London.
Goodfellow, M., Williams, S.T., 1983. Ecology of actinomycetes. Annu. Rev. Microbiol. 37 (1), 189–216.
Goodfellow, M., Williams, E., 1986. New strategies for the selective isolation of industrially important bacteria. Biotechnol. Genet. Eng. Rev. 4 (1), 213–262.
Gordon, R.E., Hagan, W.A., 1936. A study of some acid-fast actinomycetes from soil with special reference to pathogenicity for animals. J. Infect. Dis. 59 (2), 200–206.
Gregory, P.H., Lacey, M.E., 1963. Mycological examination of dust from mouldy hay associated with farmer's lung disease. Microbiology 30 (1), 75–88.
Hanka, L.J., Rueckert, P.W., Cross, T., 1985. A method for isolating strains of the genus *Streptoverticillium* from soil. FEMS Microbiol. Lett. 30, 20–23.
Hayakawa, M., 2003. Selective isolation of rare actinomycete genera using pretreatment techniques. In: Kurtböke, D.I. (Ed.), Selective Isolation of Rare Actinomycetes. Queensland Complete Printing Services: Nambour, Australia, pp. 55–81.
Hayakawa, M., Nonomura, H., 1987. Humic acid-vitamin agar, a new medium for the selective isolation of soil actinomycetes. J. Ferment. Technol. 65 (5), 501–509.
Hayakawa, M., Ishizawa, K., Nonomura, H., 1988. Distribution of rare actinomycetes in Japanese soils. J. Ferment. Technol. 66 (4), 367–373.
Higashide, E., 1988. Checking system on taxonomic stability of the International Streptomyces Project (ISP) strains in Japan. In: Okami, Y., Beppu, T., Ogawara, H. (Eds.), Biology of Actinomycetes'88. Japan Scientific Societies Press, Tokyo, Japan, pp. 210–215.
Hirsch, P.R., Mauchline, T.H., Clark, I.M., 2010. Culture-independent molecular techniques for soil microbial ecology. Soil Biol. Biochem. 42 (6), 878–887.
Hopwood, D.A., 2007. Streptomyces in Nature and Medicine: The Antibiotic Makers. Oxford University Press, New York.
Hotta, K., Okami, Y., 1996. Diversity in aminoglycoside antibiotic resistance of actinomycetes and its exploitation in the search for novel antibiotics. J. Ind. Microbiol. 17 (5–6), 352–358.
Hotta, K., Saito, N., Okami, Y., 1980. Studies on new aminoglycoside antibiotics, istamycins, from an actinomycete isolated from a marine environment. I. The use of plasmid profiles in screening antibiotic producing Streptomyces. J. Antibiot. 33, 1502–1509.
Hotta, K., Ichihara, M., Saito, N., Mizuno, S., 1992. Cross gradient plate method for differential isolation of actinomycetes with varied physiological properties. Actinomycetologica 6 (1), 33–36.

Ivantiskaya, L.P., Singal, S.M., Bibikova, M.V., Vostrov, S.N., 1978. Direct isolation of *Micromonospora* on selective media with gentamicin. Antibiotiki 23, 690–692.

Janssen, P.H., 2006. Identifying the dominant soil bacterial taxa in libraries of 16S rRNA and 16S rRNA genes. Appl. Environ. Microbiol. 72 (3), 1719–1728.

Janssen, P.H., Yates, P.S., Grinton, B.E., Taylor, P.M., Sait, M., 2002. Improved culturability of soil bacteria and isolation in pure culture of novel members of the divisions Acidobacteria, Actinobacteria, Proteobacteria, and Verrucomicrobia. Appl. Environ. Microbiol. 68 (5), 2391–2396.

Jensen, P.R., Lauro, F.M., 2008. An assessment of actinobacterial diversity in the marine environment. Antonie Van Leeuwenhoek 94 (1), 51–62.

Joint, I., Mühling, M., Querellou, J., 2010. Culturing marine bacteria—an essential prerequisite for Biodiscovery. Microb. Biotechnol. 3 (5), 564–575.

Joseph, S.J., Hugenholtz, P., Sangwan, P., Osborne, C.A., Janssen, P.H., 2003. Laboratory cultivation of widespread and previously uncultured soil bacteria. Appl. Environ. Microbiol. 69 (12), 7210–7215.

Kaeberlein, T., Lewis, K., Epstein, S.S., 2002. Isolating" uncultivable" microorganisms in pure culture in a simulated natural environment. Science 296 (5570), 1127–1129.

Kämpfer, P., 2014. Continuing importance of the "phenotype" in the genomic era. Methods Microbiol. 41, 307–320.

Kämpfer, P., Glaeser, S.P., 2012. Prokaryotic taxonomy in the sequencing era—the polyphasic approach revisited. Environ. Microbiol. 14 (2), 291–317.

Kane, W.D., 1966. A new genus of *Actinoplanaceae, Pilimelia*, with a description of two species, *Pilimelia terevasa* and *Pilimelia anulata*. J. Elisha Mitchell Sci. Soc. 82, 220–230.

Keep, N.H., Ward, J.M., Robertson, G., Cohen-Gonsaud, M., Henderson, B., 2006. Bacterial resuscitation factors: revival of viable but non-culturable bacteria. Cell. Mol. Life Sci. 63 (22), 2555–2559.

Kroppenstedt, R.M., 1985. Fatty acid and menaquinone analysis of actinomycetes and related organisms. In: Goodfellow, M., Minnikin, D.E. (Eds.), Chemical Methods in Bacterial Systematics. Academic Press, London, pp. 173–199.

Kurtböke, D.I., Murphy, N.E., Sivasithamparam, K., 1993. Use of bacteriophage for the selective isolation of thermophilic actinomycetes from composted eucalyptus bark. Can. J. Microbiol. 39, 46–51.

Kurtböke, D.I., 1996. Exploitation of host-phage interactions for the selective isolation of the industrially important bacteria. Med. Chem. Res. 6, 248–255.

Kurtböke, D.I., 2003. Use of bacteriophages for the selective isolation of rare actinomycetes. In: Kurtböke, D.I. (Ed.), Selective Isolation of Rare Actinomycetes. Queensland Complete Printing Services: Nambour, Australia, pp. 9–54.

Kurtböke, D.I., 2004. Actino-rush on the sunshine coast: prospects for bioprospecting. In: Watanabe, M.M., Suzuki, K., Seki, T., (Eds.). The Proceedings of the Tenth International Congress on Culture Collections, Tsukuba, Japan, 2004, pp. 319–322., Published by the Japan Society for Culture Collections, World Federation of Culture, Collections.

Kurtböke, D.I., 2005. Actinophages as indicators of actinomycete taxa in marine environments. Antonie van Leeuwenhoek 87, 19–28.

Kurtböke, D.I., 2009. Use of phage-battery to isolate industrially important rare actinomycetes. In: Adams, H.T. (Ed.), Contemporary Trends in Bacteriophage Research. NOVA Science Publishers, New York, pp. 119–149.

Kurtböke, D.I., 2011. Exploitation of phage battery in the search for bioactive actinomycetes. Appl. Microbiol. Biotechnol. 89 (4), 931–937.

Kurtböke, D.I., 2012. Biodiscovery from rare actinomycetes: an eco-taxonomical perspective. Appl. Microbiol. Biotechnol. 93 (5), 1843–1852.

Kurtböke, D.I., 2016. Actinomycetes in biodiscovery: genomic advances and new horizons. In: Gupta, V.K., Sharma, G.D., Tuohy, M.G., Gaur, R. (Eds.), The Handbook of Microbial Resources. CAB International Publications, Oxfordshire, UK, pp. 567–590 (Chapter 35).

Kurtböke, D.I., French, J.R.J., 2007. Use of phage battery to investigate the actinofloral layers of termite-gut microflora. J. Appl. Microbiol. 103 (3), 722–734.

Kurtböke, D.I., Chen, C.-F., Williams, S.T., 1992. Use of polyvalent phage for reduction of streptomycetes on soil dilution plates. J. Appl. Bacteriol. 72, 103–111.

Kurtböke, D.I., Grkovic, T., Quinn, R.J., 2015. Marine actinomycetes in biodiscovery. In: Kim, S.-K. (Ed.), Springer Handbook of Marine Biotechnology. Springer-Verlag, Berlin, pp. 663–676.

Kurtböke, D.I., Okazaki, T., Vobis, G., 2017. Actinobacteria in marine environments: from terrigenous origin to adapted functional diversity. In: Kim, S-K., (Ed.), Encyclopedia of Marine Biotechnology, Wiley-Blackwell.

Kurtböke, D.I, Evans-Illidge, L., Hill, R., Mancuso-Nichols, C.A., Sanderson, K., McMeekin, T.A., Wildman, H.G., 1999. Accessing Australian Diversity for Pharmaceutical purposes: Towards an improved isolation of actinomycetes. Proceedings of the Biotechnology, Biodiversity and Biobusiness Conference, Perth, Western Australia, November 1999, pp. 46–52.

Küster, E., Williams, S.T., 1964. Selection of media for isolation of streptomycetes. Nature (London) 202, 928–929.

Labeda, D.P., 1987. New actinomycetes of commercial importance. In: Megusar, F., Gantar, M. (Eds.), Perspectives in Microbial Ecology. Slovene Society for Microbiology, Ljubljana, pp. 271–276.

Lacey, J., Dutkiewicz, J., 1976. Isolation of actinomycetes and fungi from mouldy hay using a sedimentation chamber. J. Appl. Bacteriol. 41 (2), 315–319.

Lakshmanan, V., Selvaraj, G., Bais, H.P., 2014. Functional soil microbiome: belowground solutions to an aboveground problem. Plant Physiol. 166 (2), 689–700.

Lavrova, N.V., 1971. Selective media with streptomycin for isolation of new antibiotic producers. Antibiotiki 16, 781–786.

Lavrova, N.V., Preobrazhenskaya, T.P., Sveshnikova, M.A., 1972. Isolation of soil actinomycetes on selective media with rubomycin. Antibiotiki 17 (11), 965–970.

Lawrence, C.H., 1956. A method of isolating actinomycetes from scabby potato tissue and soil with minimal contamination. Can. J. Bot. 34 (1), 44–47.

Lechevalier, M.P., 1988. Actinomycetes in agriculture and forestry. In: Goodfellow, M., Williams, S.T., Mordarski, M. (Eds.), Actinomycetes in Biotechnology. Academic Press, London, pp. 327–358.

Lechevalier, H.A., Lechevalier, M.P., 1985. Biology of actinomycetes not belonging to the genus Streptomyces. In: Demain, A.L., Solomon, N.A. (Eds.), Biology of Industrial Microroganisms. Benjamin/Cummings Publishing Company Inc, California, pp. 315–318.

Lechevalier, M.P., Prauser, H., Labeda, D.P., Ruan, J.S., 1986. Two new genera of nocardioform actinomycetes: *Amycolata* gen. nov. and *Amycolatopsis* gen. nov. Int. J. Syst. Evol. Microbiol. 36 (1), 29–37.

Lewis, K., Epstein, S., D'Onofrio, A., Ling, L.L., 2010. Uncultured microorganisms as a source of secondary metabolites. J. Antibiot. 63, 468–476.

Lingappa, Y., Lockwood, J.L., 1962. Chitin media for selective isolation and culture of actinomycetes. Phytopathology 52, 317–323.

Luedemann, G.M., Brodsky, B.C., 1964. Taxonomy of gentamicin-producing *Micromonospora*. Antimicrob. Agents Chemother. 1963, 116–124.

Luedemann, G.M., Brodsky, B.C., 1965. *Micromonospora carboneceae* sp. n. an everninomicin producing organism. Antimicrob. Agents Chemother. 1964, 47–52.

Mackay, S.J., 1977. Improved enumeration of *Streptomyces* spp. on a starch casein salt medium. Appl. Environ. Microbiol. 33 (2), 227–230.

Makkar, N.S., Cross, T., 1982. Actinoplanetes in soil and on plant litter from freshwater habitats. J. Appl. Bacteriol. 52 (2), 209–218.

Mayfield, C.I., Williams, S.T., Ruddick, S.M., Hatfield, H.L., 1972. Studies on the ecology of actinomycetes in soil IV. Observations on the form and growth of streptomycetes in soil. Soil Biol. Biochem. 4 (1), 79–91.

Meiklejohn, J., 1957. Numbers of bacteria and actinomycetes in a Kenya soil. J. Soil Sci. 8 (2), 240–247.

Mertz, F.P., 1994. *Planomonospora alba* sp. nov. and *Planomonospora sphaerica* sp. nov., two new species isolated from soil by baiting techniques. Int. J. Syst. Evol. Microbiol. 44 (2), 274–281.

McCarthy, A.J., 1987. Lignocellulose-degrading actinomycetes. FEMS Microbiol. Rev. 3 (2), 145–163.

McCarthy, A.J., Cross, T., 1981. A note on a selective isolation medium for the thermophilic actinomycete *Thermomonospora chromogena*. J. Appl. Bacteriol. 51 (2), 299–302.

McCarthy, A.J., Cross, T., 1984a. A taxonomic study of *Thermomonospora* and other monosporic actinomycetes. Microbiology 130 (1), 5–25.

McCarthy, A.J., Cross, T., 1984b. Taxonomy of thermomonospora and related oligosporic actinomycetes. In: Ortiz-ortiz, L., Bojalil, J.F., Yakoleff, V. (Eds.), Biological, Biochemical and Biomedical Aspects of Actinomycetes. Academic Press, Orlando, pp. 521–526.

McCarthy, A.J., Paterson, A., Broda, P.M.A., 1986. Grass lignin degradation by actinomycetes. In: Szabo, G., Biro, S., Goodfellow, M. (Eds.), Biological, Biochemical and Biomedical Aspects of Actinomycetes. Kiado Press, Budapest, pp. 761–763.

Mulyukin, A.L., Demkina, E.V., Kryazhevskikh, N.A., Suzina, N.E., Vorob'eva, L.I., Duda, V.I., Galchenko, V.F., El-Registan, G.I., 2009. Dormant forms of *Micrococcus luteus* and *Arthrobacter globiformis* not platable on standard media. Microbiology 78 (4), 407–418.

Nesterenko, O.A., Kasumova, S.A., Kvasnikov, E.I., 1977. Microorganisms of the genus *Nocardia* and the *rhodochrous*" group in the soils of the Ukrainian SSR. Mikrobiologia 47 (5), 866–870.

Nisbet, L.J., 1982. Current strategies in the search for bioactive microbial metabolites. J. Chem. Technol. Biotechnol. 32 (1), 251–270.

Nonomura, H., 1969. Distribution of actinomycetes in soil IV. A culture method effective for both preferential isolation and enumeration of *Microbispora* and *Streptosporangium* strains in soil (Part 1). J. Ferment. Technol. 47, 463–469.

Nonomura, H., Ohara, Y., 1971a. Distribution of actinomycetes in soil VIII. Green-spore group of *Microtetraspora*, its preferential isolation and taxonomic characteristics. J. Ferment. Technol. 49, 1–7.

Nonomura, H., Ohara, Y., 1971b. Distribution of actinomycetes in soil IX. New species of the genera *Microbspora* and *Microtetraspora* and their isolation method. J. Ferment. Technol. 49, 887–894.

Nonomura, H., Ohara, Y., 1971c. Distribution of actinomycetes in soil X. New genus and species of monosporic actinomycetes. J. Ferment. Technol. 49, 885–903.

Nonomura, H., Ohara, Y., 1971d. Distribution of actinomycetes in soil. XI. Some new species of Actinomadura Lechevalier et al. J. Ferment. Technol. 49, 904–912.

Normand, P., Orso, S., Cournoyer, B., Jeannin, P., Chapelon, C., Dawson, J., Evtushenko, L., Misra, A.K., 1996. Molecular phylogeny of the genus Frankia and related genera and emendation of the family Frankiaceae. Int. J. Syst. Evol. Microbiol. 46 (1), 1–9.

Normand, P., Lapierre, P., Tisa, L.S., Gogarten, J.P., Alloisio, N., Bagnarol, E., Bassi, C.A., Berry, A.M., Bickhart, D.M., Choisne, N., Couloux, A., 2007. Genome characteristics of facultatively symbiotic *Frankia* sp. strains reflect host range and host plant biogeography. Genome Res. 17 (1), 7–15.

Okamura, K., Koki, A., Sakamoto, M., Kubo, K., Mutoh, Y., Fukagawa, Y., Kuono, K., Shimauchi, Y., Ishikura, T., 1979. Microorganisms producing a new beta-lactam antibiotic. J. Ferment. Technol. 57, 265–272.

Orchard, V.A., Goodfellow, M., 1974. The selective isolation of *Nocardia* from soil using antibiotics. Microbiology 85 (1), 160–162.

Orchard, V.A., Goodfellow, M., 1980. Numerical classification of some named strains of Nocardia asteroides and related isolates from soil. Microbiology 118 (2), 295–312.

Orchard, V.A., Goodfellow, M., Williams, S.T., 1977. Selective isolation and occurrence of nocardiae in soil. Soil Biol. Biochem. 9 (4), 233–238.

Ottow, J.C., Glathe, H., 1968. Rose bengal-malt extract-agar, a simple medium for the simultaneous isolation and enumeration of fungi and actinomycetes from soil. Appl. Microbiol. 16 (1), 170–171.

Palleroni, N.J., 1980. A chemotactic method for the isolation of *Actinoplanaceae*. Arch. Microbiol. 128 (1), 53–55.

Parenti, F., Beretta, G., Berti, M., Arioli, V., 1978. Teichomycins, new antibiotics from *Actinoplanes teichomyceticus* nov. sp. I. Description of the producer strain, fermentation studies and biological properties. J. Antibiot. 31 (4), 276–283.

Parenti, F., Pagani, H., Beretta, G., 1975. Lipiarmycin, a new antibiotic from *Actinoplanes*. I. Description of the producer strain and fermentation studies. J. Antibiot. 28 (4), 247–252.

Parenti, F., Pagani, H., Beretta, G., 1976. Gardimycin, a new antibiotic from *Actinoplanes*. I. Description of the producer strain and fermentation studies. J. Antibiot. 29 (5), 501–506.

Parks, D.H., Mankowski, T., Zangooei, S., Porter, M.S., Armanini, D.G., Baird, D.J., Langille, M.G., Beiko, R.G., 2013. GenGIS 2: Geospatial analysis of traditional and genetic biodiversity, with new gradient algorithms and an extensible plugin framework. PloS One 8 (7), e69885.

Panthier, J.J., Diem, H.G., Dommergues, Y., 1979. Rapid method to enumerate and isolate soil actinomycetes antagonistic towards rhizobia. Soil Biol. Biochem. 11 (4), 443–445.

Phillips, B.J., Kaplan, W., 1976. Effect of cetylpyridinium chloride on pathogenic fungi and *Nocardia asteroides* in sputum. J. Clin. Microbiol. 3 (3), 272–276.

Porter, J.N., Wilhelm, J.J., Tresner, H.D., 1960. Method for the preferential isolation of actinomycetes from soils. Appl. Microbiol. 8 (3), 174.

Preobrazhenskaya, T.P., Lavrova, N.V., Ukholina, R.S., Nechaeva, N.P., 1975. Isolation of new species of *Actinomadura* on selective media with streptomycin and bruneomycin. Antibiotiki 20, 404–408.

Pridham, T.G., Anderson, P., Foley, C., Lindenfelser, L.A., Hesseltine, C.W., Benedict, R.G., 1957. A selection of media for maintenance and taxonomic study of streptomycetes. Antibiot. Annu. 1956–1957, 947–953.

Rajendhran, J., Gunasekaran, P., 2008. Strategies for accessing soil metagenome for desired applications. Biotechnol. Adv. 26 (6), 576–590.

Řeháček, Z., 1956. Determination of number of spores of sporulating actinomycetes in soil and their isolation. Cˇeskoslovenská Mikrobiologie 1 (3), 129–134.

Reibach, P.H., Mask, P.L., Streeter, J.G., 1981. A rapid one-step method for the isolation of bacteroids from root nodules of soybean plants, utilizing self-generating percoll gradients. Can. J. Microbiol. 27 (5), 491–495.

Roberts, A.A., Schultz, A.W., Kersten, R.D., Dorrestein, P.C., Moore, B.S., 2012. Iron acquisition in the marine actinomycete genus *Salinispora* is controlled by the desferrioxamine family of siderophores. FEMS Microbiol. Lett. 335 (2), 95–103.

Rodríguez-Valera, F., 2004. Environmental genomics, the big picture? FEMS Microbiol. Lett. 231 (2), 153–158.

Romano, G., Costantini, M., Sansone, C., Lauritano, C., Ruocco, N., Ianora, A., 2016. Marine microorganisms as a promising and sustainable source of bioactive molecules. Marine Environ. Res, Available from: http://dx.doi.org/10.1016/j.marenvres.2016.05.002.

Rovenich, H., Boshoven, J.C., Thomma, B.P., 2014. Filamentous pathogen effector functions: of pathogens, hosts and microbiomes. Curr. Opin. Plant Biol. 20, 96–103.

Rowbotham, T.J., Cross, T., 1977. Ecology of *Rhodococcus coprophilus* and associated actinomycetes in fresh water and agricultural habitats. Microbiology 100 (2), 231–240.

Sait, M., Hugenholtz, P., Janssen, P.H., 2002. Cultivation of globally distributed soil bacteria from phylogenetic lineages previously only detected in cultivation-independent surveys. Environ. Microbiol. 4 (11), 654–666.

Sandrak, N.A., 1977. Degradation of cellulose by micromonospores. Mikrobiologiia 46 (3), 478–481.

Schall, K., Beaman, B.L., 1984. Clinical significance of actinomycetes. In: Goodfellow, M., Mordarski, M., Williams, S.T. (Eds.), The Biology of Actinomycetes. Academic Press, London, pp. 389–424.

Schmeisser, C., Steele, H., Streit, W.R., 2007. Metagenomics, biotechnology with non-culturable microbes. Appl. Microbiol. Biotechnol. 75 (5), 955–962.

Schwintzer, C.R., Tjepkema, J.D., 1983. Seasonal pattern of energy use, respiration, and nitrogenase activity in root nodules of Myrica gale. Can. J. Bot. 61 (11), 2937–2942.

Sneath, P.H.A., 1980. Basic program for the most diagnostic properties of groups from an identification matrix of percent positive characters. Computers and Geosciences 6, 21–26.

Speer, J.R., Lynch, D.L., 1969. Isolation of Actinomycetes from soils. Illinois State Academy of Science, Transactions 62, 265–272.

Stackebrandt, E., 2011. Molecular taxonomic parameters. Microbiol. Australia 32 (2), 59–61.

Stackebrandt, E., Schleifer, K.H., 1984. Molecular systematicsof actinomycetes and related organisms. In: Ortiz-ortiz, L., Bojalil, J.F., Yakoleff, V. (Eds.), Biological, Biochemical and Biomedical Aspects of Actinomycetes. Academic Press, Orlando, pp. 485–504.

Stackebrandt, E., Woese, C.R., 1981. The evolution of prokaryotes. In: Carlile, M.J., Collins, J.F., Moseley, B.E.B. (Eds.), Molecular and Cellular, Aspects of Microbial Evolution. Cambridge University Press, Cambridge, pp. 1–31.

Stoner, D.L., Geary, M.C., White, L.J., Lee, R.D., Brizzee, J.A., Rodman, A.C., Rope, R.C., 2001. Mapping microbial biodiversity. Appl. Environ. Microbiol. 67 (9), 4324–4328.

Suzuki, S.I., Okuda, T., Komatsubara, S., 1999. Selective isolation and distribution ofsporichthya strains in soil. Appl. Environ. Microbiol. 65 (5), 1930–1935.

Terekhova, L.P., 2003. Isolation of actinomycetes with the use of microwaves and electric pulses. In: Kurtböke, D.I. (Ed.), Selective Isolation of Rare Actinomycetes. Queensland Complete Printing Services, Nambour, Australia, pp. 82–101.

Thiemann, J.E., Pagani, H., Beretta, G., 1967. A new genus of the *Actinoplanaceae*: *Dactylosporangium*, gen. nov. Archiv für Mikrobiologie 58 (1), 42–52.

Tomita, K., Hoshino, Y., Sasahira, T., Hasegawa, K., Akiyama, M., Tsukiura, H., Kawaguchi, H., 1980. Taxonomy of the antibiotic Bu-2313-producing organism, *Microtetraspora caesia* sp. nov. J. Antibiot. 33 (12), 1491–1501.

Tiwari, K., Gupta, R.K., 2012. Rare actinomycetes: a potential storehouse for novel antibiotics. Crit. Rev. Biotechnol. 32 (2), 108–132.

Tiwari, K., Gupta, R.K., 2013. Diversity and isolation of rare actinomycetes: an overview. Crit. Rev. Microbiol. 39 (3), 256–294.

Tresner, H.D., Backus, E.J., 1963. System of color wheels for streptomycete taxonomy. Appl. Microbiol. 11 (4), 335–338.

Tsukamura, M., Mizuno, S., Tsukamura, S., Tsukamura, J., 1979. Comprehensive numerical classification of 369 strains of *Mycobacterium, Rhodococcus, and Nocardia*. Int. J. Syst. Bacteriol. 29 (2), 110–129.

Vickers, J.C., Williams, S.T., Ross, G.W., 1984. A taxonomic approach to selective isolation of streptomycetes from soil. In: Ortiz-ortiz, L., Bojalil, J.F., Yakoleff, V. (Eds.), Biological, Biochemical and Biomedical Aspects of Actinomycetes. Academic Press, Orlando, pp. 553–561.

Vickers, J.C., Williams, S.T., 1987. An assessment of plate inoculation procedures for the enumeration and isolation of soil streptomycetes. Microbios Lett. 35 (139–140), 113–117.

Vizcaino, M.I., Guo, X., Crawford, J.M., 2014. Merging chemical ecology with bacterial genome mining for secondary metabolite discovery. J. Ind. Microbiol. Biotechnol. 41 (2), 285–299.

Wakisaka, Y., Kawamura, Y., Yasuda, Y., Koizumi, K., Nishimoto, Y., 1982. A selective isolation procedure for *Micromonospora*. J. Antibiot. 35 (7), 822–836.

Waksman, S.A., 1961. The Actinomycetes. Classification, Identification and Descriptions of Genera and SpeciesWilliams and Wilkins, Baltimore.

Wellington, E.M.H., Cross, T., 1983. Taxonomy of antibiotic producing actinomycetes and new approaches to their selective isolation. In: Bushell, M.E. (Ed.), Progress in Industrial Microbiology. Elsevier Scientific Publication Company, Amsterdam, pp. 7–36, 17.

Williams, S.T., 1978. Streptomycetes in the soil ecosystem. Zentralblatt für Bakteriologie, Parasitenkunde und Infektionskrankheiten und hygiene. Abteilung I, Supplement 6, 137–144.

Williams, S.T., 1982. Are antibiotics produced in soil? Pedobiologia 23, 427–435.

Williams, S.T., 1986. Actinomycete ecology—A critical evaluation. In: Ortiz-ortiz, L., Bojalil, J.F., Yakoleff, V. (Eds.), Biological, Biochemical and Biomedical Aspects of Actinomycetes. Academic Press, Orlando, pp. 673–700.

Williams, S.T., Wellington, E.M.H., 1982a. Actinomycetes. In: Page, A.L., Miller, R.H., Keeney, D.R. (Eds.), Methods of Soil Analysis, Part 2, Chemical, Microbiological Properties, second ed., pp. 969–987. American Society of Agronomy/Soil Science Society of America, Madison.

Williams, S.T., Wellington, E.M.H., 1982b. Principles and problems of selective isolation of microbes. In: Bu'lock, J.D., Nisbet, L.J., Withstanley, D.J. (Eds.), Bioactive Microbial Products: Search and Discovery. Academic Press, London, pp. 9–26.

Williams, S.T., Goodfellow, M., Vickers, J.C., 1984. New microbes from old habitats? In: Kelly, D.P., Carr, N.G. (Eds.), The Microbe 1984, I.I., Prokaryotes, Eukaryotes, pp. 219–256, Cambridge University Press, Cambridge.

Williams, S.T., Vickers, J.C., Goodfellow, M., 1985. Application of new theoretical concepts to the identification of streptomycetes. In: Goodfellow, M., Jones, D., Priest, F.G. (Eds.), Computer-Assisted Bacterial Systematics. Academic Press, London, pp. 289–306.

Williams, S.T., Shameemullah, M., Watson, E.T., Mayfield, C.I., 1972. Studies on the ecology of actinomycetes in soil-VI. The influence of moisture tension on growth and survival. Soil Biol. Biochem. 4 (2), 215–225.

Williams, S.T., Wellington, E.M.H., Goodfellow, M., Alderson, G., Sackin, M.J., Sneath, P.H.A., 1981. The genus *Streptomyces*-a taxonomic enigma. Zentralblatt für Bakteriologie, Mikrobiologie und Hygiene, Supplement 11, 45–57.

Williams, S.T., Goodfellow, M., Alderson, G., Wellington, E.M.H., Sneath, P.H.A., Sackin, M.J., 1983a. Numerical classification of *Streptomyces* and related taxa. J. Gen. Microbiol. 129, 1743–1813.

Williams, S.T., Goodfellow, M., Wellington, E.M.H., Vickers, J.C., Alderson, G., Sneath, P.H.A., Sackin, M.J., Mortimer, A.M., 1983b. A probability matrix for identification of streptomycetes. J. Gen. Microbiol. 129, 1815–1830.

Williams, S.T., Locci, R., Beswick, A., Kurtböke, D.I., Kuznetsov, V.D., Le Monnier, F.J., Long, P.F., Maycroft, K.A., Palmit, R.A., Petrolini, B., Quaroni, S., Todd, J.I., West, M., 1993. Detection and identification of novel actinomycetes. Res. Microbiol. 144, 653–656.

Willoughby, L.G., 1971. Observations on some aquatic actinomycetes of streams and rivers. Freshwater Biol. 1 (1), 23–27.

Watve, M.G., Tickoo, R., Jog, M.M., Bhole, B.D., 2001. How many antibiotics are produced by the genus Streptomyces? Arch. Microbiol. 176, 386–390.

Yamamoto, H., Hotta, K., Okami, Y., Umezawa, H., 1981. Self-resistance of a *Streptomyces* which produces istamycins. J. Antibiot. 34 (7), 824–829.

Yoshioka, H., 1952. A new rapid isolation procedure of soil streptomycetes. J. Antibiot. 5, 559–561.

Zotchev, S.B., Sekurova, O.N., Kurtböke, D.I., 2016. Metagenomics of marine actinomycetes: from functional gene diversity to biodiscovery. In: Kim, S-K., (Ed.), Marine OMICS: Principles and Applications, CRC Press, Taylor and Francis Group (Chapter 9).

Chapter 4

Microbial Resources for Global Sustainability

Jim Philp* and Ronald Atlas**
**Science and Technology Policy Division, Paris, France; **University of Louisville, Louisville, KY, United States*

INTRODUCTION

Global sustainability is a controversial issue, as it might imply that we have to use renewable materials for industrial production instead of (nonrenewable) fossil resources, particularly coal, oil, and gas to reduce global greenhouse gas (GHG) emissions. High political priority has been given to biofuels and substantial progress has been made in developing ethanol, biodiesel, and jet aviation fuels at production scale levels (http://www.doegenomestolife.org/biofuels/). It is foreseeable that one day the internal combustion engine (ICE) can be replaced for light transport vehicles but it is unlikely that heavy shipping fuels can be easily replaced with biofuels although cost will likely remain a factor.

For the foreseeable future it will be much harder to replace many of the chemicals produced today from fossil fuels with biologically produced ones. Thus, here lies the real challenges for the future—producing biofuels at lower costs and chemicals using microbial resources so as to achieve global sustainability. Nevertheless, there is a real potential for biotechnology to contribute to our reduced dependence on fossil fuels.

This is at the heart of the political construct called the bioeconomy (OECD, 2009). In this chapter we examine the role of microorganisms in the production of biofuels and chemicals for a future bioeconomy. One of those is, of course, ethanol, although the authors will give an overview beyond biofuels as the greater opportunities for economic growth lie with chemicals.

Microbial Resources. http://dx.doi.org/10.1016/B978-0-12-804765-1.00004-7

REPLACING FOSSIL FUELS WITH BIOFUELS

Biomass represents an abundant carbon-neutral renewable resource for the production of bioenergy and biotechnological advances offer the potential for the development of sustainable biopower (Ragauskas et al., 2006). Renewable biofuels are needed to displace petroleum-derived transport fuels, which contribute to global warming (Chisti, 2008). Biodiesel and bioethanol are the two potential renewable fuels that have attracted the most attention. Ethanol yields 25% more energy than the energy invested in its production and biodiesel yields 93% more energy than the energy invested in its production (Hill et al., 2006).

Estimates of the potential for bioenergy by 2050 range from 2 to 20 EJ/year or 17%–30% of projected total energy requirements (Hall and House, 1995). In Brazil, 44% of the energy matrix is renewable and 13.5% is derived from sugarcane (Soccol et al., 2010). First-generation biofuels, with the exception of sugar cane ethanol, will likely have a limited role in the future transport fuel mix (Chisti, 2008). Even dedicating all US corn and soybean production to bioethanol production would meet only 12% of gasoline demand and 6% of diesel demand (Hill et al., 2006). To achieve sustainable energy production, it will be necessary to replace easily fermentable sugars with lignocellulosic biomass that can be converted to ethanol (Balat, 2011; Himmel et al., 2007). By turning to lignocellulosic biomass it has been estimated that bioethanol production could replace 32% of the global gasoline consumption when bioethanol is used in E85 fuel for mid-size passenger vehicles (Kim and Dale, 2004). Low-cost thermochemical pretreatment, highly effective cellulases and hemicellulases and efficient and robust fermentative microorganisms will make cost-effective bioethanol production from lignocellulose biomass possible (Gray et al., 2006).

Bokinsky et al. (2011) incorporated both biomass-degrading and biofuel-producing capabilities into a single organism (an *Escherichia coli* chassis organism) through metabolic engineering techniques. They engineered strains that express cellulase, xylanase, beta-glucosidase, and xylobiosidase enzymes under control of native *E. coli* promoters selected to optimize growth on model cellulosic and hemicellulosic substrates. One engineered strain carries out pinene synthesis, butanol synthesis, and fatty acid ethyl ester synthesis. These three metabolic pathways form the basis for fuel substitutes or precursors suitable for gasoline, diesel, and jet engines directly from ionic liquid-treated switchgrass. Although unproven at scale, this nevertheless demonstrates the economic potential of the techniques that come under the umbrella of synthetic biology. The authors stated that with improvements in both biofuel synthesis pathways and biomass digestion capabilities, this approach could provide an economical route to the production of advanced biofuels.

Besides providing renewable fuel resources to augment petroleum as a source of liquid fuels, biofuels can greatly reduce GHG emissions. Relative to the fossil fuels, GHG gas emissions are reduced 12% by using bioethanol and 41% by biodiesel (Hill et al., 2006). As an example of what can be achieved, the addition of 25% ethanol to gasoline in Brazil reduced the import of 550 million barrels oil and also reduced the CO_2 emissions by 110 million tons (Soccol et al., 2010).

Hydrogen is also a potential biofuel that could greatly reduce GHG emissions. Japan has been especially interested in the biotechnological production of hydrogen to reduce GHG emissions for some time (OECD, 1994). Current Japanese Prime Minister Shinzo Abe has dubbed hydrogen the "energy of the future," and hopes it will help Tokyo meet the modest emissions targets it has set and hopes to be able to see hydrogen powered vehicles on Japanese roadways by 2020 (http://phys.org/news/2015-11-japan-lofty-hydrogen-society-vision.html#jCp). The Japanese efforts are coordinated across the academic and private sectors by the Ministry of International Trade and Industry (MITI). Much of the effort is focused on engineering algae to photosynthetically produce hydrogen in sufficient quantities to enable industrial production of hydrogen as an automotive fuel. The basic advantages of biological hydrogen production from algae over other "green" energy sources are that it does not compete for agricultural land use, and it does not pollute, as water is the only by-product of the combustion. The microalga *Chlamydomonas reinhardtii* is considered one of the most promising algal hydrogen producers and bioengineering is able to increase hydrogen production by suppressing oxygenic photosynthesis (Dubin and Ghiradi, 2015; Esquível et al., 2011; Kruse et al., 2005; Surzycki et al., 2007). This alga produces increased amounts of hydrogen when deprived of sulfur and the sulfur-deprivation process may eventually become commercially viable (Torzillo et al., 2015). In the end it may be the cost of fuel cells compared to electric batteries in automobiles that limits the use of hydrogen rather than the ability to produce sufficient amounts of hydrogen through biotechnology.

Algae and cyanobacteria may also become an important source of sustainable biomass for biofuel production (Chisti, 2008). Macroalgae-based biofuels present similar but slightly different opportunities and challenges from microalgae. Macroalgae are a readily accessible source of biomass, their culture and production methods are well developed, and the supply chain is well established. However, the physical footprint of macroalgae is larger than that of microalgae and, because it is more amenable to growth in the ocean, there are challenges in terms of containment and interaction with facilities, such as fisheries and wind farms. Challenges in terms of biofuel extraction and waste are being addressed. Bio Architecture Laboratories (BAL) (http://www.biofuelsdigest.com/bdigest/2013/05/17/nlacm-leads-to-changing-times-at-bio-architecture-lab/) has recently engineered a microbe to degrade and a pathway to metabolize alginate, the most abundant sugar in seaweed (Wargacki et al., 2012). Alginic acid/alginates constitute 20%–30% of the total dry matter content of brown seaweeds. The BAL platform can convert seaweed carbohydrates into a renewable chemical intermediate that is scalable and can be used to produce both fuels and a variety of chemicals for green plastics, surfactants, agrochemicals, synthetic fibers, and nutraceuticals.

One of the great problems with biodiesel production from algae is the extraction of the oil, which produces a lot of unwanted biomass that is difficult and expensive to dry. A direct process described by Robertson et al. (2011) combines an engineered cyanobacterial organism designed to produce and secrete an alkane diesel product continuously. The process is closed and uses industrial waste CO_2 at concentrations 50–100 times higher than atmospheric. The organism is further engineered to provide a switchable control between carbon partitioning for biomass or product.

Fatty acids composed of long alkyl chains and are akin to a natural "petroleum," being primary metabolites used by cells for both chemical and energy storage functions. These energy-rich molecules are today isolated from plant and animal oils for a diverse set of products ranging from fuels to oleochemicals. A more scalable, controllable, and economic route to this important class of chemicals would be through the microbial conversion of renewable feedstocks, such as biomass-derived carbohydrates. Steen et al. (2010) demonstrated the engineering of *E. coli* to produce structurally tailored fatty esters (biodiesel), fatty alcohols, and waxes directly from simple sugars. Furthermore, they demonstrated the engineering of biodiesel-producing cells to express hemicellulases, a step toward producing these compounds directly from hemicellulose, a major component of plant-derived biomass. Combining this natural fatty acid synthetic ability of *E. coli* with new biochemical reactions that can be realized through synthetic biology has provided a means to divert fatty acid metabolism directly toward fuel and chemical products of interest.

Besides bioethanol and biodiesel bioaviation fuels are being produced from plant biomass through chemical processes. United Airlines, for example, is purchasing bio-aviation fuel from a California-based refinery that converts sustainable feedstocks, like nonedible natural oils and agricultural wastes, into low-carbon, renewable jet fuel. This biofuel is price-competitive with traditional, petroleum-based jet fuel, but achieves a 50% reduction in carbon dioxide emissions on a life cycle basis when compared to traditional jet fuel. United Airlines announced it plans to purchase up to 15 million gallons of sustainable aviation biofuel between 2015 and 2018 and to use a 30:70 blend ratio (http://www.united.com/web/en-US/content/company/globalcitizenship/environment/alternative-fuels.aspx).

REPLACING THE OIL BARREL FOR CHEMICAL PRODUCTION

Over two decades ago, Frost and Lievense (1994) recognized that "Although aromatics are currently synthesized from benzene, toluene, and xylene derived from fossil feedstocks, environmental considerations and the scarcity of petroleum will necessitate development of other industrial routes to these molecules in nations such as the United States." Enlightened in its day, the article went on to describe the derivation of aromatic chemicals by biocatalytic routes that employ water instead of aromatic solvents, and that use low reaction temperatures and pressures. In 2013, Milken Institute (2013) (http://www.milkeninstitute.org) stated that 96% of all manufactured goods in the United States use some sort of chemical product. To replace fossil carbon in the future will need a massive source of sustainable carbon and the only possible source is renewable biomass.

To replace the oil barrel in terms of chemicals production involves hundreds if not thousands of chemicals (e.g., the short chain alkenes). Therefore, this is going to be a long-term process. Ultimately, if we cease burning fossil fuels in engines, then we could conserve crude oil to make chemicals for many generations, although the GHG gas emissions would still be a concern. Fig. 4.1 shows bio-based chemicals that have been made at least at the laboratory level through metabolic engineering in microorganisms. Some have reached commercialization, others may never do.

Most bio-based chemicals are currently more expensive than their petrochemical equivalents. Bio-based chemicals are usually cost-competitive when replacing high-value petrochemicals. However, for GHG savings, it is the high volume, low-value commodity chemicals that need to be targeted. Therefore, the task of estimating how much of the oil barrel can be replaced by renewable chemicals (and when) is fraught with difficulty. Nevertheless, Saygin et al. (2014) selected the seven most important bio-based chemicals (Table 4.1) that could *technically* replace half of petrochemical polymers and fiber consumption worldwide. A major US initiative aims to take us closer to the reality of replacing the oil barrel (Box 4.1).

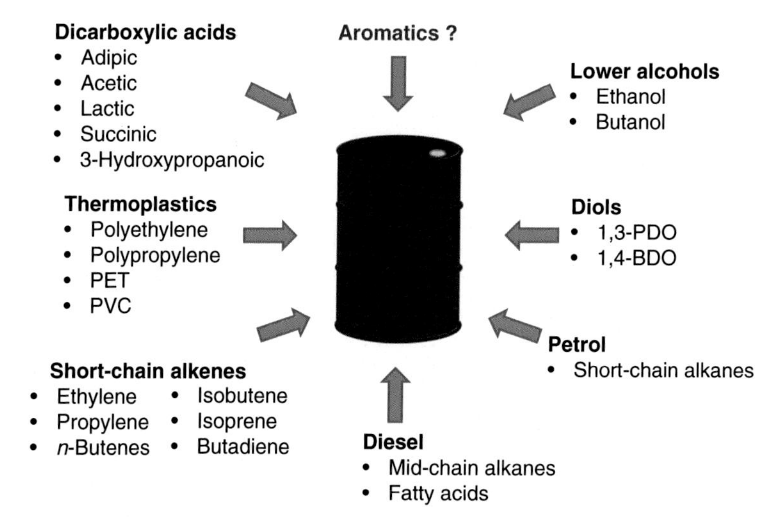

FIGURE 4.1 **Chemicals that have been made through metabolic engineering of microorganisms (OECD, 2014a).**

TABLE 4.1 The Total of Seven Polymers Could *Technically* Replace Half of the Total Polymers Production

Polymer	CO_2 emission savings (tons CO_2/ton)
Bioethylene	1.9–5.3
Biopolyethylene	2.4–4.2
Biopolyethylene terephthalate	1.9–2.5
Polyhydroxyalkanoate	1.4–4.0
Polytrimethylene terephthalate	1.1–1.9
Poly lactic acid	1.2–2.1
Starch polymers	1.7–3.6

Based on these polymers, Saygin et al. (2014) estimated a technical CO_2 emission reduction potential of 0.3–0.7 Gt CO_2 could be achieved by 2030.

Box 4.1 Living Foundries

The US Defense Advanced Research Projects Agency's (DARPA) Living Foundries programme is working with many companies, national laboratories, and universities to develop new tools to enable rapid engineering of biology. It is tackling "impossible today" industrial projects that could become "possible" if we enable, scale, and rapidly prototype genetic designs and operating systems never before accessible for industrial production. And its most recent large-scale initiative, the *1,000 Molecules Project*, seeks nothing short of a fundamental disruption of traditional chemicals and materials industries and processes by developing 1,000 new chemical building blocks for entirely new materials at the molecular scale and nanoscale in the next 3–5 years

Source: National Academy of Sciences, 2015

Short-Chain Alkenes, the Building Blocks of the Modern Petrochemical Industry

The short-chain olefins (alkenes) are extremely important due to their pivotal position in the modern petrochemicals industry (Table 4.2).

In 2010, Global Bioenergies (Evry, France) (http://www.global-bioenergies.com/) generated an artificial metabolic pathway to isobutene from glucose. Isobutene is a gas, and its recovery from a fermenter is therefore simplified. There is no product toxicity, no feedback-inhibition, and the downstream processing is inexpensive. In mid-2015, the company delivered its first batch of renewable gasoline, derived from bio-based isobutene, to Audi. The company is replicating its achievement with bio-based isobutene now with bio-based propylene and butadiene, another two members of the key gaseous olefins.

TABLE 4.2 Short Chain Alkenes and Their Uses

Olefin	Existing market $billion	Main applications
Ethylene	144	Polyethylene (60%)
Propylene	88	Polypropylene (65%)
Linear butenes	37–74	Comonomers in various plastics
Isobutene	29	Tires, organic glass, PET, fuels
Butadiene	14.6	Tires, nylon, coating polymers
Isoprene	2	Tires, adhesives

Bio-Based 1,3-Propanediol (PDO): A Metabolic Engineering Classic

1,3-PDO has appealing properties for many synthetic reactions, such as polycondensation, and for uses as solvents, adhesives, resins, detergents, and cosmetics. It is especially well known as a monomer for the synthesis of polytrimethylene terephthalate (PTT), a polyester with excellent properties for fibers, textiles, carpets, and coatings (Zeng and Sabra, 2011). Globally, the 1,3-PDO market will grow from an estimated $ 157 million in 2012 to $ 560 million in 2019 with a compound annual growth rate (CAGR) of 19.9% during the period 2012–19 (OECD, 2014a).

One of the considerations for working in *E. coli* is strains based on the *E. coli*, K12 strain are eligible for favorable regulatory status in the United States. The engineered strain relies on a carbon pathway that diverts carbon from dihydroxyacetone phosphate (DHAP), a major "pipeline" in central carbon metabolism, to 1,3-PDO.

The two most fundamental changes described were (Fig. 4.2):

1. To remove a theoretical yield limitation, the phosphotransferase (PTS) system was replaced with a synthetic system comprising galactose permease (*galP*) and glucokinase (*glk*); both genes are endogenous to *E. coli*.
2. Triosephosphate isomerase (*tpi*) was deleted in an early construct [part (A) in Fig. 4.2. But this also imposed a yield limitation. To overcome this, *gap* (glyceraldehyde 3-phosphate dehydrogenase) was downregulated, which, along with the reinstatement of *tpi* (part (B) in Fig. 4.2], provided an improved flux control point.

Along with other changes, the end result was a metabolically engineered organism that produced 1,3-PDO at a titre of 135 g/L, compared to the typical anaerobic titre of 78 g/L.

Bio-Based 1,4-Butanediol (BDO) Production: An Unnatural Molecule in a Bacterium

1,4-Butanediol (BDO) is an important commodity chemical used to manufacture over 2.5 million tons of valuable polymers annually. Currently its

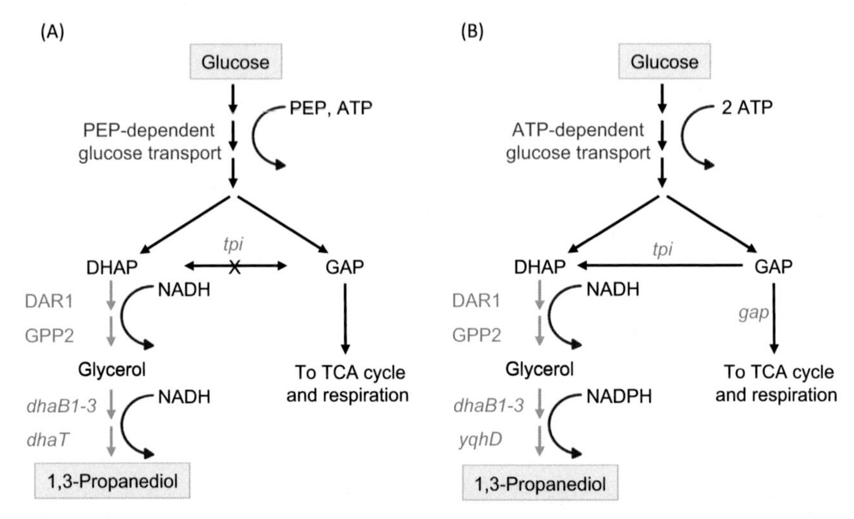

FIGURE 4.2 **Metabolic engineering for the production of 1,3-propanediol (Nakamura and Whited, 2003).** (A) an early construct; (B) a later construct with improved yield. *GAP*, glyceraldehyde 3-phosphate; *DHAP*, dihydroxyacetone phosphate.

production is entirely through petrochemistry. Unlike the other diols described here, there is no natural 1,4-BDO produced in any organism. Additionally, like many, commodity chemicals, it is highly reduced compared to carbohydrates, which makes its biosynthesis even more improbable.

As a result, Yim et al. (2011) were required to use metabolic pathway models to identify and rank the potential pathways from *E. coli* central metabolites to BDO. The forced ranking eliminated most of the potential pathways, and those remaining were prioritized according to several criteria, including the number of nonnative steps required. On the experimental side, the host *E. coli* strain required to be metabolically engineered to channel carbon and energy resources into the pathway, followed by optimization of the downstream activity. In addition, they examined nonpurified feedstocks, such as crude, mixed sugars, and biomass hydrolysates.

These challenges are likely to be similar in the production of any reduced commodity chemical in a biological process. While a commercial process would require at least a three- to fivefold increase in yield of BDO, this work represents a breakthrough as there was no prior example of high TCA flux toward a reduced compound in the literature. It is a strategy that can be applied to the design of other biocatalysts. A commercial production route from sugar to 1,4-BDO has now been described (Burgard et al., 2016).

Sugar to Plastic Through Metabolic Engineering and Fermentation

Polylactic acid (PLA) has been considered a good alternative to petroleum-based plastic because it possesses several desirable properties, such as biodegradability and biocompatibility. The major driver for its production is for large

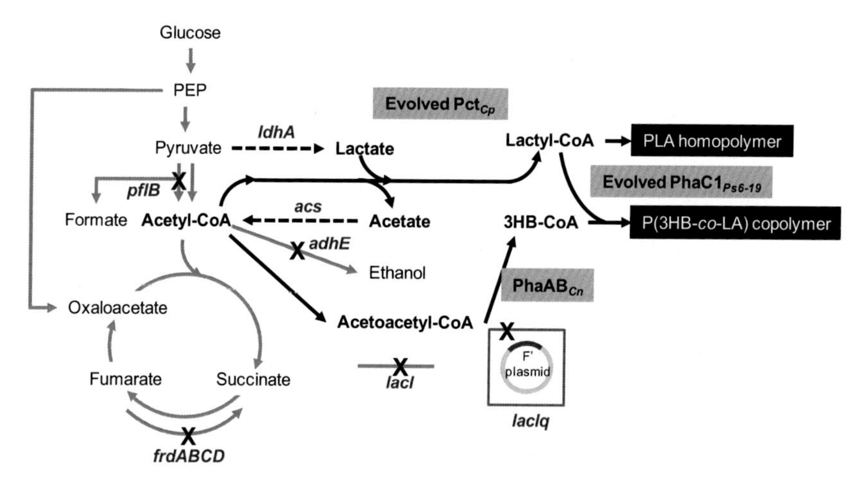

FIGURE 4.3 **Direct fermentation of glucose to PLA in *E. coli*, replacing the chemical polymerisation step.**

scale use in fibers. For example, it is being used in car interiors, replacing plastics with greater GHG emissions. Current manufacturing consists of fermentation to produce lactic acid followed by one of two major chemical routes to the polymer, both of which are difficult and either use high temperatures and solvents or heavy metal catalysts (Mehta et al., 2005). But there is no existing natural bacterial route to PLA. However, Jung and Lee (2011) described efficient production of PLA by a direct fermentation of glucose without a chemical step (Fig. 4.3) in a metabolically engineered *E. coli* chassis strain.

The overall metabolic network is shown in blue together with the introduced metabolic pathways shown in black for the production of the PLA homopolymer and the P(3HB-co-LA) copolymer in *E. coli*. The genes with cross marks shown in black represent the chromosomal gene inactivation and the elimination of F' plasmid shown in the box, and the genes with dashed arrows shown in black represent the over-expression of the genes by chromosomal promoter replacement.

TECHNICAL BARRIERS TO BIO-BASED PRODUCTION

Major investment in the development and deployment of efficient biomass conversion technologies is necessary (Hellsmark et al., 2016). There are considerable technical barriers to overcome before a significant bio-based production industry can be achieved. Some relate to the recalcitrance of cellulose and lignocellulose in the preferred feedstocks (essentially waste materials) for second generation ethanol and bio-based materials production. Another relates to the fact that microorganisms did not evolve for the operation in bioreactors at high substrate concentrations and the "extreme" conditions of industrial operations, for example, pH extremes, solvent tolerance (Burk and Van Dien, 2016).

Pretreatment of Biomass

One of the more significant challenges in utilizing the vast global lignocellulose resource is the need for large quantities of glycoside hydrolase (GH) enzymes to efficiently convert lignocellulose, hemicellulose, and cellulose into fermentable sugars (Table 4.3) (Chandel et al., 2012; van Zyl et al., 2007). The presence of lignin and hemicellulose reduce the efficiency of the hydrolysis (Sun and Cheng, 2002), but much progress has been made in the last decade in modifying enzymes. These enzymes represent the second highest contribution to raw material cost after the feedstock itself (Klein-Marcuschamer et al., 2010).

It seems likely that an efficient biomass degradation system requires a large number of enzymes to act in a coordinated fashion, and yet the individual and collective actions of these enzymes are poorly understood. An alternative approach is to combine these enzyme activities with the machinery for making bio-based products within a single bacterial biocatalyst. This is the consolidated bioprocessing (CBP) approach. The complexity of biomass conversion and the rational design approach of synthetic biology makes what has been termed "a match made in heaven" (French, 2009). CBP could potentially improve bioprocess economics (Lynd et al., 2005) by avoiding the costs of a dedicated enzyme generation step. The US Department of Energy (http://energy.gov) endorsed the view that CBP technology is widely considered the ultimate low-cost configuration for cellulose hydrolysis and fermentation (US DoE, 2006). Moreover, in the CBP strategy, cellulosic and hemicellulosic materials should be fermented simultaneously (see review by Hasunuma et al., 2013). In a recent

TABLE 4.3 Composition of Different Lignocellulosic Materials (Jørgensen et al., 2007)

	Glucose	Xylose	Arabinose	Mannose	Lignin
Hardwood					
Birch	38.2	18.5	—	1.2	22.8
Willow	43.0	24.9	1.2	3.2	24.2
Softwood					
Spruce	43.4	4.9	1.1	12.0	28.1
Pine	46.4	8.8	2.4	11.7	29.4
Grasses					
Wheat straw	38.2	21.2	2.5	0.3	23.4
Rice straw	34.2	24.5	n/d	n/d	11.9
Corn stover	35.6	18.9	2.9	0.3	12.3

Figures are percentage of total dry weight, Glucose is mainly derived from cellulose; xylose, arabinose, and mannose from hemicellulose. Lignin is comprised mainly of phenolics. *n/d*, not determined, —, below detection limit.

example, an *E. coli* strain was engineered to express recombinant xylanases and polyhydroxyalkanoate (PHA)-producing enzymes for the biosynthesis of the copolymer poly(lactate-*co*-3-hydroxybutyrate) [P(LA-*co*-3HB)] from xylan as a consolidated bioprocess (Salamanca-Cardona et al., 2016).

Inhibitory Compounds in CBP

CBP-enabling microorganisms encounter a variety of toxic compounds produced during biomass pretreatment that inhibit microbial growth and ethanol yield (Hasunuma and Kondo, 2012). However, the harsh conditions used in the pretreatment of the raw material release fermentation inhibitors including weak organic acids (particularly acetic and formic acids), furan derivatives, and phenolic compounds (there are several reviews, e.g., Almeida et al., 2007). Indeed, ethanol itself is inhibitory to xylose utilization, which is considered to be critical for the production of fuels from biomass hydrolysates (Zhang et al., 2016).

To improve fermentation ability of industrial yeast strains for ethanol production, for example, several strategies have been applied to overcome the effect of inhibitors, including: controlling inhibitor concentrations during the fermentation (Martin et al., 2007); a mutagenesis and genome shuffling approach (Zheng et al., 2011); and the overexpression of genes encoding enzymes that confer resistance toward specific inhibitors (Hasunuma and Kondo, 2012).

Titre, Yield, and Productivity

Most natural microbial processes are incompatible with an industrial process as the product titres (gram per liter of product), yields (gram product per gram substrate, often glucose), and productivity (gram per liter per hour) rates are often too low to be scalable (e.g., Harder et al., 2016). The required economic yield, titre, and productivity of a microbial process depend on whether the product is a bulk or niche chemical. The higher the value of the chemical, the more that low titre, yield, and productivity can be tolerated. For low-value, bulk chemicals, however, these factors make or break a bioprocess (Box 4.2).

Box 4.2 Glucaric Acid

Glucaric acid is an example that highlights central issues. The biggest bulk applications are its potential use as a building block for a number of polymers, including new nylons and hyper-branched polyesters (Moon et al., 2010). For example, it can be converted to adipic acid for nylon production (US Patent Application, 2010). For these and other reasons, D-glucaric acid has been identified as a *"top value-added chemical from biomass"* (Werpy and Petersen, 2004).

As a result, researchers have been exploring metabolic engineering routes to production with a view to maximising product titre. There is a natural biochemical route in mammals but it has too many steps for an industrial production, and therefore most effort has been focused on designing a microbial pathway, typically in *E. coli.*

Metabolic engineering publications often demonstrate huge potential for improvement in titres and yields. In 2009, Dueber et al. (2009) reported a 200-fold increase in glucaric acid titre, but still to only 1.7 g/L. Raman et al. (2014) achieved a 22-fold increase over their *E. coli* parent strain; however, absolute production of glucaric acid remained substantially lower (1.2 mg/L) than previously reported titres. Despite much elegant metabolic engineering in *E. coli*, available yields through microbial processes are still way too low: Schiue (2014) further improved titres, but a variety of strategies never achieved more than 5 g/L. This is a long way from the "relatively low conversion" in the glucose chemical oxidation process.

A review indicates that production titres for organic acids range from 29 to 771 g/L (Sauer et al., 2008). For comparative purposes, the yeast *Saccharomyces cerevisiae* has an outstanding capacity to produce ethanol and CO_2 from sugars with high productivity, titre, and yield: some wine strains can tolerate 15% ethanol or more. Also an oleaginous microorganism is one that can accumulate greater than 20% of its dry body mass as oil in the form of lipids (Ratledge and Wynn, 2002). A recent review of metabolic engineering studies to synthesize lactic acid (Upadhyaya et al., 2014) cited many studies with titres well over 100 g/L and yields in excess of 90%. Despite its identification as a top value-added biochemical from biomass in 2004, more than a decade later a commercially viable glucaric acid bioprocess is still elusive.

THE CONVERGENCE WITH GREEN CHEMISTRY

The greatest likelihood is that many successful academic metabolic engineering studies will not make it through to commercialization. However, the convergence of industrial biotechnology with green chemistry may increase the success rate. For example, Gerbaud et al. (2016) have proposed the computer aided molecular design (CAMD) approach for designing bio-based commodity molecules.

Green chemistry is taken to mean designing environmentally "benign" chemical processes:

- that lead to the manufacture of chemicals with lesser environmental footprint, that is, that use less energy in their production (lower temperatures and pressures);
- that use reduced levels of solvents or none at all, renewable, recyclable catalysts; and
- that lead to products with lower GHG emissions than chemicals produced from fossil raw materials and supply chains.

Increasingly there will be examples of how green chemistry and industrial biotechnology converge to solve a particular bio-based chemical synthesis. For example, a novel efficient green chemical process to lactide from fermentation-derived lactic acid was described recently (Dusselier et al., 2015). A forward-looking example is given here—artificial photosynthesis (Box 4.3).

Box 4.3 Artificial Photosynthesis

A long-term goal of basic science has been an understanding at the molecular level of how photosynthesis operates naturally and achieves the photosplitting of water. The applied science and technology implications have not been lost on practitioners—an artificial catalyst that mimics the small inorganic oxygen evolution complex (OEC) cluster within the much larger photosystem II (PSII) enzyme could be used to create fuels, such as hydrogen from water via sunlight (Hammarström et al., 1998).

In nature, the OEC in PSII catalysis the photosplitting of water into oxygen by plants, algae, and cyanobacteria. The resulting electrons and protons are then ultimately used to create adenosine triphosphate (ATP) to convert CO_2 into organic compounds (Sun, 2015). Over the years a tremendous effort has been involved in trying to produce an artificial structure that mimics OEC that can be used for engineering purposes. The closest to a working model, for example, Zhang et al. (2015) still has weaknesses as a functioning catalyst. Making the synthetic models work as real water oxidation catalysts will require consideration of other structural features. However, significant progress has been made along the way to expropriating natural photosynthesis to exploit solar energy to replace fuels and petrochemicals.

GENE AND GENOME EDITING IN PRODUCTION STRAINS-THE FUTURE FOR BIO-BASED PRODUCTION?

For several decades, antibiotic resistance markers have been used in genetic engineering as a means of selection and maintenance of recombinant plasmids in various industrial microorganisms. Plasmid instability and loss in a bacterial culture during cell growth is mainly caused by factors leading to an uneven distribution of plasmids to the daughter cells during cell division. Loss of the metabolic burden of carrying a plasmid leads to production run failure as the desired product is usually encoded on the plasmid and the plasmid loss represents a major industrial problem (Hägg et al., 2004).

The use of some antibiotic resistance genes on the plasmid followed by adding the corresponding antibiotic to the growth medium represents the most conventional selection pressure for plasmid maintenance. Moreover, antibiotic-based selection has long been used in production systems, both in fermentation and in finished products. This has proven very efficacious in the prevention of microbial contamination in large-scale fermentation (Sodoyer et al., 2012). Antibiotic resistance markers have come under scrutiny for various reasons, including biosafety. There is tentative evidence, for example, that resistance genes deriving from synthetic plasmids released into the environment could represent an unrecognized source of antibiotic resistance (Chen et al., 2012).

Several strategies for the production of strains free of antibiotic markers have been devised, all of them with problems. In future, chromosomal integration is likely to be more prevalent to provide genetic and phenotypic stability

(Wiles et al., 2003). To avoid the use of recombinant plasmids and antibiotic markers altogether, Lemuth et al. (2011) created a strain harbouring the biosynthetic genes that are required for the formation of astaxanthin in *E. coli* stably inserted into the chromosome. Such a strategy also has drawbacks, particularly a limitation of gene dosage, but this can be at least partially overcome. Nevertheless, this may have applications with very high-value bio-based chemicals. Synthetic astaxanthin is a high-value chemical used to create the pink color in farmed salmon. Biological sources are still limited and extremely expensive. However, the efficiency of integration decreases when native genes or promoters are present on the fragment to be integrated, or in the case of multiple simultaneous integrations (Ronda et al., 2015). Increasing the efficiency of targeted integration without selection is therefore important for accelerating and potentially automating the strain engineering process.

Despite recent advances, the sheer size of even the smallest bacterial genomes renders serial modification of limited utility for truly genome-scale engineering endeavours. Targeted genome editing and engineering have till recently been laborious and costly. Efficient methods enabling multiplex genome editing are urgently needed (Esvelt and Wang, 2013). Here, progress has been rapid even in the last 3 years. In combination with more sophisticated metabolic modeling tools, such new techniques will substantially accelerate metabolic engineering (Sandoval et al., 2012) and could transform the costs involved in making new strains for bio-based production. Chromosomal insertion and gene editing have particular potential in *Saccharomyces cerevisiae*, the "original" industrial microbe used for ethanol production (Box 4.4) (Fig. 4.4).

Box 4.4 *Saccharomyces cerevisiae* and gene editing: the ultimate bio-based production tools?

Saccharomyces cerevisiae has many advantages as an industrial microorganism. In-depth knowledge of its genetics, genetic engineering, physiology, and biochemistry has been accumulated, and industrial-scale fermentation technologies are readily available (for a review, see Nevoigt, 2008). "Traditional" brewing techniques utilize *Saccharomyces cerevisiae* strains because they have high ethanol yield, high productivity, and their tolerance to high ethanol concentrations keeps distillation costs low (Kasavi et al., 2012). As a eukaryote, it has the sub-cellular machinery for performing posttranslational protein modification. And of course, it plays a central role in modern industrial biotechnology in bioethanol production as a biofuel, both first and second generation.

A particular advantage in synthetic biology is its extraordinary efficiency of homologous recombination Kuijpers et al. (2013). In vivo recombination of overlapping DNA fragments for assembly of large DNA constructs in this yeast holds great potential for pathway engineering for automated high-throughput strain construction. Kuijpers et al. (2013) described a robust method for in vivo assembly of plasmids for use in *S. cerevisiae* strain engineering.

But its efficiency of homologous recombination has also facilitated targeted manipulations within chromosomes (Klinner and Schäfer, 2004). Amyris, to avoid problems with plasmids, takes advantage of this homologous recombination efficiency to work on chromosomal insertion. Since the adoption of methods using their patented standardized linker system and an automated assembly process, Amyris yeast DNA construction costs have dropped more than 90%, its capacity has increased at least 10-fold, and the fabrication success rates are greater than 95% (Gardner and Hawkins, 2013).

Another extremely important attribute is the fortuitous positive and negative selection system based on *URA3*, a gene on chromosome 5 in *Saccharomyces*. It encodes orotidine-5′phosphate decarboxylase, required for the biosynthesis of uracil. Positive selection is carried out by auxotrophic complementation of *ura3* mutations, whereas highly discriminating negative selection is based on the specific inhibitor 5-fluoro-orotic acid (5-FOA) that prevents growth of the prototrophic strains but allows growth of the *ura3* mutants. This works because 5-FOA appears to be converted to the toxic compound 5-fluorouracil by the action of decarboxylase. In other words, cells transformed with plasmids that contain the wild-type genes and a *ura* marker can be isolated by selecting for growth without uracil, and plasmid-free cells can be recovered by growth with 5-FOA (Forsburg, 2001). Because of this negative selection and its small size, *URA3* is the most widely used yeast marker in yeast vectors, freeing strain engineering from the need for antibiotic resistance markers.

The very recent development of CRISPR/Cas9 genome editing tools may be the ultimate solution for making *S. cerevisiae* production strains (and bacterial strains, for that matter). DiCarlo et al. (2013) first described the high specificity of CRISPR/Cas9 for *S. cerevisiae* gene editing. The CrEdit (CRISPR/Cas9 mediated genome Editing) (Ronda et al., 2015) combines CRISPR/Cas9 with the convenient genome engineering tool EasyClone (Jensen et al., 2014). They achieved highly efficient and accurate simultaneous genomic integration of multiple pathway gene expression cassettes in different loci in the genome of *S. cerevisiae*. In the same paper they describe integration of three pathway genes involved in the production of β-carotene (Fig. 4.4), a high-value food supplement (Borowitzka, 2013), at three different integration sites located on three different *S. cerevisiae* chromosomes.

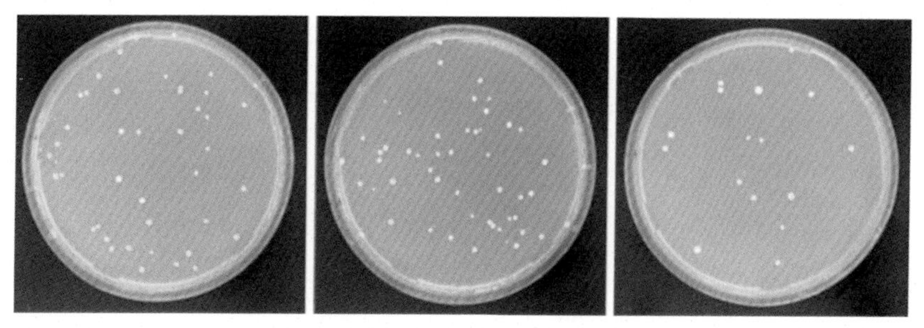

FIGURE 4.4 **Single-step integration of the beta-carotenoid pathway created *S. cerevisiae* colonies that accumulate β-carotene, resulting in an orange pigmentation.**

SHORTENING THE INNOVATION CYCLE TIME: THE ULTIMATE CONTRIBUTION OF SYNTHETIC BIOLOGY?

There is a large number of ways in which synthetic biology can improve the production of bio-based products. As an overall contribution, perhaps the greatest will be reducing the innovation cycle time. Currently it takes on average 7.4 years to launch a bio-based product (Il Bioeconomista, 2015). The development of a viable product containing synthetic parts remains a Herculean effort. It took DuPont and Genencor approximately 15 years and 575 person years to develop and produce 1,3-PDO (Hodgman and Jewett, 2012). It cost $ 25 million to genetically engineer *E. coli* and yeast to produce the chemical precursor to the antimalarial drug artemisinin (National Academy of Sciences, 2013). However, it took 5 years to commercialize BDO (Yim et al., 2011). Anecdotal evidence from industry insiders suggests that the innovation cycle is constantly dropping as better hardware, better software, and more in-house experience is gained in synthetic biology and strain construction.

FUTURE PROSPECTS: OVERCOMING THE BOTTLENECKS IN BIO-BASED PRODUCTION

Synthetic biologists who work in industry pushed for standards that would simplify commercial decisions. For instance, organisms that make products well in the lab do not always work when grown to bigger scales, Hayden (2015).

Despite the fact that at least 26 bio-based chemicals have been identified at technology readiness level (TRL) 8–9 (European Commission, 2015), bio-based market capitalization is still low. Why the deluge of metabolically engineered, bio-based drop-in fuels and chemicals has not already occurred? Partly the reason is the low oil price. The more fundamental technical answers relate to the engineering design cycle (Fig. 4.5), and essential differences between the scientific method (test a hypothesis through experimentation) and engineering design (design a solution to a problem and test the outcome) (http://research.uc.edu/sciencefair/resources-forms/topic-suggestions/

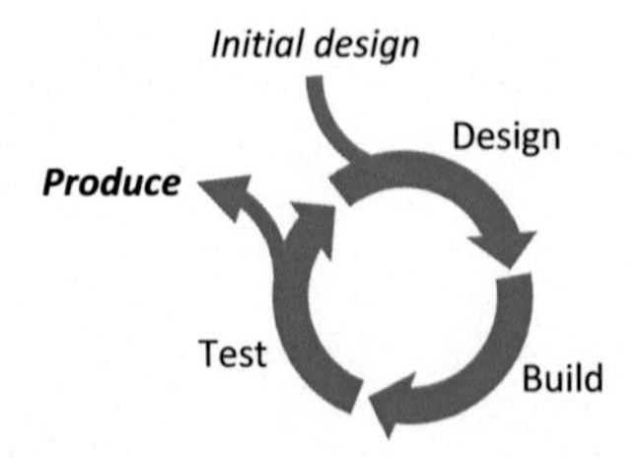

FIGURE 4.5 **The engineering design cycle.** Many variants on the above exist, but this shows the basic elements of engineering design through a phase of initial design, building and testing of a part/system/device. Nobody expects an optimal design at the first attempt. Therefore, the process is iterated as often as necessary to meet the engineering specifications.

scientific-method-v-engineering-design-procedures.aspx). Concepts, such as interoperability, separation of design from manufacture, standardization of parts and systems, all of which are central to engineering disciplines, have been largely absent from biotechnology (OECD, 2014b).

In metabolic engineering, the design cycle is restrained by the sheer number of variants that require to be tested before the optimum variant is achieved. As the rate of product development is directly related to the throughput of the design cycle, it is necessary to identify where the bottlenecks are and overcome them.

Despite the huge advances achieved in DNA sequencing and forward engineering, metabolic engineering design is still a flawed process. Many more designs have to be evaluated than in, say, a civil engineering design process. A huge number of designs for evaluation is at odds with an overly strict interpretation of the *one factor at a time* (OFAT) approach taught in the scientific method. The build phase is no longer the challenge that it was even 2 years ago. DNA synthesis prices have dropped to the point that ordering several genes is trivial for many laboratories, although there is still some way to go. Combined with other tools, genome engineering has made the build process routine: construction of billions of genomic variants is now possible in a single day.

The Test Phase is the Current Bottleneck

Phenotype evaluation is a major rate-limiting step in metabolic engineering. When constructing production strains for biofuels or bio-based chemicals, design success will be measured in the amount of product formed. If this requires the separation of individual strains into, say, 96 well plates and the determination of the concentration of the chemical of interest in each, then the process of multiplexing in design and build has been defeated: in effect, this results in demultiplexing (Rogers and Church, 2016). The throughput is limited to hundreds or thousands of design evaluations per day. Improving this throughput by mechanical or electronic automation is going to be limited as the orders of magnitude of improvement needed are so high. The advances necessary have to come from biology itself.

Thorough characterization of biological parts is the bedrock for abstraction and complexity in engineered biological systems. Biosensors are of primary importance as they provide channels of communication into and out of the cell. Measurement of intracellular metabolites is often desirable, but is typically a slow and destructive process. Rogers et al. (2015) proposed the use of fluorescent biosensors because when intracellular metabolite concentration is transduced to fluorescence, high-throughput single cell measurements become possible. When coupled to intracellular metabolite production, each cell expresses a fluorescent protein at a rate proportional to its ability to produce the target compound. Therefore such coupled biosensors reporting on gene expression enable rapid and multiplexed phenotype evaluation, facilitating faster design-build-test cycles (Rogers et al., 2015).

However, it is well recognized that nongenetic variation is known to arise in microbial cultures as a result of naturally inherent factors. These factors are known to generate huge variation in protein and metabolite concentrations, regardless of plasmid-based or chromosome-based gene expression (e.g., Zenobi, 2013). This suggests the existence of both high- and low-performance variants in all cultures. Xiao et al. (2016) developed and validated a biosensor tool termed in vivo PopQC (population quality control) to exploit nongenetic variation for enhanced biosynthetic performance. They found that most cells in an *E. coli* population exhibit low biosynthetic performance and that only a small fraction of the population generates a majority of product (in this case free fatty acid). PopQC uses an intracellular product–responsive biosensor to regulate the expression of a survival gene, which continuously selects for high-performing, nongenetic variants (Fig. 4.6). If applicable to a wide range of metabolically engineered systems, the test process in bio-based chemical production would be simplified due to a larger fraction of the population already being high performers.

Small-Scale Fermentation Models

Fermenters are the ultimate arbiters of process optimization, but they are costly to run and typically require expert supervision. While multiplexing, the design and test process should drastically reduce the number of strains to be tested in a fermenter, the fermenter is still necessary to make sure that it is truly the best

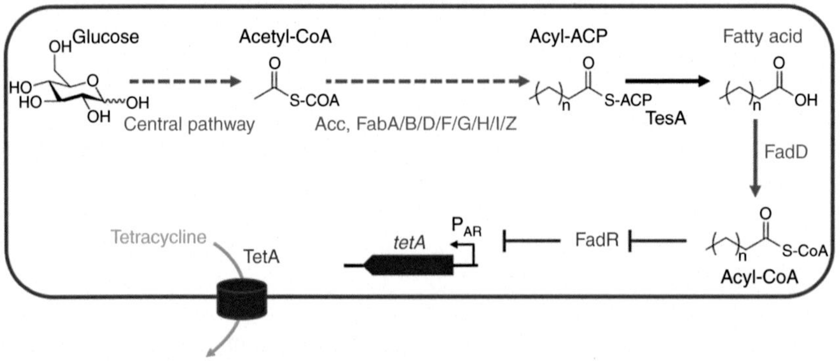

FIGURE 4.6 **PopQC confers a growth advantage to non-genetic high-performing cells, increasing their proportion in the whole population.** The FFA-responsive transcription factor FadR, which responds to FFA via acyl-CoA, and the synthetic promoter P_{AR} are used to regulate the expression of a tetracycline efflux protein (encoded by *tetA*). In the presence of tetracycline, high-performance cells can out-compete low-performance cells and dominate the population. Native enzymes and pathways (*blue*), engineered enzymes (*black*) and sensing and selection genes (*purple*) are shown. *Acc*, Acetyl-CoA carboxylase; *FabA/B/D/F/G/H/I/Z*, multisubunit fatty acid synthase; *TesA*, acyl-ACP thioesterase (with leader sequence deleted); *FadD*, acyl-CoA synthetase; *FadR*, FFA-responsive transcription factor (Xiao et al., 2016).

strains that are chosen for industrial production. Small-scale fermentation is lacking in a number of areas, such as pH and aeration control and the ability to sample frequently, and it is hoped that the solutions will lie in microfluidics (Burk and Van Dien, 2016).

Reproducibility is a Continuing Problem

Early in the history of synthetic biology, reproducibility was highlighted as a challenge (Kwok, 2010) and this remains so (Beal et al., 2016; Hayden, 2015). This has to be conquered for bio-based manufacturing to take its place as a credible manufacturing platform of the future.

Many researchers have called for completely new computational languages for biotechnology. It is argued that variants of natural languages, such as English are too imprecise and ambiguous to be useful in tackling the highly complex systems of biology and biotechnology. Antha is perhaps the first *bona fide* attempt to create a programming language for general purpose computation in biology (Sadowski et al., 2016). It is built on Google's Go programming language, but incorporates domain-specific features, such as liquid handling planning. It is claimed to enable experiments of an entirely new level of complexity. It embraces the departure from OFAT by enabling detection of interactions between different experimental factors.

The creator of Antha, Synthace of London, exemplified the challenge. Synthace worked with Merck to create a new microbial manufacturing platform for biotherapeutics. They examined the interactions between 27 factors in order to integrate strain construction with process development. This is far too large a dimensional space to address with a screening approach. Even screening a billion assays a second would require billions of years to investigate every permutation of these 27 genetic and process factors. Using multifactorial methods navigated this space, revealing key factor interactions in, of course, a small fraction of the time.

It is hoped that the combination of automated execution of highly complex experimental designs facilitated by novel programming languages will end the problem of irreproducibility.

Conquering These Challenges Will Move the Bottleneck to Data Analysis and Storage

A fully multiplexed design-build-test cycle that links phenotype to DNA sequence will enable the evaluation of millions of designs per cycle. However, this will create an unprecedented amount of data, which may move the bottleneck to data storage. It will also necessitate a new data analysis pipeline that simplifies the interrogation of phenotype-sequence relationships. In the age of machine learning, ultimately the data should inform the next iteration of design without lumpen human intervention (Rogers and Church, 2016). For example,

AutoBioCAD promises to design genetic circuits for *E. coli* with virtually no human user input (Rodrigo and Jaramillo, 2013).

CONCLUDING REMARKS

Bio-based production needs to be placed into the wider context of the sustainable agenda (El-Chichakli et al., 2016). The essence of the bioeconomy is how to reconcile the future demands for the trinity of food, energy, and chemicals. Much of the world has embarked on energy reform. It is clearly a long, expensive process. It is harder to foresee a society with drastic reductions in chemical products without large lifestyle changes, for example, a large reduction in the use of plastics. All three currently are highly dependent on fossil fuel resources, and that will ultimately change. Nevertheless, the futures of food, energy, and chemicals are interlinked, which is essentially the over-arching challenge for policymakers and society.

At the end of 2015 the European Commission announced that: *The shift to a European bio-economy is now irreversible and this transition will now accelerate after the COP21*. A central part of this vision for a European bioeconomy is bio-based production. There are now over 40 countries for which key objectives for a bioeconomy are now embedded in their strategic activities (Bioökonomierat, 2015). Each varies in its approach, such as Finland, which has 29 biorefineries either in operation or under construction.

Major changes at the socioeconomic level will require societal acceptance. There are many aspects to this. Negative perceptions of genetic modification are likely to cause the bioeconomy to develop unevenly: Asia and the Americas have seen much less resistance to genetic modification than Europe. Communication targeting public understanding of the value of scientific advancements is a key factor. It is difficult to convey a need for a switch to bio-based production of, say, ethylene if the public fail to realize the central role of ethylene in our society. On the other hand, the same technology could be used to *make morphine production as simple as brewing beer* (Ehrenberg, 2015), opening the possibility for the perversion of the technology for illegal drug production.

Conveying the benefits of a technology can therefore be difficult, much more so than simply seeing the negatives impacting public perception. Better economic data are needed on bioeconomy development, to properly inform policy making (Carlson, 2016), and to instill confidence in society at large. Technology developers and governments should clearly illustrate the overweighing benefits of technological applications for the advancement of mankind.

DISCLAIMER STATEMENT

The views expressed are those of the authors and not necessarily those of the OECD or of the governments of its member countries.

REFERENCES

Almeida, J.R.M., Modig, T., Petersson, A., Hahn-Hägerdal, B., Lidén, G., Gorwa-Grauslund, M.F., 2007. Increased tolerance and conversion of inhibitors in lignocellulosic hydrolysates by *Saccharomyces cerevisiae*. J. Chem. Technol. Biotechnol. 82, 340–349.

Balat, M., 2011. Production of bioethanol from lignocellulosic materials via the biochemical pathway: a review. Energ. Convers. Manage. 52, 858–875.

Beal, J., Haddock-Angelli, T., Gershater, M., de Mora, K., Lizarazo, M., Hollenhorst, J., Rettberg, R., 2016. Reproducibility of fluorescent expression from engineered biological constructs in *E. coli*. PLoS ONE 11 (3), e0150182.

Bioökonomierat, 2015. Bioeconomy Policy Synopsis and Analysis of Strategies in the G7. The Office of the Bioeconomy Council, German Federal Government, Berlin, Germany.

Bokinsky, G., Peralta-Yahya, P.P., George, A., Holmes, B.M., Steen, E.J., Dietrich, J., Lee, T.S., Tullman-Ercek, D., Voigt, C.A., Simmons, B.A., Keasling, J.D., 2011. Synthesis of three advanced biofuels from ionic liquid-pretreated switchgrass using engineered *Escherichia coli*. Proc. Natl. Acad. Sci. USA 108, 19949–19954.

Borowitzka, M.A., 2013. High-value products from microalgae—their development and commercialisation. J. Appl. Phycol. 25 (3), 743–756.

Burgard, A., Burk, M.J., Osterhout, R., Van Dien, S., Yim, H., 2016. Development of a commercial scale process for production of 1,4-butanediol from sugar. Curr. Opin. Biotechnol. 42, 118–125.

Burk, M.J., Van Dien, S., 2016. Biotechnology and chemical production: challenges and opportunities. Trends Biotechnol. 34, 187–190.

Carlson, R., 2016. Estimating the biotech sector's contribution to the US economy. Nat. Biotechnol. 34, 247–255.

Chandel, A.K., Chandrasekhar, G., Silva, M.B., Silvério da Silva, S., 2012. The realm of cellulases in biorefinery development. Crit. Rev. Biotechnol. 32, 187–202.

Chen, J., Jin, M., Qiu, Z.G., Guo, C., Chen, Z.L., Shen, Z.Q., Wang, X.W., Li, J.W., 2012. A survey of drug resistance *bla* genes originating from synthetic plasmid vectors in six Chinese rivers. Environ. Sci. Technol. 46, 13448–13454.

Chisti, Y., 2008. Biodiesel from microalgae beats bioethanol. Trends Biotechnol. 26, 126–131.

DiCarlo, J.E., Norville, J.E., Mali, P., Rios, X., Aach, J., Church, G.M., 2013. Genome engineering in *Saccharomyces cerevisiae* using CRISPR-Cas systems. Nucleic Acids Res. 41, 4336–4343.

Dubin, A., Ghiradi, M.L., 2015. Engineering photosynthetic organisms for the production of biohydrogen. Photosynth. Res. 123, 241–253.

Dueber, J.E., Wu, G.C., Malmirchegini, G.R., Moon, T.S., Petzold, C.J., Ullal, A.V., Prather, K.L.J., Keasling, J.D., 2009. Synthetic protein scaffolds provide modular control over metabolic flux. Nat. Biotechnol. 27, 753–759.

Dusselier, M., Van Wouwe, P., Dewaele, A., Jacobs, P.A., Sels, B.F., 2015. Shape-selective zeolite catalysis for bioplastics production. Science 349, 78–80.

Ehrenberg, R., 2015. Engineered yeast paves way for home-brew heroin. Nature 521, 267–268.

El-Chichakli, B., Von Braun, J., Lang, C., Barben, D., Philp, J., 2016. Five cornerstones of a global bioeconomy. Nature 535, 221–223.

Esquível, M.G., Amaro, H.M., Pinto, T.S., Fevereiro, P.S., Malcata, F.X., 2011. Efficient H_2 production via *Chlamydomonas reinhardtii*. Trends Biotechnol. 29 (12), 595–600.

Esvelt, K.M., Wang, H.H., 2013. Genome-scale engineering for systems and synthetic biology. Mol. Syst. Biol. 9, 641.

European Commission, 2015. From the sugar platform to biofuels and biochemical. Final report for the European Commission Directorate-General Energy. ENER/C2/423-2012/SI2.673791, April 2015.

Forsburg, S.L., 2001. The art and design of genetic screens: yeast. Nat. Rev. Genet. 2, 659–668.

French, C.E., 2009. Synthetic biology and biomass conversion: a match made in heaven? J. R. Soc. Interface 6, S547–S558.

Frost, J.W., Lievense, J., 1994. Prospects for biocatalytic synthesis of aromatics in the 21st century. New J. Chem. 18, 341–348.

Gardner, T.S., Hawkins, K., 2013. Synthetic Biology: evolution or revolution? A co-founder's perspective. Curr. Opin. Chem. Biol. 17, 1–7.

Gerbaud, V., Dos Santos, M.T., Pandya, N., Aubry, J.M., 2016. Computer aided framework for designing bio-based commodity molecules with enhanced properties. Chem. Eng. Sci. 159, 177–193.

Gray, K.A., Zhao, L., Emptage, M., 2006. Bioethanol. Curr. Opin. Chem. Biol. 10, 141–146.

Hägg, P., De Pohl, J.W., Abdulkarim, F., Isaksson, L.A., 2004. A host/plasmid system that is not dependent on antibiotics and antibiotic resistance genes for stable plasmid maintenance in *Escherichia coli*. J. Biotechnol. 111, 17–30.

Hall, D.O., House, J.I., 1995. Biomass energy in Western Europe to 2050. Land Use Policy 12, 37–48.

Hammarström, L., Sun, L., Åkermark, B., Styring, S., 1998. Artificial photosynthesis: towards functional mimics of photosystem II? Biochim. Biophys. Acta 1365, 193–199.

Harder, B.-J., Bettenbrock, K., Klamt, S., 2016. Model-based metabolic engineering enables high yield itaconic acid production by *Escherichia coli*. Metabol. Eng. 38, 29–37.

Hasunuma, T., Kondo, A., 2012. Development of yeast cell factories for consolidated bioprocessing of lignocellulose to bioethanol through cell surface engineering. Biotechnol. Adv. 30, 1207–1218.

Hasunuma, T., Okazaki, F., Okai, N., Hara, K.Y., Ishii, J., Kondo, A., 2013. A review of enzymes and microorganisms for lignocellulosic biorefinery and the possibility of their application to consolidated bioprocessing technology. Bioresour. Technol. 135, 513–522.

Hayden, E.C., 2015. Synthetic biology called to order. Meeting launches effort to develop standards for fast-moving field. Nature 520, 141–142.

Hellsmark, H., Mossberg, J., Söderholm, P., Frishammar, J., 2016. Innovation system strengths and weaknesses in progressing sustainable technology: the case of Swedish biorefinery development. J. Clean. Prod. 131, 702–715.

Hill, J., Nelson, E., Tilman, D., Polasky, S., Tiffany, D., 2006. Environmental, economic, and energetic costs and benefits of biodiesel and ethanol biofuels. Proc. Natl. Acad. Sci. 103, 11206–11210.

Himmel, M., Himmel, E., Ding, S.Y., Johnson, D.K., Adney, W.S., Nimlos, M.R., Brady, J.W., Foust, T.D., 2007. Biomass recalcitrance: engineering plants and enzymes for biofuels production. Science 315, 804–807.

Hodgman, C.E., Jewett, M.C., 2012. Cell-free synthetic biology: thinking outside the cell. Metabol. Eng. 14, 261–269.

Il Bioeconomista, 2015. Synbio start-ups need an average of 7.4 years to launch first chemical product. Available from: https://ilbioeconomista.com/2015/06/10/synbio-start-ups-need-an-average-of-7-4-years-to-launch-first-chemical-product/

Jensen, N.B., Strucko, T., Kildegaard, K.R., David, F., Maury, J., Mortensen, U.H., Forster, J., Nielsen, J., Borodina, I., 2014. EasyClone: method for iterative chromosomal integration of multiple genes *Saccharomyces cerevisiae*. FEMS Yeast Res. 14, 238–248.

Jørgensen, H., Kristensen, J.B., Felby, C., 2007. Enzymatic conversion of lignocellulose into fermentable sugars: challenges and opportunities. Biofuel. Bioprod. Bior. 1, 119–134.

Jung, Y.K., Lee, S.Y., 2011. Efficient production of polylactic acid and its copolymers by metabolically engineered *Escherichia coli*. J. Biotechnol. 151, 94–101.

Kasavi, C., Finore, I., Lama, L., Nicolaus, B., Oliver, S.G., Oner, E.T., Kirdar, B., 2012. Evaluation of industrial *Saccharomyces cerevisiae* strains for ethanol production from biomass. Biomass Bioenerg. 45, 230–238.

Kim, S., Dale, B.E., 2004. Global potential bioethanol production from wasted crops and crop residues. Biomass Bioenerg. 26, 361–375.

Klein-Marcuschamer, D., Oleskowicz-Popiel, P., Simmons, B.A., Blanch, H.W., 2010. Technoeconomic analysis of biofuels: a wiki-based platform for lignocellulosic biorefineries. Biomass Bioenerg. 34, 1914–1921.

Klinner, U., Schäfer, B., 2004. Genetic aspects of targeted insertion mutagenesis in yeasts. FEMS Microbiol. Rev. 28, 201–223.

Kruse, O., Rupprecht, J., Bader, K., Thomas-Hall, S., Schenk, P., Finazzi, G., Hankamer, B., 2005. Improved photobiological H_2 production in engineered green algal cells. J. Biol. Chem. 280, 34170–34177.

Kuijpers, N.G., Solis-Escalante, D., Bosman, L., van den Broek, M., Pronk, J.T., Daran, J.M., Daran-Lapujade, P., 2013. A versatile, efficient strategy for assembly of multi-fragment expression vectors in Saccharomyces cerevisiae using 60 bp synthetic recombination sequences. Microb. Cell Fact. 12, 47.

Kwok, R., 2010. Five hard truths for synthetic biology. Nature 463, 288–290.

Lemuth, K., Steuer, K., Albermann, C., 2011. Engineering of a plasmid-free Escherichia coli strain for improved in vivo biosynthesis of astaxanthin. Microb. Cell Fact. 10, 29, 1–12.

Lynd, L.R., van Zyl, W.H., McBride, J.E., Laser, M., 2005. Consolidated bioprocessing of cellulosic biomass: an update. Curr. Opin. Biotechnol. 16, 577–583.

Martin, C., Alriksson, B., Sjöde, A., Nilvebrant, N.O., Jönsson, L.J., 2007. Dilute sulfuric acid pretreatment of agricultural and agro-industrial residues for ethanol production. Appl. Biochem. Biotechnol. 137, 339–352.

Mehta, R., Kumar, V., Bhunia, H., Upadhyaya, S.N., 2005. Synthesis of poly(lactic acid): a review. J. Macromol. Sci. C Polymer Rev. 45, 325–349.

Milken Institute, 2013. Unleashing the power of the bio-economy. Santa Monica, CA, US.

Moon, T.S., Dueber, J.E., Shiue, E., Prather, K.L., 2010. Use of modular, synthetic scaffolds for improved production of glucaric acid in engineered *E. coli*. Metabol. Eng. 12, 298–305.

Nakamura, C.E., Whited, G.M., 2003. Metabolic engineering for the microbial production of 1,3-propanediol. Curr. Opin. Biotechnol. 14, 454–459.

National Academy of Sciences, 2013. Positioning Synthetic Biology to Meet the Challenges of the 21st Century: Summary Report of a Six Academies Symposium Series. National Academies Press, Washington DC.

National Academy of Sciences, 2015. Industrialization of Biology: A Roadmap to Accelerate the Advanced Manufacturing of Chemicals. National Academies Press, Washington DC.

Nevoigt, E., 2008. Progress in metabolic engineering of *Saccharomyces cerevisiae*. Microbiol. Mol. Biol. Rev. 72, 379–412.

OECD, 1994. Biotechnology for a Clean Environment. OECD Publishing, Paris.

OECD, 2009. The Bio-Economy to 2030—Designing a Policy Agenda. OECD Publishing, Paris.

OECD, 2014a. Biobased Chemicals and Plastics. Finding the Right Policy Balance. OECD Science, Technology and Industry Policy Papers No. 17. OECD Publishing, Paris.

OECD, 2014b. Emerging Policy Issues in Synthetic Biology. OECD Publishing, Paris.

Ragauskas, A.J., Williams, C.K., Davison, B.H., Britovsek, G., Cairney, J., Eckert, C.A., Frederick, Jr., W.J., Hallett, J.P., Leak, D.J., Liotta, C.L., Mielenz, J.R., Murphy, R., Templer, R., Tschaplinski, T., 2006. The path forward for biofuels and biomaterials. Science 311, 484–489.

Raman, S., Rogers, J.K., Taylor, N.D., Church, G.M., 2014. Evolution-guided optimization of biosynthetic pathways. Proc. Natl. Acad. Sci. 111, 17803–17808.

Ratledge, C., Wynn, J.P., 2002. The biochemistry and molecular biology of lipid accumulation in oleaginous microorganisms. Adv. Appl. Microbiol. 51, 1–51.

Robertson, D.E., Jacobson, S.A., Morgan, F., Berry, D., Church, G.M., Afeyan, N.B., 2011. A new dawn for industrial photosynthesis. Photosynth. Res. 107, 269–277.

Rodrigo, G., Jaramillo, A., 2013. AutoBioCAD: full biodesign automation of genetic circuits. ACS Synth. Biol. 2, 230–236.

Rogers, J.K., Church, G.M., 2016. Multiplexed engineering in biology. Trends Biotechnol. 34, 198–206.

Rogers, J.K., Guzman, C.D., Taylor, N.D., Raman, S., Anderson, K., Church, G.M., 2015. Synthetic biosensors for precise gene control and real-time monitoring of metabolites. Nucleic Acids Res. 43, 7648–7660.

Ronda, C., Maury, J., Jakocˇiu¯nas, T., Jacobsen, S.A.B., Germann, S.M., Harrison, S.J., Borodina, I., Keasling, J.D., Jensen, M.K., Nielsen, A.T., 2015. CrEdit: CRISPR mediated multi-loci gene integration in *Saccharomyces cerevisiae*. Microb. Cell Fact. 14, 97.

Sadowski, M.I., Grant, C., Fell, T.S., 2016. Harnessing QbD, programming languages, and automation for reproducible biology. Trends Biotechnol. 34, 214–227.

Salamanca-Cardona, L., Scheel, R.A., Bergey, N.S., Stipanovic, A.J., Matsumoto, K., Taguchi, S. Nomura, C.T., 2016. Consolidated bioprocessing of poly(lactate-co-3-hydroxybutyrate) from xylan as a sole feedstock by genetically-engineered *Escherichia coli*. J. Biosci. Bioeng. 122(4), 406–414.

Sandoval, N.R., Kim, J.Y., Glebes, T.Y., Reeder, P.J., Aucoin, H.R., Warner, J.R., Gill, R.T., 2012. Strategy for directing combinatorial genome engineering in *Escherichia coli*. Proc. Natl. Acad. Sci. 109, 10540–10545.

Sauer, M., Porro, D., Mattanovich, D., Branduardi, P., 2008. Microbial production of organic acids: expanding the markets. Trends Biotechnol. 26, 100–108.

Saygin, D., Gielen, D.J., Draeck, M., Worrell, E., Patel, M.K., 2014. Assessment of the technical and economic potentials of biomass use for the production of steam, chemicals and polymers. Renew. Sust. Energ. Rev. 40, 1153–1167.

Schiue, E.C.J., 2014. Improvement of D-glucaric acid production in *Escherichia coli*. PhD Thesis, Massachusetts Institute of Technology.

Soccol, C.R., de Souza Vandenberghe, L.P., Pedroni Medeiros, A.B., Karp, S.G., Buckeridge, M., Ramos, L.P., Pitarelo, A.P., Ferreira-Leitão, V., Fortes Gottschalk, L.M., Ferrara, M.A., da Silva Bon, E.P., Pepe de Moraes, L.M., de Amorim Araújo, J., Gonçalves Torres, F.A., 2010. Bioethanol from lignocelluloses: status and perspectives in Brazil. Bioresour. Technol. 101, 4820–4825.

Sodoyer, R., Mignon, C., Peubez, I., Courtois, V., 2012. Antibiotic-Free Selection for Bio-Production: Moving Towards a New "Gold Standard". INTECH Open Access Publisher, Croatia.

Steen, E.J., Kang, Y., Bokinsky, G., Hu, Z., Schirmer, A., McClure, A., Del Cardayre, S.B., Keasling, J.D., 2010. Microbial production of fatty-acid-derived fuels and chemicals from plant biomass. Nature 463, 559–562.

Sun, L., 2015. A closer mimic of the oxygen evolution complex of photosystem II. Science 348, 6235–6236.

Sun, Y., Cheng, J., 2002. Hydrolysis of lignocellulosic materials for ethanol production: a review. Biores. Technol. 83, 1–11.

Surzycki, R., Cournac, L., Peltiert, G., Rochaix, J., 2007. Potential for hydrogen production with inducible chloroplast gene expression in *Chlamydomonas*. Proc. Natl. Acad. Sci. 104, 17548–17553.

Torzillo, G., Scoma, A., Faraloni, C., Giannelli, L., 2015. Advances in the biotechnology of hydrogen production with the microalga *Chlamydomonas reinhardtii*. Crit. Rev. Biotechnol. 35, 485–496.

Upadhyaya, B.P., DeVeaux, L.C., Christopher, L.P., 2014. Metabolic engineering as a tool for enhanced lactic acid production. Trends Biotechnol. 32, 637–644.

US DoE, 2006. Breaking the biological barriers to cellulosic ethanol: a joint research agenda. DOE/SC-0095. US Department of Energy Office of Science and Office of Energy Efficiency and Renewable Energy. Available from: http://www.doegenomestolife.org/biofuels/

US Patent Application, 2010. Production of adipic acid and derivatives from carbohydrate-containing materials. Publication number: US 20100317823 A1. Original assignee: Rennovia, Inc.

van Zyl, W.H., Lynd, L.R., den Haan, R., McBride, J.E., 2007. Consolidated bioprocessing for bioethanol production using *Saccharomyces cerevisiae*. Adv. Biochem. Eng. Biotechnol. 108, 205–235.

Wargacki, A.J., Leonard, E., Win, M.N., Regitsky, D.D., Santos, C.N.S., Kim, P.B., Cooper, S.R., Raisner, R.M., Herman, A., Sivitz, A.B., Lakshmanaswamy, A., Kashiyama, Y., Baker, D., Yoshikuni, Y., 2012. An engineered microbial platform for direct biofuel production from brown macroalgae. Science 335, 308–313.

Werpy, T., Petersen, G., 2004. Top Value Added Chemicals from Biomass. Volume I: Results of Screening for Potential Candidates from Sugars and Synthesis Gas. US Department of Energy, Washington, DC.

Wiles, S., Whiteley, A.S., Philp, J.C., Bailey, M.J., 2003. Development of bespoke bioluminescent reporters with the potential for in situ deployment within a phenolic-remediating wastewater treatment system. J. Microbiol. Methods 55, 667–677.

Xiao, Y., Bowen, C.H., Liu, D., Zhang, F., 2016. Exploiting nongenetic cell-to-cell variation for enhanced biosynthesis. Nat. Chem. Biol. 12, 339–344.

Yim, H., Haselbeck, R., Niu, W., Pujol-Baxley, C., Burgard, A., Boldt, J., Khandurina, J., Trawick, J.D., Osterhout, R.E., Stephen, R., Estadilla, J., Teisan, S., Schreyer, H.B., Andrae, S., Yang, T.H., Lee, S.Y., Burk, M.J., Van Dien, S., 2011. Metabolic engineering of *Escherichia coli* for direct production of 1,4-butanediol. Nat. Chem. Biol. 7, 445–452.

Zeng, A.P., Sabra, W., 2011. Microbial production of diols as platform chemicals: recent progresses. Curr. Opin. Biotechnol. 22, 749–757.

Zenobi, R., 2013. Single-cell metabolomics: analytical and biological perspectives. Science 342, 1243259.

Zhang, C., Chen, C., Dong, H., Shen, J.-R., Dau, H., Zhao, J., 2015. A synthetic Mn_4Ca-cluster mimicking the oxygen-evolving center of photosynthesis. Science 348, 690–693.

Zhang, B., Sun, H., Li, J., Wan, Y., Li, Y., Zhang, Y., 2016. High-titer-ethanol production from cellulosic hydrolysate by an engineered strain of *Saccharomyces cerevisiae* during an in situ removal process reducing the inhibition of ethanol on xylose metabolism. Process Biochem. 51, 967–972.

Zheng, D.Q., Wu, X.C., Tao, X.L., Wang, P.M., Li, P., Chi, X.Q., Li, Y.D., Yan, Q.F., Zhao, Y.H., 2011. Screening and construction of *Saccharomyces cerevisiae* strains with improved multi-tolerance and bioethanol fermentation performance. Bioresour. Technol. 102, 3020–3027.

Chapter 5

Modern Natural Products Drug Discovery and Its Relevance to Biodiversity Conservation

C. Benjamin Naman, Christopher A. Leber and William H. Gerwick
Center for Marine Biotechnology and Biomedicine, Scripps Institution of Oceanography, University of California, San Diego, CA, United States

INTRODUCTION

Natural products (NP) drug discovery research is inherently and intricately intertwined with biodiversity conservation (Cragg et al., 2012; Kingston, 2011; Kursar et al., 2006; Tan et al., 2006). NP research relies explicitly on the availability of biodiversity as the source material from which new drugs, pharmacological probes, and interesting chemical leads may be obtained. The fundamental goal of NP drug discovery is to isolate and determine the structures of small molecule NP chemicals, frequently referred to as secondary metabolites, and to aid in their development as useful societal materials (Kinghorn et al., 2011; Kingston, 2011; Newman and Cragg, 2014; Tan et al., 2006).

As a general principle, different life forms tend to elaborate different types of NPs. For example, the NPs of terrestrial and marine microorganisms, fungi, invertebrates, and higher plants each possess distinctive and unique structural features. This results from the production of NPs via enzymatic reactions, catalyzed by proteins programmed by DNA sequences that differ greatly between the source organisms. These differences in DNA reflect the disparate evolutionary forces acting on these organisms. Furthermore, geographically diverse samples tend to also be chemically diverse, and thus, the examination of the same or similar organisms from different locations has been a productive approach to the discovery of novel chemical entities (Bravo-Monzón et al., 2014; Forner et al., 2013; Li et al., 2006; Singh and Pelaez, 2008). As NP bioprospecting has become worldwide in scope, this has necessarily given rise to collaborative interactions between scientists in different countries as well as government involvement in the regulation of international science.

Microbial Resources. http://dx.doi.org/10.1016/B978-0-12-804765-1.00005-9

There have been many benefits for conducting NP drug discovery research on a worldwide scope; however, this has necessitated international policies and protocols so as to prevent potential harm and other problems. For example, harmful microorganisms, insects, or other invasive species might be transported along with scientific samples, and cause significant environmental damage. In another sense, international agreements, such as the Convention on Biological Diversity (CBD) (https://www.cbd.int/) have been enacted along with subsequent accords (Cartagena in 2003 and Nagoya in 2014) to ensure that the proper Memorandum of Understanding and benefit sharing agreements are in place, in the event that an economically valuable compound be discovered or developed from indigenous life forms, or "genetic resources," of a foreign country. Microorganisms, which are especially easily transported, have been the focus of many industrial drug discovery research programs due in large part to the scalability of their growth and compound production. These latter aspects of NP drug discovery are addressed in another chapter of this book (see Chapter 8) and are not further discussed here.

Studies of the biosynthesis of NPs in microorganisms over the last two decades have revealed that the genes encoding the biosynthetic enzymes tend to be clustered, and this has greatly accelerated their discovery, characterization, and downstream production by genetic methods. This has also enabled genomics-based NP drug discovery once the DNA sequence of an organism has been determined. DNA sequence information is not only a helpful tool for NP drug discovery, but also for the taxonomic description of biodiversity itself. Development of a genetic basis for determining the taxonomy of a given organism, such as by the 16S rRNA gene sequence for prokaryotes, has greatly improved our ability to accurately identify NP producing organisms, as well as appreciate that there is much greater diversity in microbial life forms than was previously thought. Ultimately, this information is a critical part of the data portfolio needed to better understand the value and the benefits of a biodiverse planet.

NATURAL PRODUCT DRUG DISCOVERY

Historically, NP drug discovery began with research on traditional medicines and naturally produced toxins. Morphine is a well-known example of a naturally produced substance with a long history of utilization in its crude form. In the early 1800s, it was subsequently purified and utilized for medicinal purposes as a pure substance. This discovery spurred great interest into the chemical composition of plant pharmaceuticals, and this interest later expanded to include fungi, microorganisms, and most recently, marine life forms. In their purified forms and not accounting for existing systems of traditional medicine, unmodified NPs account for approximately 5% of the drugs currently approved worldwide. Furthermore, approximately another 50% of such approved drug agents represent chemicals that were either synthetically modified from a NP or designed with inspiration from the knowledge of a NP's molecular structure.

These modifications have typically been made to improve beneficial drug properties for the final product, such as increased efficacy, reduced toxicity, more desirable routes of delivery, or for production cost effectiveness. Sometimes, the structure of the final drug product has been changed so significantly from the original lead molecule that it is difficult to discern their relationship.

Some early success stories from NP drug discovery have entered the common vernacular. For example, the alkaloidal substance quinine, which is added to "tonic waters" worldwide and is responsible for the bitter taste, was first administered to prophylactically combat malaria (Greenwood, 1992; Treadgold, 1918). More recent discoveries, made possible by technological developments that have allowed for the bioprospecting of microbial organisms, have significantly contributed in increasing the average human lifespan on a global level. In particular, the "golden age of antibiotics drug discovery" during the 1940s to 1960s included the discovery of the cephalosporin, penicillin, tetracycline, and vancomycin antibiotics. A current strategy for drug discovery and delivery involves the linking of highly efficacious NP drugs and drug candidates to designer antibodies, forming antibody-drug conjugates (ADCs), which enable highly specific drug targeting, increased efficacy, and reduced toxicity for patients. The anticancer agents, trastuzumab emtansine and brentuximab vedotin are ADCs that utilize derivatives of terrestrial and marine microbial NPs as "warhead" toxins to enact their therapeutic effects. Emtansine is a simple modification of the NP toxin maytansine whereas brentuximab vedotin possesses a derivative of the cyanobacterial metabolite dolastatin 10 (Fig. 5.1). Still, unmodified NPs, such as ecteinascidin (ET-743 or trabectedin) and homoharringtonine, and various NP chemical derivatives, such as the halichondrin A analog, eribulin, also continue to be approved as new molecular entities for use in the clinic (Fig. 5.2).

The advent of the *Self-Contained Underwater Breathing Apparatus* (SCUBA) enabled chemists, beginning in the 1960s, to begin to investigate the unique chemical constituents of marine organisms. Notable early successes include the isolation of prostaglandins in gram quantities from the Caribbean Sea Whip, *Plexaura homomalla*, characterization of the exceptionally potent seafood toxin saxitoxin, and discovery that marine algae of the phylum Rhodophyta produce an abundance of halogenated terpenoid and polyketide NPs. As modern technologies and tools have improved over the years, investigations of marine life have risen to the forefront of NP research. Many different and innovative approaches have emerged, including sourcing marine organisms from new and unique niches, conducting ecology-driven investigations, developing novel laboratory culture conditions, and utilizing genomic approaches.

Cyanobacteria have been of particular interest for drug discovery due to their abundant biosynthetic capacities to elaborate structurally diverse NPs with potential biomedical application. This has been particularly true in the search for new cancer chemotherapeutic agents. For example, the potential anticancer depsipeptide "coibamide A" was purified from a Panamanian

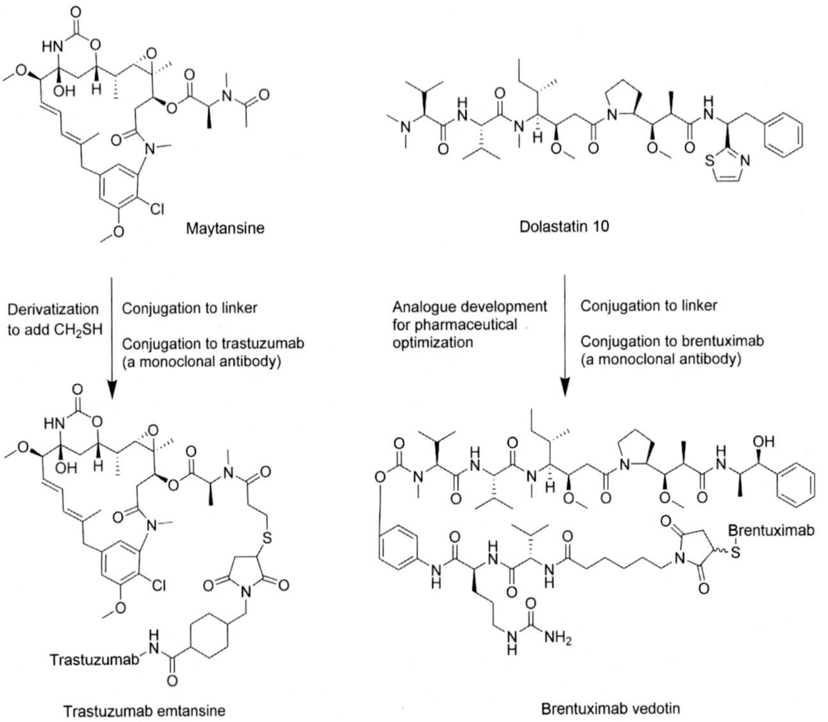

Maytansine

Dolastatin 10

Derivatization to add CH_2SH

Conjugation to linker

Conjugation to trastuzumab (a monoclonal antibody)

Analogue development for pharmaceutical optimization

Conjugation to linker

Conjugation to brentuximab (a monoclonal antibody)

Trastuzumab emtansine

Brentuximab vedotin

FIGURE 5.1 Chemical structure of the NP maytansine and antibody-drug conjugate trastuzumab emtansine along with the NP dolastatin 10 and antibody-drug conjugate brentuximab vedotin.

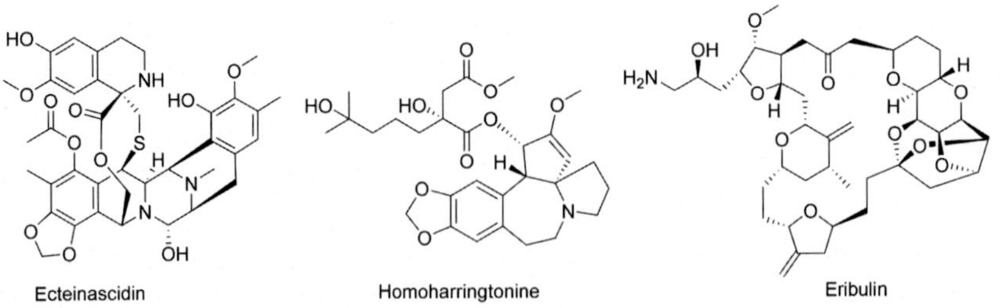

Ecteinascidin

Homoharringtonine

Eribulin

FIGURE 5.2 Chemical structures of the NP drugs ecteinascidin and homoharringtonine, and NP derivative drug eribulin.

cyanobacterium collected using SCUBA from a reef pinnacle in the Coiba National Park. Coibamide A is a potent cancer cell cytotoxin and represents an exciting lead molecule, as described later in this chapter. Another potential anticancer lead molecule is the alkaloid curacin A, which was isolated from a sample of *Moorea producens* (formerly *Lyngbya majuscula*) collected in Curaçao. Curacin A is a potent cytotoxic agent, and its discovery sparked significant

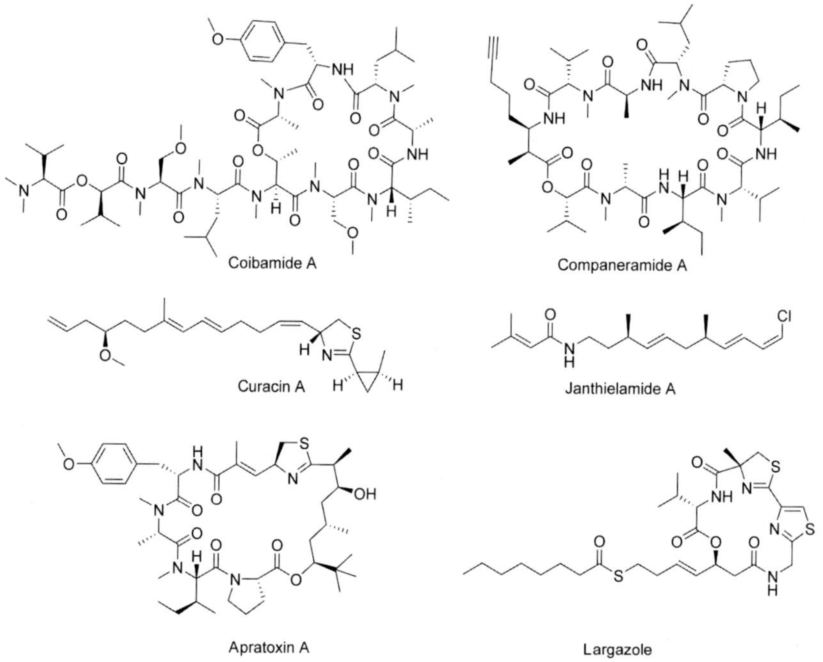

FIGURE 5.3 **Chemical structures of the biologically active cyanobacterial NPs coibamide A, companeramide A, curacin A, janthielamide A, apratoxin A, and largazole.**

conservation efforts in the island country for more than a decade. Many more examples of biologically active cyanobacterial NPs have been reported, such as apratoxin A and largazole, and there are several ongoing research programs actively exploring their potential development. The structures of these molecules are presented in Fig. 5.3.

POLICY REGARDING BIOPROSPECTING

Several sets of international conventions and policy agreements are in current enforcement that relate to the preservation and appropriate usage of biodiversity. An early example is the Convention on the International Trade of Endangered Species of Wild Flora and Fauna (CITES). Devised by the International Union for the Conservation of Nature (IUCN) in the 1960s and enacted in the early 1970s, CITES put in place regulations on the trade of endangered species to prevent their extinction and to slow the rate of biodiversity decline worldwide. The list of protected plant and animal species is maintained and periodically updated by IUCN, and also applies to products made from those organisms. Furthermore, participating countries have the opportunity to request protection of indigenous wildlife to prevent its unauthorized collection and distribution to other nations. Although CITES does not currently regulate the trade of microbial organisms, it has been successful in slowing the rate of eukaryotic species extinction, and has served as an important mechanism by which to raise societal awareness of biodiversity and its importance.

The subsequent CBD was presented at the 1992 United Nations Conference on Environment and Development. This conference was held in Rio de Janeiro at the so-called "Rio Earth Summit"; the convention itself has often been referred to as the "Rio Convention" (https://www.cbd.int/). Several issues relevant to bioprospecting were raised by the CBD, such as the "need to share costs and benefits between developed and developing countries" along with "ways and means to support innovation by local people." Prior informed consent is required in order for foreign investigators to gain access to the sovereign genetic resources of a state. In practice, this has led to intellectual property agreements that ensure the equitable sharing of profits, and are negotiated in advance of bioprospecting research programs. The CBD has become the most widespread policy adopted internationally that relates to biodiversity and bioprospecting. Unfortunately, in a few cases an unintended consequence of the CBD has been the creation of unreasonable expectations for the monetary rewards of bioprospecting, thereby inhibiting research on the NPs of organisms from those countries. In fact, the rate at which newly discovered NP molecules gain regulatory approval and clinical use is notoriously low, and this process is coupled to an immense financial outlay. The different outcomes expected by some developed and developing countries during advanced negotiations has been cited in the past as a reason that strategic partnerships fail to be formed with higher frequency. The United States signed but never ratified the CBD, effectively meaning that federally funded researchers are required to uphold the policies of the CBD whereas private industries are not. Nevertheless, many productive and mutually beneficial partnerships have been created between developed and developing countries in NP drug discovery research, and examples of two of these will be presented later in this chapter.

A supplementary agreement was added to the CBD in 2010, called the "Nagoya Protocol on Access to Genetic Resources and the Fair and Equitable Sharing of Benefits Arising from their Utilization" ("Nagoya Protocol" or ABS). The terms of the ABS did not go into effect until late 2014, after it had been signed and ratified by 50 member countries or "parties." Some of the major issues addressed by the ABS were the lack of requirement for transparency of research progress, the need to raise awareness of the importance of biodiversity in the countries both sharing and accessing genetic resources, as well as the necessity for the provision of a clear and consistent framework for benefit-sharing and prior informed consent. Some members of the scientific community have voiced concerns that the rigidity of the ABS guidelines could hamper conservation and research efforts. One such criticism of this agreement is that it requires the establishment of national focal points and competent national authorities through which all access and benefit-sharing agreements are handled. The intent of this particular article of ABS appears to be prevention of back-room dealings among individuals wishing to circumvent benefit sharing, as well as to unify the procedures for accessing biodiversity in countries both large and small. However, the fear and likely reality is that this requirement will actually lead to further delays in relationship building, as well as overwhelm national focal point offices in biodiversity hotspots. Through the end of 2015, the United States had not signed or ratified the new ABS agreement. The long-term effect of implementing these new ABS articles is not yet clear. The development of benefit-sharing agreements has always been a long-term and committed process, and not enough time has passed to yet know these consequences. However, the Nagoya Protocol Implementation Fund, a sizeable trust that offers grant money to research programs with partnerships in the private sector, has begun to provide sizeable research funding to encourage ratification and implementation of the Nagoya Protocol (https://www.thegef.org/gef/Press_release/GEF_establishes_NPIF).

BIODIVERSITY ESTIMATES

Biodiversity can be practically defined in a number of different ways: number of species or species richness, relative species abundance or evenness, concentration of endemic or rare species, as well as considerations of alternative taxonomic levels or focus on particular taxonomic clades. Each measure of biodiversity provides its own insights and shortcomings, but a universally shared limitation in quantifying biodiversity is the uncertainty associated with our knowledge of what is actually present. Our planet is host to tremendous amounts of biodiversity, but the great majority of this has not yet been studied or even discovered. There are accepted taxonomic names and descriptions for about 300,000 land plants and approximately 1.9 million species of animals. In comparison, the total number of plant species on Earth is estimated to be greater than 450,000, and the number of insect species alone is thought to range between 5 and 6 million! Estimates for total eukaryote diversity, as suggested by various models, range from 2 to 100 million, with some even suggesting that no estimate is realistic given our incomplete state of knowledge, especially in regards to fungal and insect diversity.

We have even less knowledge of prokaryote biodiversity, though the initiation of metagenomic research and the development of next-generation sequencing technology have greatly expanded our appreciation of the magnitude and ecological importance of the diverse global microbiome. Late in the 20th century, when the study of bacteria relied heavily on traditional culture methods, bacteria were classified into 11 main phyla. As of 2004, after approximately 20 years of metagenomics research, the number of detected bacterial phyla had risen to 53. The advent of next-generation sequencing has expanded our understanding of prokaryotic diversity even further. As of 2014, over 4 million 16S rRNA gene sequences have been deposited in publicly available databases, and sequence-based rarefaction curves estimate the number of prokaryote phyla to range from about 1,100 to about 1,350. Remarkably, estimates of species diversity in prokaryotes extend from approximately 30,000 to some eight orders of magnitude greater (e.g., on the order of a trillion) (Keller and Zengler, 2004; Yarza et al., 2014), and some even question the appropriateness of the "species concept" when applied to prokaryotic lifeforms (Oren and Garrity, 2014).

The continued development in understanding prokaryotic genetic diversity has been closely accompanied by increases in knowledge of the diversity of prokaryote ecological function. Prokaryotes are ubiquitous throughout Earth's ecosystems, and they serve myriad ecological roles of tremendous importance. These roles include primary production via photosynthesis and chemosynthesis, nitrogen fixation, methanogenesis and methane oxidation, nutrient remineralization, and initiation of countless other oxidation and reduction reactions that transform and significantly impact the environments in which eukaryotes (and other prokaryotes) live.

Eukaryote biodiversity is not uniformly distributed across the surface of the Earth, but is instead geographically concentrated. Abundant species tend to inhabit larger geographic ranges, whereas less abundant species tend to be more geographically concentrated. These rarer species of limited abundance and geographic range (= endemic species) often cooccur with other rare species so as to create areas of high biodiversity known as biological "hotspots." Just as eukaryotes have biogeographic patterns, there is pronounced spatial and temporal variability in microbial communities. Heterogeneous environments present diverse microhabitats and niches that support distinct and specific microbiomes. For example, the microbial community that inhabits a particular swath of oxic ocean sediment will greatly differ from the microbial communities in the suboxic and anoxic sediment layers below, and the ocean water above. Similarly, microbial communities differ greatly between such niches as sponges, corals, algal surfaces, fish intestines, and wood boring worms (i.e., shipworms). Considering the number of such microhabitats, each with its specific microbiome, makes more reasonable the estimates of the enormous number of prokaryotic life forms.

The vast biodiversity that is hosted by our planet, especially the microbial diversity, holds great promise for the continued productivity and success of NP research. The variety of prokaryotic ecological functions combined with the diversity of environments they inhabit, provides some indication of the likely extent of metabolic capabilities that these organisms possess. For this reason, each newly investigated species has significant potential to yield biologically useful NPs.

THREATS TO BIODIVERSITY

There is considerable interest to describe Earth's complete biodiversity in advance of any large-scale extinction events. Anthropogenic impacts including, most notably, habitat destruction and over-hunting and fishing have been particularly efficient in eroding biodiversity. It has been estimated that the current extinction rate for eukaryotes is at least 1000 times greater than the background extinction rate. There is grave concern that global climate change will exacerbate this situation with resulting nonlinear increases in extinction rates.

Loss of biodiversity is of great concern to ecologists and NP researchers alike in that each extinction of a species represents an irretrievable loss to the planet as well as to society. With each eukaryotic extinction, a piece of an ecosystem is lost, permanently altering how that ecosystem is structured and how it functions. Coupled with each eukaryotic extinction is the destruction of any number of microbiomes that exist within that organism, or otherwise rely on it for continued existence (e.g., coprophilous fungal microbiomes). NP researchers are thus implicit stakeholders in the preservation of biodiversity, making partnerships between NP drug discovery and biodiversity conservation both logical and necessary.

CLASSIFICATION OF BIODIVERSITY RESULTING FROM BIOPROSPECTING; CYANOBACTERIA AS AN EXAMPLE

The phylum Cyanobacteria represents a broad genetic group; until recently, proper classification systems were insufficient to appropriately delineate phylogenetic relationships in this group. Bioprospecting efforts have significantly contributed to developing a better understanding of these relationships. Previously, morphological and culture condition characteristics were used to define genera and species of cyanobacteria; however, 16S rRNA gene sequencing as well as multilocus sequence typing have revealed much greater genetic diversity in the cyanobacteria, and completely altered our understanding of their evolutionary relationships. For example, the genus *Lyngbya* was previously thought to be responsible for over 40% of the total number of NPs produced by cyanobacteria. It was not until specimens of this genus were examined phylogenetically that *Lyngbya* was revealed to be polyphyletic. It is interesting to note that when the 16S rRNA gene sequences were evaluated using the Bayesian method, these morphologically similar groups formed distinct and distant clades (Fig. 5.4). Efforts to correct this misclassification resulted in the formation of a new genus, *Moorea*.

In addition to the reclassification of previously described taxa, phylogenetic investigations can also lead to more accurate initial descriptions of newly isolated taxa. Due to systemic structural issues in the funding of scientific research, it is common that phylogenetic studies are only conducted on organisms once their potential value from bioprospecting research has been established. Consequently, deeper insights into the extent of cyanobacterial biodiversity as well as their capacity for NP biosynthesis have been hindered.

For example, a cf. *Symploca* sp. collected in Curaçao yielded a new NP, janthielamide A (Fig. 5.3). Initial taxonomic assignment was based solely on morphological characteristics; however, examination of its 16S rRNA gene

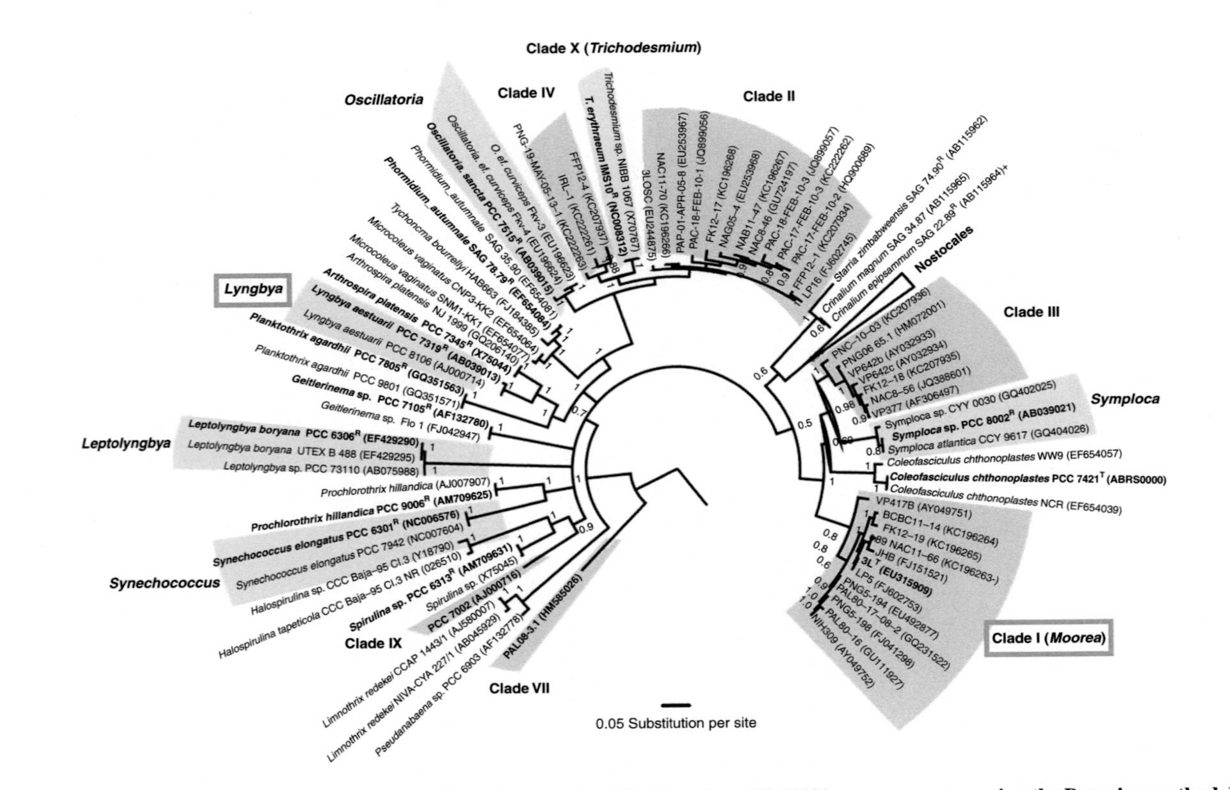

FIGURE 5.4 Evolutionary tree of marine cyanobacteria known to produce NPs based on 16S rRNA gene sequences using the Bayesian method. NAC11-66 corresponds to *Moorea producens* (formerly *Lyngbya majuscula*). The genera *Lyngbya* and *Moorea* are outlined in *orange boxes*. *(Reproduced with permission from Engene, N., Gunasekera, S.P., Gerwick, W.H., Paul, V.J., 2013. Phylogenetic inferences reveal a large extent of novel biodiversity in chemically rich tropical marine cyanobacteria. Appl. Environ. Microbiol. 79(6), 1882–1888.)*

sequence suggested that it belonged to an evolutionarily distant and undescribed genus. Similarly, a Panamanian cyanobacterium that yielded the antiplasmodial NP companeramide A (Fig. 5.3) was initially described as a *Leptolyngbya* sp. However, phylogenetic analysis indicated that it was more similar to, but significantly distinct from, the genus *Symploca*; a new genus, *Hyalidium*, has been proposed for this organism. Another Panamanian example comes from the compound coibacin A, a selective antileishmanial and antiinflammatory agent produced by a cyanobacterium thought to be of the genus *Oscillatoria*, based on its morphology. However, phylogenetic studies revealed that it is distinct and likely a representative of a cyanobacterial genus that is currently undescribed. Thus, it is unclear how many species of cyanobacteria have been misidentified in the past and have yet to be reclassified. Nonetheless, it is clear that bioprospecting and biodiversity studies are contributing significantly to these efforts.

BIODIVERSITY PARTNERSHIPS, BIOPROSPECTING, AND CONSERVATION EFFORTS

Much of Earth's biodiversity is concentrated in tropical terrestrial and shallow water subtidal environments. These same geographical locations are comprised of mainly economically-disadvantaged countries that struggle to meet the basic infrastructural, medical, and educational needs of their citizens. As a result, it can be difficult to allocate resources for the preservation of biodiversity because this heritage is perhaps not fully appreciated or necessarily prioritized. However, by identifying and developing the economic value of biodiversity in specific and concrete terms, the goals of economic development and biodiversity preservation can become synergistic. Such has been the case in Panama as a result of a two decade long effort that involved an in-country renaissance of support for the natural sciences that synergized with an NIH-Fogarty International Center-sponsored project, namely the Panama International Cooperative Biodiversity Group (Panama ICBG).

The overall ICBG program was established in 1992 as a cooperative effort between the US National Institutes of Health, National Science Foundation, and Agency for International Development, and has been coordinated by the Fogarty International Center. The aim of this program has been to foster incentives for biodiversity conservation and sustainable economic growth through scientific training and research, particularly focused on NP drug discovery. Participants in the ICBG programs have included not only the US agencies and university researchers, but also the government agencies and research centers in the partner countries, and a great exchange of scientific capability and training has accompanied the benefit sharing agreements of these efforts. The first ICBG programs were initiated in partnership with Cameroon and Nigeria, Costa Rica, Peru, Suriname, and cooperatively between Argentina, Chile, and Mexico. Several additional ICBG programs were subsequently organized, with the Panama ICBG among them, and a few are ongoing with funding secured through at least 2019.

The diverse experiences of the extended ICBG effort in Panama, which had the central focus of discovering novel therapeutic agents from tropical marine and terrestrial life forms, especially against neglected parasitic diseases and cancer, resulted in a number of discoveries which have had a positive impact on the country's preservation of biodiversity. In hindsight, it was the remarkable insight and inspiration of the ICBG program itself that sowed the seeds for much of this success. The ICBG programs were designed to intimately link the exploration of biodiversity for potentially useful products along with its characterization and preservation. Additional goals were to enhance economic development, training, and infrastructure development in the host country. The embodiment of this within the Panama ICBG was to locate much of the actual NP drug discovery research and conservation efforts in Panama, thereby ensuring the training of Panamanian students and faculty, development of scientific infrastructure, and greatly increasing the country's appreciation of the value of its biodiversity.

To a significant degree, the ability to conduct substantial research activities within Panama was made possible as a result of the purchase and placement of several key pieces of scientific instrumentation in the host country. For example, the NCI, the Fogarty International Center, and the Smithsonian Tropical Research Institute collectively purchased for the Panama ICBG effort a research-grade 300 MHz NMR instrument. This was placed in the Smithsonian Tropical Research Institute's Tupper Campus in Panama City, and enabled high quality NP research to occur in-country. Subsequently, a second high field instrument, this time operating at 400 MHz, was purchased with Panamanian government funds and placed at their nascent City of Knowledge research facilities. These were paradigm-shifting events with wide ranging consequences. Panamanians receiving graduate and postdoctoral education in NP and organic chemistry elsewhere in the world could now return to Panama and practice their disciplines to the level that they were trained, making their return home a more viable and attractive career path. Upon their return, they could not only conduct world-class NP science, but also train and develop the next generation of such scientists within their own nation state and its academic programs. This has had the consequence of opening up academic science to an increasingly large cadre of young and well-trained Panamanian chemists, growing the in-country capacity and appreciation for science and its aspirations.

A major geographic focus of efforts in Panama centered on the island of Coiba, a roughly 200 square mile island some 25 miles off the West Coast of the country. From 1919 to 2003, Coiba served as home to a notorious penal colony, the fearsome nature of which had the unintended consequence of keeping the majority of the island free from habitation or resource utilization. When the prison closed, it was thus prime real estate for potential development. At about this time, the Panama ICBG project began explorations of the island's exceptionally rich terrestrial and marine flora in search of unique NPs and with a special interest in compounds active in antiparasitic and anticancer screens. One organism of interest, collected from several subtidal environments around the island and appearing as gelatinous red strands, was identified as a *Leptolyngbya* species of cyanobacterium. Exceptionally potent cancer cell toxicity was noted in the crude extract, and this was ultimately traced to a novel cyclic depsipeptide, given the trivial name "coibamide A." Coibamide A became a very high priority anticancer lead compound as a result of its novel chemical structure, very potent cancer cell toxicity, and unique though incompletely characterized mechanism of action. Total chemical synthesis of coibamide A has been hampered by its propensity for epimerization, decomposition, and conformational flexibility. However, recent mechanistic studies have revealed that it induces cancer cell death via mTOR independent autophagy, and its specific molecular target is the Sec61 protein, an endoplasmic reticulum membrane protein translocator.

At about the same time as the discovery of coibamide A and several other notable NPs from Coiba's unique flora and fauna, it was becoming clear that its biodiversity required protection. If steps were not taken to protect the island, it would have fallen prey to commercial interests and been developed as a tourist destination with resort hotels and their associated facilities. Several key individuals from the Panama ICBG program became deeply invested in the preservation of Coiba, namely Drs. Todd Capson and Alicia Ibanez. As a result of their efforts, along with many others in nongovernmental organizations and the Panamanian National Government, Coiba Island was named a World Heritage Site by UNESCO in 2005. Management of the park, especially in regards to the prevention of illegal activities, such as commercial fishing and timber harvests, has been difficult to oversee, but with steady progress continuing to be made. The Coiba National Park thus stands as one of a number of lasting testaments and accomplishments of the Panama ICBG program, and part of its natural and biodiverse beauty can be appreciated in Fig. 5.5.

Another example of natural product driven conservation efforts derives from our previous work in the Southern Caribbean island of Curaçao. This low lying arid island has played an important role as an access point between much of South America, especially Venezuela, and the rest of the world. Early on, its exceptional deep-water port made it a practical site for trade and commerce. With the discovery of oil in Venezuela's nearby Maracaibo Basin, Curaçao became the logical location for a major oil refinery, which has now been in operation for roughly 100 years. The storage and refinement of oil on Curaçao made it a location of considerable strategic importance during the Second World War. Moreover, an exceptional mountainous deposit of calcium carbonate exists in the southern portion of the island, and this has been mined and shipped around the world for many decades. Whereas these industries have certainly contributed to the economy of the island country, they have also left their indelible anthropogenic mark on the island's once natural landscape. Furthermore, the vegetation of Curaçao has been enormously damaged by the free-range pasture of goats, resulting in a desert-like environment populated by cacti, thorny

FIGURE 5.5 **Photographs of the natural beauty and biodiversity of Coiba National Park in the Gulf of Chiriqui on the west side of Panama.** *(Reproduced with permission from the copyright holder, Alicia Ibáñez.)*

bushes, and other goat-tolerant vegetation. Curaçao is thus a curious mixture of the relatively pristine and the environmentally degraded; these have existed side-by-side for many years. However, a slow but steady economic resurgence is evident, with hotels, private villas, and destination resorts being introduced to the island. This has initiated a new round of environmental pressure on the island, as bay-front and beachside touristic facilities and housing have recently appeared. One such conflict arose in the Barbara Beach region of the Spanish Bay on Curaçao's Southwest coast.

In 1993, our laboratory initiated collections of marine algae and cyanobacteria from the island of Curaçao. Hosted through the increasingly sophisticated marine laboratory of CARMABI (Caribbean Research and Management of Biodiversity), we made collections of shallow water cyanobacteria from various locations along the leeward side of the island, from Playa Kalki in the Northwest to Punt Kanon in the Southeast. One such collection was made of trellises of red filaments descending from mangrove roots into the shallow waters of Spanish Bay. These trellises were comprised of unicyanobacterial growths, some as much as 40 cm long, which accumulated into soft tufts easily collected by hand. This cyanobacterium is now known to be *Moorea producens*, with the common name of "Mermaid's Hair." Coexisting in this unique habitat were stinging hydrozoans, scorpion fish, *Cassiopeia spp.*, the occasional young barracuda, and a diversity of other juvenile fish species. At the time of first collection, this region of Spanish Bay was only reachable by a nearly 30 min swim from the closest access at Barbara Beach, and lacked any significant point of interest to anyone other than marine scientists.

Subsequently, a cell toxicity assay run as part of a screen for new antiviral agents at the then Syntex Corporation in Palo Alto revealed the extract of these trellises to possess phenomenal levels of cancer cell cytotoxicity. We proceeded to isolate the active compound, named "curacin A," and determined its planar structure, which was first published in 1993 in the *Journal of Organic Chemistry*. Ensuing mechanistic studies revealed curacin A to be an exquisitely potent cytotoxin that acted at the colchicine site of microtubules, inhibiting their polymerization and thereby disrupting the ordered separation of chromosomes during mitosis. With determination of the absolute configuration of curacin A by a combination of chemical degradation and synthetic chemistry efforts, along with the development of total synthetic routes to the molecule, its status as an anticancer lead molecule was established in the field.

Advancement of curacin A to this status had a number of consequences, including that aspects of its biology, ecology, and biosynthesis became of great interest to the scientific community. It also helped to accentuate the point that this forgotten bay in Curaçao, with its fringe of mangroves and tufts of cyanobacteria, was a habitat with a specific and very tangible value to society. The average person broadly appreciates the need for new, more efficacious medicines to treat diseases, such as cancer. So through seminars given to the general public in Willemstad, television interviews and newspaper stories, and other communications provided by the CARMABI research station in Curaçao, a greater understanding and appreciation of the value of the local unique marine biodiversity spread throughout the island's populace.

In the late 1990s, interest in commercial development of the Barbara Beach area grew. Plans were laid for a resort-style hotel, as well as an array of up-scale residences, complete with individual boat dock facilities and swimming pools. These plans involved radical alteration of the waterfront vegetation in a portion of Spanish Bay, and possibly included the complete removal of the narrow fringe of mangroves that had provided the unique habitat from which the Mermaid's Hair was collected. Legislation was introduced in Curaçao to block this development, and this was successfully upheld for a number of years. The argument most easily understood and appreciated in this situation was that the natural resource, specifically *M. producens*, potentially held a real and tangible value to society, and so was worthy of preservation. However, after a number of years in the absence of realization of this value in monetary or material terms, the arguments became less convincing, and so the developers were ultimately successful in overcoming these environmental impact concerns (Fig. 5.6). Today, a luxury hotel, fringing docks, speedboats, a golf course and villa-style homes occupy this area, which was previously only the domain of cacti, lizards, mangroves, and Mermaid's Hair.

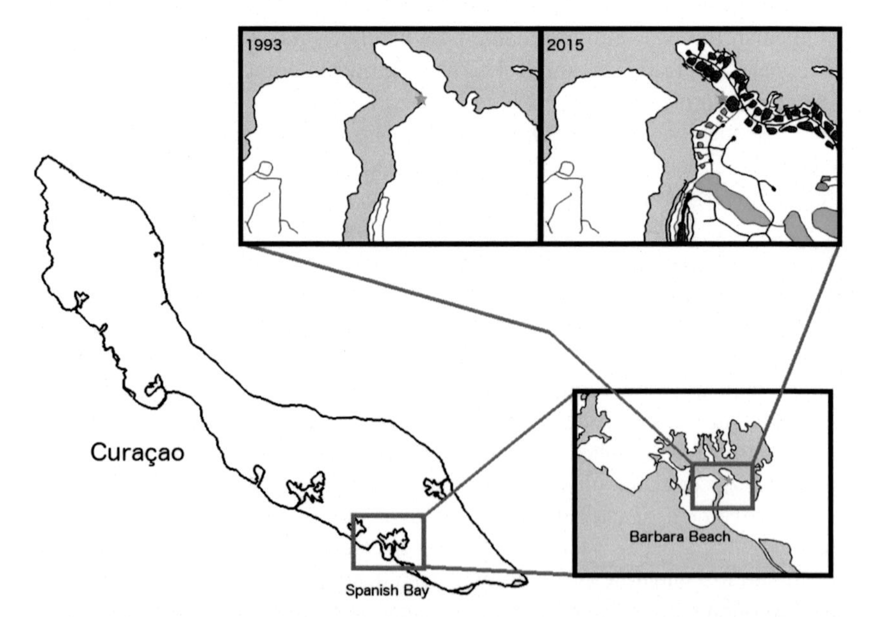

FIGURE 5.6 **Outline of the island of Curaçao with expansions of the Spanish Bay and Barbara Beach regions where the curacin A producer, *Moorea producens*, was collected in 1993.** *Orange star* denotes the specific site of the mangrove from where *M. producens* was collected. *Red blocks* represent housing developments, *green patches* signify the Santa Barbara golf course, and *pink blocks* are prospective development sites.

FUTURE OUTLOOKS

The efficacy of NP research in discovering compounds worthy of pharmaceutical utilization is supported by a history of successes and the current importance of NPs and associated derivatives in drug markets worldwide. The profound impact of discoveries in this field has been well-recognized, including the joint awarding of the 2015 Nobel Prize in physiology or medicine to the three lead researchers responsible for the late 20th century discovery of the naturally occurring antiparasitic drugs avermectin and artemisinin. Paired with the preponderance of Earth's biodiversity that is yet to be explored, the field of NP research has tremendous potential for continued advancement, as well as further improvements to global human health through the provision of new and more efficacious pharmaceutical agents. International policies, such as CBD, the Cartagena and Nagoya accords, and even yet unimagined agreements, are important in protecting the rights of developing nations as well as encouraging their interest in biodiversity exploration and conservation efforts. These policies also serve to facilitate related worldwide research efforts and scientific training programs, ultimately benefitting society as a whole by many direct and indirect avenues. However, as global climate change begins to compound the already burgeoning anthropogenic threats to biodiversity, the available opportunities for discovery leading to sustainable economic growth and usage of biodiversity is progressively lessened. Faced with this grim reality, it is absolutely imperative that conservationists and NP researchers unite in their efforts to protect and study biologically diverse habitats. With adequate funding, proper support from policy makers and governments worldwide, and the formation of international partnerships, significant efforts can and must be made to preserve our planet's incredible and invaluable biodiversity.

REFERENCES

Bravo-Monzón, A.E., Ríos-Vásquez, E., Delgado-Lamas, G., Espinosa-García, F.J., 2014. Chemical diversity among populations of *Mikania micrantha*: geographic mosaic structure and herbivory. Oecologia 174 (1), 195–203.

Cragg, G.M., Katz, F., Newman, D.J., Rosenthal, J., 2012. The impact of the United Nations Convention on Biological Diversity on natural products research. Nat. Prod. Rep. 29, 1407–1423.

Forner, D., Berrué, F., Correa, H., Duncan, K., Kerr, R.G., 2013. Chemical dereplication of marine actinomycetes by liquid chromatography-high resolution mass spectrometry profiling and statistical analysis. Analyt. Chim. Acta 805, 70–79.

Greenwood, D., 1992. The quinine connection. J. Antimicrob. Chemother. 30 (4), 417–427.

Keller, M., Zengler, K., 2004. Tapping into microbial diversity. Nat. Rev. Microbiol. 2 (2), 141–150.

Kinghorn, A.D., Pan, L., Fletcher, J.N., Chai, H., 2011. The relevance of higher plants in lead compound discovery programs. J. Nat. Prod. 74 (6), 1539–1555.

Kingston, D.G.I., 2011. Modern natural products drug discovery and its relevance to biodiversity conservation. J. Nat. Prod. 74 (3), 496–511.

Kursar, T.A., Caballero-George, C.C., Capson, T.L., Cubilla-Rios, L., Gerwick, W.H., Gupta, M.P., Ibañez, A., Linington, R.G., McPhail, K.L., Ortega-Barría, E., Romero, L.I., Solis, P.N., Coley, P.D., 2006. Securing economic benefits and promoting conservation through bioprospecting. Bioscience 56 (12), 1005–1012.

Li, H.-J., Jiang, Y., Li, P., 2006. Chemistry, bioactivity and geographical diversity of steroidal alkaloids from the Liliaceae family. Nat. Prod. Rep. 23 (5), 735–752.

Newman, D.J., Cragg, G.M., 2014. Natural products as sources of new drugs from 1981 to 2014. J. Nat. Prod. 79 (3), 629–661.

Oren, A., Garrity, G.M., 2014. Then and now: a systematic review of the systematics of prokaryotes in the last 80 years. Antonie van Leeuwenhoek 106, 43–56.

Singh, S.B., Pelaez, F., 2008. Biodiversity, chemical diversity and drug discovery. Prog. Drug Res. 65, 141–174.

Tan, G., Gyllenhaal, C., Soejarto, D.D., 2006. Biodiversity as a source of anticancer drugs. Curr. Drug Targets 7 (3), 265–277.

Treadgold, C.H., 1918. The prophylactic use of quinine in malaria: with special reference to experiences in Macedonia. Br. Med. J. 1 (2993), 525–529.

Yarza, P., Yilmaz, P., Pruesse, E., Glöckner, F.O., Ludwig, W., Schleifer, K., Whitmann, W.B., Euzéby, J., Amann, R., Rosselló-Móra, R., 2014. Uniting the classification of cultured and uncultured bacteria and archaea using 16S rRNA gene sequences. Nat. Rev. Microbiol. 12 (9), 635–645.

FURTHER READING

Calteau, A., Fewer, D.P., Latifi, A., Coursin, T., Laurent, T., Jokela, J., Kerfeld, C.A., Sivonen, K., Piel, J., Gugger, M., 2014. Phylum-wide comparative genomics unravel the diversity of secondary metabolism in cyanobacteria. BMC Genomics 15, 977.

Harvey, A.L., Edrada-Ebel, R., Quinn, R.J., 2015. The re-emergence of natural products for drug discovery in the genomics era. Nat. Rev. Drug Discov. 14 (2), 111–129.

McCarthy, D.P., Donald, P.F., Scharlemann, J.P.W., Buchanan, G.M., Balmford, A., Green, J.M.H., Bennun, L.A., Burgess, N.D., Fishpool, L.D.C., Garnett, S.T., Leonard, D.L., Maloney, R.F., Morling, P., Schaefer, H.M., Symes, A., Wiedenfeld, D.A., Butchart, S.H.M., 2012. Financial costs of meeting global biodiversity conservation targets: current spending and unmet needs. Science 338 (6109), 946–949.

Myers, N., Mittermeier, R.A., Mittermeier, C.G., da Fonseca, G.A.B., Kent, J., 2000. Biodiversity hotspots for conservation priorities. Nature 403, 853–858.

Pimm, S.L., Jenkins, C.N., Abell, R., Brooks, T.M., Gittleman, J.L., Joppa, L.N., 2014. The biodiversity of species and their rates of extinction, distribution, and protection. Science 344 (6187), 1246752.

Tanm, G., Gyllenhaal, C., Soejarto, D.D., 2006. Biodiversity as a source of anticancer drugs. Curr. Drug Targets 7 (3), 265–277.

Uemura, D., 2010. Exploratory research on bioactive natural products with a focus on biological phenomena. Proc. Jpn. Acad. Ser. B Phys. Biol. Sci. 86 (3), 190–201.

Chapter 6

Hydrocarbon-Oxidizing Bacteria and Their Potential in Eco-Biotechnology and Bioremediation

Irena B. Ivshina, Maria S. Kuyukina and Anastasiya V. Krivoruchko
Institute of Ecology and Genetics of Microorganisms, Ural Branch of the Russian Academy of Sciences, Perm State University, Perm, Russia

INTRODUCTION

The current ecological situation is extremely challenging, resulting from the detrimental environmental impacts of the large-scale development of natural resources, especially oil and gas. Despite the push into the generation of alternative energy sources and development of oil-replacement technologies, the energy supplier role of "fossil energy," namely petroleum hydrocarbons, will continue to be significant for the next few decades. Oil pollution, however, inevitably results from oil exploration, storage, transport, and processing and the accidental oil spills are the most dangerous ones for the environment and characterized by shock loads on bio-cenoses and their long-term destabilization (Cameron, 2012; Duarte et al., 2013). These events demonstrate the growing threat of secondary soil and water pollution stemming from lateral and radical migration of petroleum hydrocarbons in soil and water profiles, as well as the resultant current global ecological challenge of the anthropogenic hydrocarbon contamination of the natural environment.

Indigenous microorganisms able to degrade petroleum hydrocarbons contribute significantly toward the natural attenuation processes of oil-contaminated ecosystems (Van Beilen and Funhoff, 2007; Van Beilen et al., 2003; Wentzel et al., 2007). This property alone contributes toward their survival under environmental conditions detrimental for other microorganisms. Thus the rationale of studying this biogeochemically important group of microorganisms is obvious by virtue of their features (e.g., appropriate enzyme systems for hydrocarbon

Microbial Resources. http://dx.doi.org/10.1016/B978-0-12-804765-1.00006-0

degradation, and mechanisms for tolerance to their toxic effects). They contribute considerably to the global carbon cycle and are key players in preventing the accumulation of hydrocarbon pollutants in the atmosphere, dominate the hydrocarbon-contaminated biotopes. Some of them are able to assimilate gaseous *n*-alkanes in addition to liquid hydrocarbons as sole carbon sources. As a result, the role of these microorganisms in bioremediation and restoration of soil fertility is significant.

Recently, there has been a renewed interest in hydrocarbon-oxidizing microorganisms propelled by the advances in genomics, proteomics, and synthetic biology. This has been accompanied by the increased attention toward the in-depth studies of the metabolism of hydrocarbon-oxidizing bacteria (HOB), in particular the hydrocarbon transport into cells and molecular mechanisms of their oxidation (Galvão et al., 2005; Suenaga et al., 2007). Recent advances in functional metagenomics and mass-spectrometry facilitated the discovery of novel highly active versatile enzymes produced by bacteria and involved in biodegradation of hydrocarbon pollutants for their potential applications in environmental biotechnology (De Vasconcellos et al., 2010; Dellagnezze et al., 2014; Sierra-García et al., 2014; Ufarté et al., 2015). However, the taxonomic and functional diversity of hydrocarbon-oxidizing microorganisms is still not fully revealed, and there are many unclear issues concerning their adaptation mechanisms to anthropogenically disturbed habitats. Generation of sound knowledge is essential for understanding the natural processes and molecular mechanisms involved in bacterial catabolism of petroleum hydrocarbons, and to assess their potential for biotechnological applications, particularly in pollution prospecting and oil bioremediation.

Based on the previously listed recent advances, we attempted to review the diversity and biological features of bacteria in hydrocarbon-contaminated ecosystems, with selected examples related to the optimization of bacterial hydrocarbon degradation processes, as well as the potential of such bacteria for commercial applications as bioremediation agents. Analyses of the latest reviews (Abbasian et al., 2015; Kleindienst et al., 2015) on remediation of hydrocarbon pollution revealed that aerobic and anaerobic Gram-negative bacteria are most frequently studied as key bio-oxidizers of aliphatic and aromatic hydrocarbons. Nevertheless, this review focuses mainly on Gram-positive organotrophic bacteria, members of the phylum *Actinobacteria*, which produce oxygenases, are continually found in oil-contaminated soils, possess the highest metabolic activities toward petroleum hydrocarbons, and prevail among other *n*-alkane biodegraders (Larkin et al., 2006; Martínková et al., 2009; Shennan, 2006; Van Beilen and Funhoff, 2007). Aspects of biological oxidation of petroleum hydrocarbons by actinobacteria and their adaptation to assimilate hydrophobic substrates are discussed; advances and challenges of using these microorganisms for bioindication of contaminations and applications in oil-contaminated system bioremediation are also highlighted in this chapter.

BACTERIOLOGY OF HYDROCARBON DEGRADATION

Hydrocarbons (HCs) are highly reduced oil constituents, possessing solvent properties, and can affect adversely microbial survival, leading to cell respiration and growth inhibition, reduced cell viability, or lysis. Toxic effects are caused by HCs depending on their water solubility (De Carvalho et al., 2009), and when accumulated in the cell wall, more hydrophobic compounds may not be highly toxic, whereas soluble oil components exhibit a profound toxic effect (Sei et al., 2003). Therefore, microorganisms, persistent and adapting to HC substrates, are essential in self-cleaning of oil-contaminated biotopes and applied in HC pollutant detoxification. It is generally accepted that bacteria play a major role in HC degradation in soil, as evidenced by significant numbers of HOB (10 million cells/g and higher) compared to that of microfungi (Atlas and Bartha, 1992; Sorkhoh et al., 1990). Common features of soil bacteriocenoses chronically exposed to petroleum HCs are a limited diversity, formation of "specialist" ecotrophic groups of microorganisms utilizing HCs of a certain chemical structure. Despite a large variety of HC degradation enzymes and pathways, none of these microorganisms is capable of utilizing all HCs, but only a particular range (Alvarez, 2003).

Our knowledge of the distribution, taxonomic and functional diversity of HOB is still limited, and it can be assumed that a large fraction of microorganisms that provide their energy and constructive metabolism from petroleum HCs, and enzyme systems they synthesize, remain yet undiscovered. Though the phylogenetically diverse microorganisms can transform and degrade HC pollutants, the bacteria are to a greater extent studied and more often used in bioremediation. Many studies reviewed the degradative abilities of the Gram-positive aerobic heterotrophic bacteria of the genera *Actinomyces* (Etoumi, 2007), *Arthrobacter* (Plotnikova et al., 2001), *Bacillus* (Kumar et al., 2007; Sathishkumar et al., 2008), *Corynebacterium* (Sathishkumar et al., 2008; Watrinson and Morgan, 1990), *Dietzia* (Bødtked et al., 2009; Dellagnezze et al., 2014; Von der Weid et al., 2007; Wang et al., 2011), *Geobacillus* (Feng et al., 2007), *Gordonia* (Alvarez, 2003; Kotani et al., 2003), *Micrococcus* (Dellagnezze et al., 2014; Jegan et al., 2010), *Mycobacterium* (Kotani et al., 2006; Sakai et al., 2004; Van Beilen et al., 2002), *Micromonospora* (Barabás et al., 2001; De Vasconcellos et al., 2010), *Nocardia* (Alvarez, 2003; Sakai et al., 2004), *Nocardioides* (Hamamura et al., 2001), *Pseudonocardia* (Kotani et al., 2006), *Rhodococcus* (Alvarez, 2003; Alvarez and Silva, 2013; Binazadeh et al., 2009; Van Beilen et al., 2002) and *Streptomyces* (Barabás et al., 2001). Among the above listed genera, the most abundant and notable for the largest variety of degradable xenobiotics were found to be coryne- and nocardioform cluster of the genera *Arthrobacter*, *Corynebacterium*, *Dietzia*, *Gordonia*, *Mycobacterium*, *Nocardia*, and *Rhodococcus* (Alvarez, 2003; Hamamura et al., 2001; Van Beilen et al., 2003; Wentzel et al., 2007). Specifically, the genus *Rhodococcus*, which dominate the biocenoses of specific habitats, such as high salinity and/or oil-contaminated biotopes, are promising biodegraders (De Carvalho and da Fonseca, 2005; Herter et al., 2012; Martínková et al., 2009). They synthesize enzyme systems that catalyze oxidative biotransformations of organic compounds from almost all known classes (Warhurst and Fewson, 1994). The features of rhodococci, such as an oligotrophic growth, molecular nitrogen fixation, high competitiveness, cell aggregation, flexible and multifunctional oxidoreductases, biosurfactant production, lack of marked pathogenic properties [except for *R. hoagii* (formerly "*R. equi*") (Kämpfer et al., 2014; Kedlaya et al., 2001), *R. globerulus* (Cuello et al., 2002), and *R. fascians* (Goethals et al., 2001)], and finally their high catabolic activities in extreme environments are due to their remarkable genome versatility (De Carvalho and da Fonseca, 2005; Kuyukina and Ivshina, 2010; Larkin et al., 2006).

Until now, 114 *Rhodococcus* genomes have been sequenced which belong to ecologically significant species, including strains isolated from oil-contaminated soils and utilizing a wide aliphatic and aromatic HC range (Larkin et al., 2010). Large-size (5.6–10.1 Mb) genomes and plenty of functional genes have been discovered, indicating a versatile metabolism and huge catabolic capabilities of these actinobacteria. In this context, the genome of *R. ruber* IEGM 231, an efficient biodegrader of petroleum HCs (C_3-C_8 and C_{11}-C_{32}) and a biosurfactant producer, revealed 5,928 predicted genes, including those coding for propane monooxygenases, rubredoxin-dependent alkane 1-monooxygenases, cytochrome P450-dependent monooxygenases, flavin monooxygenases, dioxygenases, peroxidases, dehydrogenases, etc (Ivshina et al., 2014; Serebrennikova et al., 2014; Kuyukina et al., 2015; Catalogue of Strains, 2016). The abovementioned investigations indicate obvious technological advantages of rhodococci and provide a background for their various ecobiotechnology applications ranging from biological indication of HC deposits and pollutions to enhanced degradation of oil pollutions.

Specific ecological interest is addressed to the ability (quite a rare feature among other microorganisms) of individual rhodococcal species (*R. rhodochrous*, *R. ruber*) to utilize both liquid HCs and gaseous methane homologs (C_2-C_4). The studies on distribution of propane- and butane-oxidizing bacteria in ground water and soils of oil-bearing and nonoil-bearing areas revealed their confinement to oil-bearing structures and possible applications as HC deposit bioindicators (Ivshina et al., 1981, 1995). In oil and gas deposit areas, the indicator bacteria form a strong biological oxidative filter which completely intercepts the gas flow from generating layers. The gas-oxidizing rhodococci population was shown to be not influenced by sharp seasonal fluctuations, but is controlled by the flow rate of HC gases.

In natural environments, microorganisms organize stable multicomponent consortia that allow them to convert different HC types relatively easy. Enzyme systems synthesized by HC degraders can also oxidize compounds other than their nutrient sources. Other consortium members, namely heterotrophic microorganisms lacking a hydrocarbon-oxidizing activity, utilize the HC oxidation intermediates (Sayavedra-Soto et al., 2006). The varying abilities of microorganisms to utilize HCs result from a horizontal transfer of specific biodegradation genes in mobile genetic elements (plasmids, transposons, integrons), including transfers among phylogenetically distant bacteria (Leahy et al., 2003). There are also data on a horizontal transfer of bioemulsifiers from producer bacteria where upon such exchange the degradation of recalcitrant HCs becomes possible due to the acceptor organism's transformed activity (Ron and Rosenberg, 2001).

The carbon chain length is crucially important in relation to HC utilization. Representatives of only a few species of microorganisms oxidize gaseous *n*-alkanes (up to C_4). The most difficult to oxidize are short-chain liquid alkanes (C_5-C_{10}), and the most active growth is observed on C_{13}–C_{19}. Hard paraffins (C_{30}-C_{44}) are degraded extremely slowly. Cycloalkanes, mono- and polycyclic aromatic HCs are significantly resistant to microbial attacks (Alvarez et al., 2001; Sayavedra-Soto et al., 2006). The vast majority of bacterial HC transformations are oxidative reactions occurring most actively under aerobic conditions. The anaerobic HC biodegradation is less extensively studied (Heider and Schühle, 2013). The progress is evident in current knowledge on molecular mechanisms and pathways for aerobic HC biodegradation which are: (1) plenty of multipurpose oxygenase systems have been found that form active complexes with HC substrates and molecular oxygen; (2) several enzymes involved in the initial stage of aerobic alkane biodegradation have been identified and characterized (Coon, 2005; Funhoff et al., 2006; Kotani et al., 2003; Van Beilen and Funhoff, 2007); (3) a metagenomic approach allowed describing novel metabolic pathways for HC degradation distinct from those previously characterized in culturable representatives (Sierra-García et al., 2014); and (4) novel alkane hydroxylase (*alkB*) gene phylotypes in marine ecosystems have been detected (Smith et al., 2013; Wasmund et al., 2009). Characteristics of known enzymes involved in HC degradation are reported in an excellent review by Van Beilen and Funhoff (2007). However, it should be noted that the initial stage of aerobic HC bioconversion, which is an extremely difficult reaction, is so far studied only in general terms.

Initial alkane oxidation involving cytochrome P450 is nearly similar in bacteria and animals where the enzyme is cell membrane-bound. Bacterial cytochrome P450s are dissolved in the cytoplasm (Urlacher and Eiben, 2006). The known cytochrome P450 alkane hydroxylases of the CYP153 family (Class I) have an obligatory component iron and its valence state change is characteristic of the catalytic hydroxylase action (Fig. 6.1A) (Van Beilen et al., 2006). A purified cytochrome P450 preparation from *Mycobacterium* sp. HXN-1500 was used to hydroxylate C_5-C_{11} alkanes to corresponding 1-alkanols (Funhoff et al., 2006).

An oxygen-activating system lacking this cytochrome is typical only of prokaryotes and formed from the integral membrane-bound nonheme iron monooxygenase encoded by an *alkB* gene in most bacteria, and from electron transfer proteins, such as rubredoxin and NADH-dependent rubredoxin reductase

(A)

$NAD(P)^+$ FADH$_2$ Protein (Fe^{3+}) Cyt-P450 (Fe^{2+}) O_2 R−CH_3

NAD(P)H FAD Protein (Fe^{2+}) Cyt-P450 (Fe^{3+}) H_2O R−CH_2OH

$$R{-}CH_3 + NAD(P)H + O_2 + H^+ \rightarrow R{-}CH_2OH + NAD(P)^+ + H_2O$$

(B)

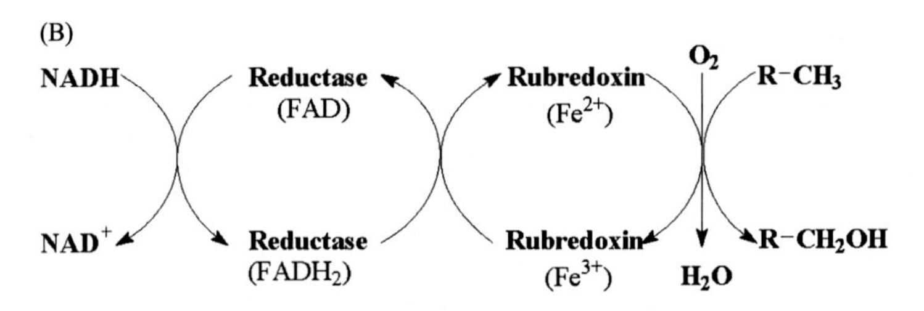

$$R{-}CH_3 + NADH + O_2 + H^+ \rightarrow R{-}CH_2OH + NAD^+ + H_2O$$

FIGURE 6.1 **Reactions catalyzed by components of the bacterial enzyme system of hydroxylation.**

encoded by *alkG* and *alkT*, respectively (Fig. 6.1B) (Cappelletti et al., 2011; Van Beilen et al., 2006). The AlkB hydroxylases are widespread in bacteria and oxidize alkanes (up to C_{16}) at the terminal position (Van Beilen and Funhoff, 2007; Wasmund et al., 2009). The LadA alkane monooxygenase found in *Geobacillus thermodenitrificans* NG80-2 and different from AlkB hydroxylases can catalyze C_{15}-C_{36} alkanes (Feng et al., 2007). Every microorganism has a specific set of inducible oxygenase systems, and the ability to degrade some HCs or others depends on the expression of corresponding oxygenases (Redmond et al., 2010).

Mechanisms of bacterial HC oxidation are covered in several reviews (Abbasian et al., 2015; Pérez-Pantoja et al., 2010; Yam et al., 2010). The microbial alkane degradation pathways are most extensively studied (Rojo, 2009; Shennan, 2006; Van Beilen and Funhoff, 2007; Wentzel et al., 2007). According to current concepts, there are three possible alkane oxidation pathways (Fig. 6.2): (1) Monoterminal oxidation of the alkane methyl group resulting in a primary alcohol further converted to an aldehyde and monocarboxylic acid, (2) Subterminal oxidation to a corresponding methyl ketone via a secondary alcohol, and (3) Diterminal oxidation when terminal methyl groups of the alkane are oxidized simultaneously or consecutively to form dicarboxylic acids similar to the parent HC molecule in carbon chain length. Monoterminal oxidation is the most widespread HC assimilation pathway. One of the possible options for

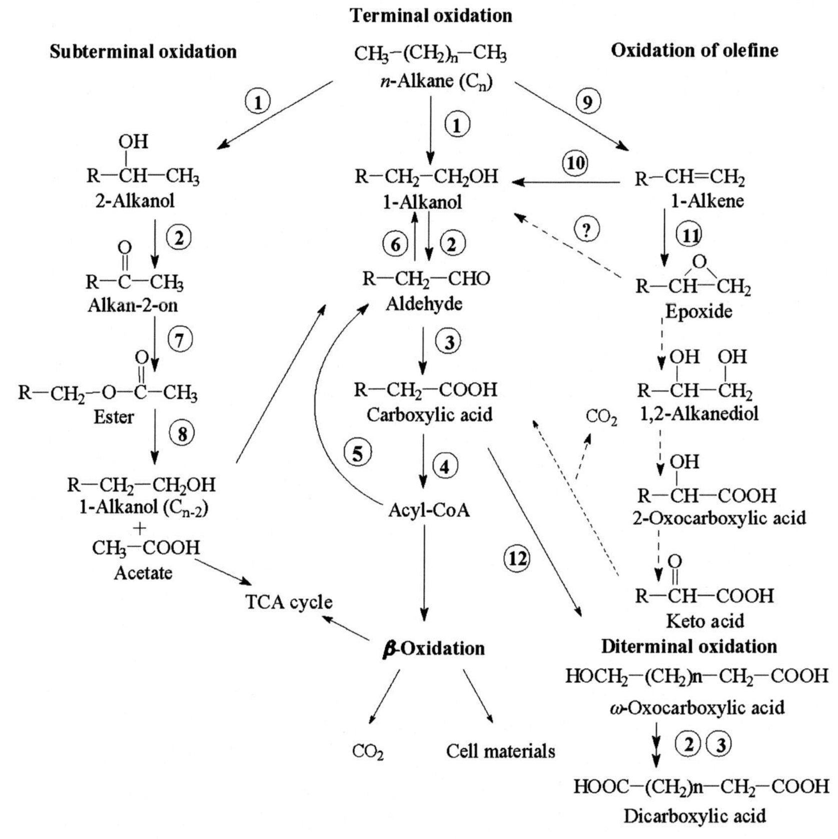

FIGURE 6.2 **Pathways for bacterial degradation of *n*-alkanes.** (1) substrate-specific terminal alkane monooxygenase/alkane hydroxylase; (2) alcohol dehydrogenase; (3) aldehyde dehydrogenase; (4) acyl-CoA synthetase; (5) fatty acyl-CoA reductase; (6) fatty aldehyde reductase; (7) Baeyer-Villiger monooxygenase; (8) esterase; (9) alkane dehydrogenase; (10) alkene dehydrogenase; (11) alkene monooxygenase; (12) ω-hydroxylase; *TCA*, tricarboxylic acids

initial HC oxidation is by oxidases, like an oxygen transferase, via hydroperoxide formation (Ji et al., 2013).

$$R\text{-}CH_2-CH_3 + O_2 \rightarrow R\text{-}CH_2\text{-}CH_2\text{-}OOH$$

$$R\text{-}CH_2\text{-}CH_2\text{-}OOH + NAD(P)H + H^+ \rightarrow R\text{-}CH_2\text{-}CH_2\text{-}OH + NAD(P)^+ + H_2O$$

Another option is the dehydration to olefins involving a NAD^+-dependent dehydrogenase (Fig. 6.2). Alkanotrophic bacteria can degrade olefins by alkene monooxygenase to form corresponding epoxides, toxic compounds (Jin et al., 2003; Lin et al., 2011). Based on the data from Abbasian et al. (2015), a scheme for further epoxide transformation is hypothesized. Fatty acids formed in HC oxidation can accumulate (Binazadeh et al., 2009) or be further oxidized by β-oxidation, namely a consecutive detachment of two-carbon fragments as active acetate that enters the tricarboxylic acid cycle (TCA) (Alvarez, 2003; Wentzel et al., 2007). β-Oxidation results in acetic acid accumulation from even-numbered HCs, and propionic acid from odd-numbered alkanes. Several minor options for further fatty acid assimilation are described: α-oxidation, acid decarboxylation, that often occurs when a subterminal carbon atom has a keto or a hydroxyl group; ω-hydroxylation resulting in ω-hydroxycarboxylic acids and then in dicarboxylic acids and their further fragmentation; and fatty acid dehydrogenation followed by a double bond cleavage by lipoxydases (Coon, 2005).

Postcultural liquids of *n*-alkane-grown bacterial cultures are reported to contain the penultimate atom-oxidized products (subterminal oxidation)-secondary alcohols and methyl ketones (Kotani et al., 2003, 2007). Actinobacteria oxidize short-chain alkanes (C_3-C_6) via secondary alcohols to corresponding ketones (Atlas, 1981; Kotani et al., 2007). α-Oxidation of *n*-alkanes dominates mainly when microorganisms utilize lower paraffins (C_2-C_5) (Kotani et al., 2003, 2006, 2007). Oxygen fixation occurs both in the initial *n*-alkane oxidation to secondary alcohol and in further methyl ketone cleavage required for their terminal group oxidation. There are microorganisms with enzyme systems capable of diterminal oxidation to form ω-oxymonocarboxylic and dicarboxylic acids (Fig. 6.2).

Routes for aromatic HC degradation are extremely diverse (Gupta et al., 2015; Yam et al., 2010). Initial aromatic HC transformations generally include hydroxylation and only then the aromatic ring cleavage via *ortho-* or *meta*-cleavage to form TCA compounds. Decomposition products are transformed into pyruvate, fumarate, succinate, and other intermediates involved in general metabolism. Naphthalene biodegradation processes are studied largely using the representative members of the genera *Arthrobacter*, *Bacillus*, *Corynebacterium*, *Geobacillus*, *Micrococcus*, *Nocardia*, *Rhodococcus*, and *Streptomyces* (Bubinas et al., 2008; Kulakov et al., 2000).

BACTERIAL ADAPTATION TO HYDROCARBON ASSIMILATION

Because of their hydrophobicity, HCs require specific transport means to enter into the microbial cell. Apparently, such entry is achieved by passive diffusion, facilitated diffusion, and active transport. Bacterial cells may contact with liquid HCs by one of the mechanisms: (1) absorption of mono-disperse aqueous-phase alkanes with $C < 10$; (2) direct cell contact with hydrocarbon droplets or films that largely eliminates substrate limitations; and (3) contact with HC microdroplets (pseudosolubilization) (De Vasconcellos et al., 2009; Hua and Wang, 2012; Lang and Philp, 1998; Wick et al., 2002).

Bacteria can utilize HCs at extremely low concentrations (e.g., minimal evaluated propane concentration is 0.001%, and for volatile HCs, the minimum is below 10 ppm). Low-molecular HCs (C_5-C_{10}) are absorbed as true solutions, whereas uptake of HCs longer than 10 carbon atoms occurs only upon direct contact with the cell. Solid paraffins may enter the cell by a solid substrate pretreatment into a drop-liquid state by specific exometabolites or by cell adsorption and biofilm formation on sold HC surfaces (Wick et al., 2002). The lipid composition of the cell wall plays a key role in HCs sorption by the cell (Alvarez, 2003). The amount of total lipids and saturated fatty acids ($C_{16:0}$, $C_{18:0}$, $C_{21:0}$) found in *n*-hexadecane-grown *R. ruber* IEGM 333 was twice as high as that found in propane- or nutrient broth-grown cells, and qualitatively diverse phospholipids have been also found (Kuyukina et al., 2000). Members of *Corynebacterium*, *Dietzia*, *Gordonia*, *Mycobacterium*, *Nocardia*, *Pseudonocardia*, and *Rhodococcus* possess a unique tough lypophilic cell wall containing mycolic acids, long-chain 3-hydroxy-2-branched acids (C_{20}-C_{90}). These acids are present not only as part of complex cell wall lipids (peptidolipids, glycolipids, and peptidoglycolipids) but may be in a free form, thus enhancing the hydrophobic cell surface properties (Sutcliffe, 1998). They play a major role in HC binding, and their long-chain aliphatic radicals form HC-harbored globules inside the cell wall. It should be noted that in rhodococci, for example, passive HC diffusion occurs across the cell wall where they accumulate and remain nonoxidized in considerable amounts, that is, up to 50%-80% HC absorbed (Koronelli, 1996). Moreover, the adhesive properties of rhodococci are also related to their elevated hydrophobicity (Ivshina et al., 2013). Along with diffusion across the cell wall, HCs may penetrate through special ultramicroscopic pores filled with a lipopolysaccharide that often comes partially outside, contacts the alkane, and facilitates its introduction into the cell wall and dissolution in lipids. Such formations are produced by hexadecane-grown arthrobacteria and dietziae that also hydrophobize the cell wall synthesizing triglycerides and waxes as internal store material (Alvarez, 2003; Koronelli, 1996).

Microorganisms have also developed another adaptive ability, namely to synthesize biosurfactants (Christofi and Ivshina, 2002; Kuyukina and Ivshina, 2010) that are usually glycolipids or lipopeptides possessing high surface and interfacial activities (Philp et al., 2002). Rhodococci, as an example, can synthesize biosurfactants that decrease the surface tension of water from 72 to 25 mN/m and exhibit high emulsifying and oil-washing activities (Christofi and Ivshina, 2005; Ivshina et al., 1998, 2001). Bacterial cell adaptations developed to assimilate HCs are realized in their cytology. Upon rhodococcal shift from carbohydrates to HC nutrition source, the electron-microscopy reveals the structural–functional changes that clearly correlate with the chain length of a HC substrate (Fig. 6.3). Most typical *Rhodococcus* cell responses to *n*-alkanes are fibrillar, less frequently homogenous, polysaccharide capsules (Fig. 6.3E); poly-β-oxibutyrate and polyphosphate granules (Fig. 6.3A); the increased size of the latter (up to 100 nm), and also lipid inclusions (Fig. 6.3A and E); a developed intracellular membrane system formed at the cytoplasm periphery (Fig. 6.3F); expressed flexibility of the cell wall in propane- and

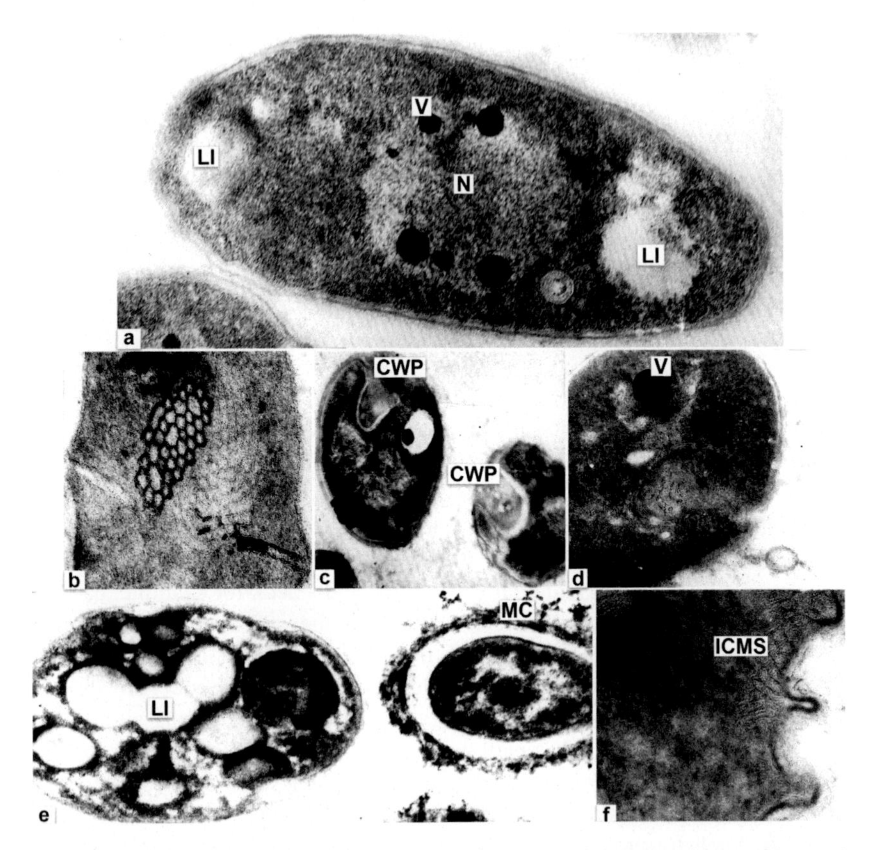

FIGURE 6.3 Ultrathin sections of *R. ruber* cells grown on the mineral medium supplemented with *n*-alkanes: propane (A, C, F), *n*-pentane (B), *n*-octane (D), and *n*-hexadecane (E). (A) IEGM 73, × 100,000; (B) IEGM 79, × 85,000; (C) IEGM 342, × 50,000; (D) IEGM 87, × 80,000; (E) IEGM 333, × 85,000; (F) IEGM 342, × 100,000. *CWP*, proliferation of the cell wall; *ICMS*, intracytoplasmic membrane system; *LI*, lipid inclusions; *MC*, microcapsule; *V*, volutin.

n-butane-cultivated *Rhodococcus* (Fig. 6.3C and F). Obviously, the excess cell wall growth, increasing its surface, is related to formation of a kind of a depot for the cells to be completely alkane-saturated that provides fixation of the hydrophobic substrate entered. These functional adaptations of rhodococci grown in the presence of *n*-alkanes (C_3-C_5, C_8, C_{16}) disappeared in ultrathin structures of the nutrient broth-incubated cells.

The next cellular stage in HC uptake is a metabolic process associated with the transmembrane HC transport involving carrier proteins localized in membrane structures. In bacteria, passive and active HC transport systems have been found (Li et al., 2014). There are interesting data indicating that HC utilization occurs by passive transport at high *n*-octadecane (C_{18}) concentration in the medium, whereas at lower HC concentrations this specific system switches to energy-dependent transport (Hua et al., 2013). Therefore,

HC biodegradation is a global and ubiquitous phenomenon, though occurs differently depending on biochemical features of a bacterial strain and HC composition. The process is determined by the presence of complex enzymes (oxygenases) and cell adaptations ensuring solubilization of HCs. Their assimilation is characterized by the primary and secondary metabolite accumulation. Two steps in HC uptake can be distinguished: (1) physicochemical uptake and diffusion through the cell wall that is possible due to specific cell adaptations ranging from hydrophobic cell wall and lypophylic channels formed in it and to bioemulsifier synthesis; and (2) active uptake due to metabolic reactions that produce an appropriate sorption gradient ensuring continuous HC input into the cell.

ENVIRONMENTAL BIOSENSORS FOR HYDROCARBON POLLUTANTS

Biosensors are analytical devices that combine biological sensing elements (cells, enzymes, DNA/RNA, antibodies, etc.) with physicochemical transducers to generate a measurable signal (usually electrical or optical) proportional to the analyte concentration. In recent years, a large number of microbial biosensors have been developed for diverse environmental applications, such as ecotoxicity testing of xenobiotics, human health risk assessment, and evaluation of bioremediation/biodegradation potential for polluted environments (Diplock et al., 2009; Harms et al., 2006; Su et al., 2011). It should be noted that biosensors can have certain advantages over the physicochemical sensing (e.g., GC-MS, HPLC-MS), such as detection of bioavailable pollutant fractions (while chromatographic techniques measure total amounts of solvent extractable pollutants), relatively low cost, and simple equipment (Tecon, 2015). Since HCs easily bind to natural sorbents (soils, sediments, clays, humic materials, and dissolved organic matters), their bioavailability and bioactivity could change dramatically depending on environmental conditions. Therefore, biosensors could better predict the real exposure of living organisms, including humans, to HC pollutants (Tecon and van der Meer, 2008).

Microorganism-based biosensors include whole microbial cells, enzyme systems, or hybrid biosensors typically consisting of either mixed cultures or cells and purified enzymes (Reshetilov et al., 2010). Whole-cell biosensors seem better suited for environmental applications as they are self-producing, easy to manipulate, and more stable under harsh environmental conditions than enzyme systems (Park et al., 2013). Furthermore, using living cells allows us to obtain functional information rather than analytical, which is, in many cases, more important for ecotoxicology and bioremediation monitoring. For example, it is essential to know whether the conditions are favorable for bioremediation and whether the pollutant availability limits biodegradation. After gaining such information on the metabolic functions of viable cells exposed to a pollutant, it is possible to detect either stimulating or inhibitory effects of this pollutant.

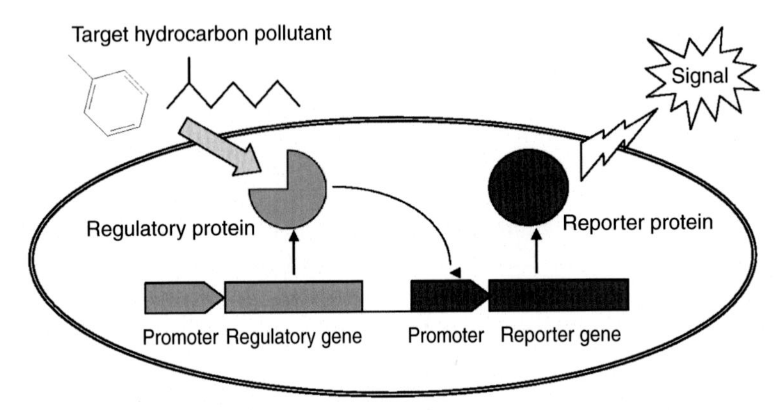

FIGURE 6.4 **Schematic illustration of a whole-cell transcriptional biosensor for HC pollutants.**

Many bacterial biosensors have been developed for the detection of petroleum HCs, such as alkanes (Sevilla et al., 2015; Sticher et al., 1997); monoaromatic HCs, such as benzene, toluene, ethylbenzene, and xylene (BTEX) (Applegate et al., 1998; Paitan et al., 2004; Stiner and Halverson, 2002; Tizzard and Lloyd-Jones, 2007), and polycyclic aromatic hydrocarbons (PAHs) (Tecon, 2015). These biosensors are based on the analysis of gene expression, typically by creating transcriptional fusions between a promoter of interest and a reporter gene, and the expression level of a reporter gene serves as a measure of the available pollutant concentration (Fig. 6.4) (Stiner and Halverson, 2002).

Originally, genetically engineered *Escherichia coli* strains carrying the fusion of bacterial bioluminescence (*lux*) genes as a reporter with selected bacterial gene promoters were developed as environmental biosensors. Depending on the promoter choice, they responded to a large variety of environmentally significant pollutants and served as "general toxicity" biosensors (Bechor et al., 2002). Although they performed well under laboratory conditions, these biosensors suffered a number of drawbacks that limited their use in detecting HC pollutants. For example, mass transfer problems, which limit the availability of HCs for bacterial cells, thus resulting in slow response and low signal to noise ratios (Sevilla et al., 2015). Therefore, many factors should be considered when choosing a host strain for biosensor constructing, such as physiology and activity, cell wall and membrane compositions, presence of uptake channels, and the capacity of the organism to influence the HC bioavailability (e.g., by excreting biosurfactants) (Harms et al., 2006). The question is, will the biosensor bacteria be able to uptake and metabolize target HCs or not? On the one hand, if the host strain is equipped with the biosensor construction, while maintaining its degradation capacity or surfactant production, the resulting bioreporter will perform similarly to a wild-type degrader (Harms et al., 2006), which is preferable for bioremediation monitoring. On the other hand, due to intracellular degradation, less HC molecules entered into the cell will be available for the detector to senses that may result in a lower signal. Hence, both biosensor types involving strains with active and inactivated biodegradation capacities are in use. One way to overcome major limitations of microbial biosensors (for example, slow response, low signal to noise ratio, fixed temperature, and pH requirements) could be the exploiting of strains adapted to toxic HC substrates and harsh environmental conditions. For example, a *p*-nitrophenol adapted *Pseudomonas* sp. biosensor was developed, which operated in a wide range of temperature and pH (Banik et al., 2008).

Microbial biosensors are potentially suitable for specific determination of various HC pollutants because of their broad selectivity. Such multireceptor behavior has been exploited for the determination of complex variables, such as total biodegradable compounds in oil-contaminated water or mutagenicity of aromatic petroleum fractions, while it could result in poor selectivity when detecting individual HCs in the presence of other contaminants (Park et al., 2013). In recent years, advanced protocols for biosensing HC pollutants were developed based on genetic/protein engineering and synthetic biology that allowed constructing microbial sensors with improved signal outputs, sensitivity, and selectivity. These biosensors depend on the analysis of gene expression typically by bioluminescent/fluorescent or electrochemical detection (Su et al., 2011). Bioluminescence has been very successful as a reporter for pollutant detection because of the sensitive instrumentation for detecting the light production (optic fiber probes, integrated circuit chips) (Banik et al., 2008; Dawson et al., 2008), and the expression of the whole *lux* operon (*luxCDABE*), beneficial of not requiring the addition of an exogenous substrate for signal production (Applegate et al., 1998).

Functional genes commonly used as indicators of aerobic HC biodegradation are those involved in the initial oxidation of HCs, such as alkane, xylene, and toluene monooxygenases (Sevilla et al., 2015; Tizzard and Lloyd-Jones, 2007; Zhang et al., 2012), benzene and naphthalene dioxygenases (Habe and Omori, 2003), or in the ring cleavage of PAH catabolic pathways, such as catechol dioxygenase genes (Vilchez-Vargas et al., 2013). The key factors in selecting a catabolic promoter are sensitivity and specificity. However, such promoters often respond to groups of compounds rather than to a specific compound, and may also behave differently in different host organisms.

A whole-cell biosensor based on *P. fluorescens* fused with a green fluorescent protein (GFP) was constructed to measure benzene, toluene, ethylbenzene, and related compounds in aqueous samples (Belkin, 2006). The biosensor was based on a plasmid carrying the toluene-benzene utilization (*tbu*) pathway transcriptional activator TbuT from *Ralstonia pickettii* PKO1 and a transcriptional fusion of its promoter *PtbuA1* with a promoterless *gfp* gene on a broad-host-range vector. The biosensor responded in an additive mode to the presence of multiple inducers and was unaffected by the presence of compounds that were not inducers, like those present in gasoline. More recently, a GFP-based sensor system built on marine HCoclastic bacteria *Alcanivorax borkumensis* containing the AlkS regulatory protein and *PalkB* promoter from *P. putida* was successfully used for the detection of alkanes in seawater in the presence of petrol (Sevilla et al., 2015).

Further advances in this field include the expansion of reporters with multicolored autofluorescent proteins, which show different spectral properties, improved brightness and photostability (Tamura and Hamachi, 2014). Most importantly, such genetically engineered cells can be incorporated into whole-cell array formats on silicon chips, optic fibers and other devices (Belkin, 2006). Future applications of such multiplex biosensor systems in environmental monitoring seem promising. Also new promoter-based, whole-cell biosensors suitable for online and in situ monitoring of HC pollutants have been constructed in which the monitoring of gene expression is electrochemical. For example, Paitan et al. (2004) developed two bacterial whole-cell electrochemical biosensors to monitor aromatic HCs. These bacterial biosensors are based on fusions of the *xylS* promoter of the TOL operon from *P. putida* with *lacZ* or *phoA* as reporter genes. These constructs reacted specifically to aromatic compounds and allowed a rapid detection of xylene and toluene in micromolar concentrations.

The use of two different reporter genes allows the future construction of a multianalyte detection system for simultaneous monitoring of several pollutants. In addition, a portative electrochemical nanobiochip for the water toxicity detection was described, where whole cells of genetically engineered *E. coli* MC1061 expressing detectable electrochemical signals in the presence of toxicants were placed on the array of nanovolume electrochemical cells, which provided rapid and sensitive real-time detection of acute toxicity in water (Popovtzer et al., 2005). More recently, another complex electrochemical biosensor, the DNA/hemin/nafion–graphene/glassy carbon electrode, was constructed to quantitatively assay DNA damage induced by benzo(a)pyrene, which can be used for the detection of this procarcinogenic exotoxicant (Ni et al., 2014). It should be noted that most biosensor developments are based on Gram-negative bacteria, particularly easily engineered *E. coli* and *Pseudomonas* strains. However, Gram-positive bacteria are ubiquitous in oil-contaminated environments and many of them can degrade HC pollutants. Among Gram-positive HC-oxidizing bacteria, members of the genus *Rhodococcus* are widely used as whole-cell biosensors for detecting halogenated alkanes (Peter et al., 1996), chlorinated phenols (Riedel et al., 1993), and nitrophenols (Emelyanova and Reshetilov, 2002). However, their biosensing potential for HCs present in oil-contaminated environments is currently unexploited. Recent advances in rhodococcal genome sequencing (Ivshina et al., 2014; Kriszt et al., 2012; McLeod et al., 2006) and functional genomics revealed novel metabolic pathways for the biodegradation of petroleum HCs (Cappelletti et al., 2011; Na et al., 2005), which could allow isolating new inducible promoters useful in biosensor constructions.

BIOREMEDIATION OF OIL-CONTAMINATED ENVIRONMENTS

Bioremediation, a cleanup approach based on metabolic activities of living organisms, is currently recognized as environmentally safe and economically reasonable (Atlas and Bartha, 1992; Megharaj et al., 2011). In this chapter, we focus on sustainable use of HOB in bioremediation and their applications in restoration of extreme, for example, cold climate soils, marine, hypersaline, and anoxic environments.

"Omics" technologies (genomics, proteomics, transcriptomics, metabolomics, metagenomics, metaproteomics, metatranscriptomics, and meta–metabolomics) and construction of microbial biofilms are indicated in literature as approaches which could guarantee bioremediation performance. "Omics" are used to study the community dynamics and relationships in contaminated sites and to reveal genes and enzymes involved in the contaminant biodegradation. A key advantage of "omics" is a possibility to identify biodegrading microorganisms without their isolation and cultivation (Paliwal et al., 2012; Sierra-García et al., 2014; Ufarté et al., 2015). Furthermore, "omics" is a basement for the construction of strains with improved abilities, such as resistance to stresses, high competitiveness, increased oxidizing activity, expressed targeted pathways, etc. A bacterial hybrid based on the hydrocarbon-oxidizing strain *Cupriavidus necator* H850 was constructed with introduction of three genetic elements to mineralize substituted polychlorinated compounds (Ramos et al., 2011). Manipulation with genes encoding global regulatory factors, such as the integration host factor, cAMP receptor protein, alternative sigma factors, the PtsN (IIANtr) protein, and the alarmone (p)ppGpp, was conducted to construct HOB resistant to environmental stresses (Paliwal et al., 2012).

Modification of HOB means the release of obtained genetically engineered microorganisms (GEMs) into the environment that is regulated (CBD, 2001, 2011). For example, in the USA, the only GEM approved for bioremediation is *P. fluorescens* HK44 possessing a naphthalene catabolic plasmid (pUTK21) with inserted *lux* genes (Megharaj et al., 2011). In most countries, even the GEM-field trials are restricted (Paliwal et al., 2012). To increase biosafety of GEMs, the number of solutions is developed. The programmed cell death based on triggering killer-anti-killer gene(s) after pollutant detoxification was used to construct "suicidal GEMs" (Megharaj et al., 2011). Genes encoding synthesis of *quorum sensing* molecules were expressed to sustain the microbial community, which was disrupted at the end of bioremediation (Paliwal et al., 2012). In situ breeding through inherent plasmids with catabolic gene or biosurfactant production operons of soil bacteria was an approach to acquire desired catabolic potential in oil-contaminated ecosystems (Paliwal et al., 2012).

The role of microbial biofilms in bioremediation has also increasingly been explored. Biofilms consisting of HOB enhance biodegradation of organic pollutants, especially long-chain alkanes and large PAHs. Adhesion of bacteria to the HC-water interface minimizes the diffusion path for HCs to the cell interior and thereby facilitates HC uptake. Cells in biofilms are robust, resistant to stresses, and metabolically stable (Abbasnezhad et al., 2011; McGenity et al., 2012). Biofilms were used to enhance biodegradation of PAHs by a marine strain *P. mendocina* NR802 (Shukla et al., 2014). *R. ruber* IEGM 231 cells immobilized on hydrophobized sawdust and formed firm biofilms were shown to oxidize 1.6–2.5 times more *n*-hexadecane than freely suspended cells (Ivshina et al., 2013). In microcosms simulating shallow aquifer conditions, colonization of sediments by anaerobic degraders provide efficient oxidation of HCs (Kleinsteuber et al., 2012).

Sustainable HOB are used in bioremediation of extreme environments. Soil bioremediation trials have been performed in Arctic, Antarctic, sub-Arctic, and high mountain regions. HOB typically inhabiting cold environments and thus applied in bioremediation included members of *Rhodococcus*, *Pseudomonas*, *Sphingomonas*, *Caulobacter*, *Acinetobacter*, *Chryseobacterium*, and *Flavobacterium*. Most strains were psychrotrophic with metabolic activities at low (10°C) temperature compared to activities of mesophilic HC degraders at 25–30°C (Kauppi et al., 2011; Margesin, 2000; Panicker et al., 2010; Yergeau et al., 2012). Ex situ treatments, where soils were excavated and treated in biopiles to stimulate aerobic bacteria, had larger effects on indigenous microbial communities than in situ methods, where soils were fertilized on place to keep the soil structure intact (Yergeau et al., 2012). It is important that the Antarctic Treaty does not permit the introduction of allochthonous microorganisms, and therefore biostimulation or bioaugmentation with indigenous microorganisms are the only feasible approaches in this area (Panicker et al., 2010). Paliwal et al. (2012) compared biostimulation and bioaugmentation in removal of diesel oil from freshly and chronically contaminated Antarctic soils. At the contamination level ≥2% and temperature 2°C, the most effective treatment was the addition of enriched microbial consortium isolated from this contaminated site. As mentioned in Koronelli (1996), introduction is essential to reach high rates of biodegradation under cold climate conditions. *Rhodococcus* actinobacteria are promising candidates for introduction into polar soils. Psychrotolerant rhodococci having a cold-shock gene *cspA* and isolated from Antarctic soils grow on a number of alkanes from hexane to eicosane at temperatures between −2 and 15°C (Bej et al., 2000). Some authors suppose that maintaining sufficient levels of aeration and moisture is more essential than a choice of the bioremediation method. Thus, the applications of wooden chips as bulking agents advanced microbial activities in biostimulated and bioaugmentated Finland soils, whereas bioremediation level was negligible in soil piles without wooden chips (Kauppi et al., 2011). Addition of sand resulted in maximum diesel removal in Newfoundland soil (Coles et al., 2009).

A scheme for bioremediation of soils under cold and temperate climate conditions was developed in our laboratory (Fig. 6.5), which tested in field trials on the territory of Russian Federation (Ivshina et al., 2013; Kuyukina et al., 2003). The 80%–90% removal of crude oil during one summer season was achieved for soil with initial level of contamination 5%–30% (w/w), and soil was restored after phytoremediation performed with permanent grasses.

During the last decade, especially after the Gulf Mexico oil spill in 2010, novel data on biodiversity of marine HOB have been propelled, and seawater biodegrading bacteria have been identified. They include members of *Alcanivorax* and *Cycloclasticus*, most abundant in temperate marine ecosystems, *Thalassolituus* diatoms, predominant in C_{12}-C_{32} *n*-alkane-enriched microcosms, alkane-degrading obligate psychrophiles *Oleispira* spp., and members of *Acinetobacter*, *Marinobacter*, *Pseudomonas*, *Rhodococcus*, and *Roseobacter* (Dash et al., 2013; McGenity et al., 2012). There are few examples of bioremediation in marine habitats. Uric acid was applied as a bioavailable nitrogen

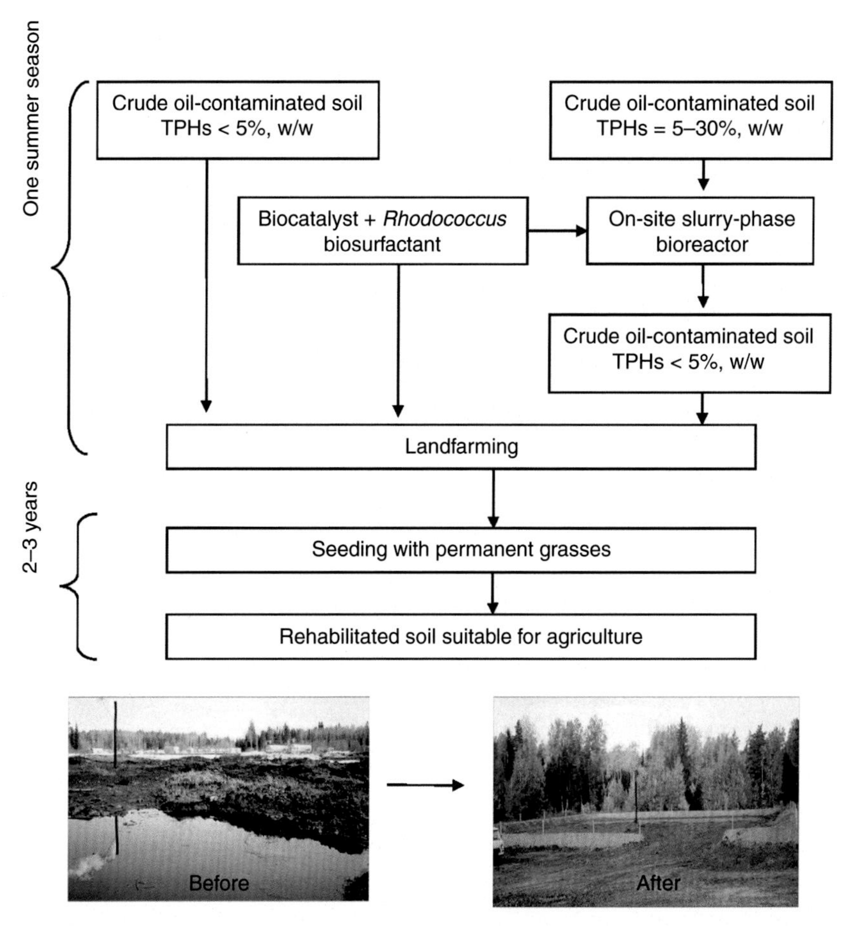

FIGURE 6.5 **Scheme for rehabilitation of crude oil-contaminated soils under cold and temperate climate conditions.** *Rhodococcus* actinobacteria immobilized on modified sawdust were used as a biocatalyst. *TPHs*, total petroleum hydrocarbons.

source for biostimulation of HOB in model seawater (Ron and Rosenberg 2014). A slurry system was constructed for containment of marine sediments with subsequent temperature regulation and provision with air and a slow-released fertilizer (Cappello et al., 2015). Association of indigenous *Cycloclasticus* bacteria supplemented with cell-protective exopolysaccharide from *R. rhodochrous* S-2 was used for bioaugmentation in PAH-contaminated seawater (Iwabuchi, 2014).

Hypersaline (>3.5% NaCl) habitats are widespread on Earth, including flooded areas around oil wells, solar salterns, saline lakes, salt marshes, groundwater, and industrial wastewater (Al-Mailem et al., 2013; Belyaev et al., 2004; Castillo-Carvajal et al., 2014). Biodegradation performance usually is decreased in the presence of high concentrations of NaCl. To overcome these limitations, halotolerant and halophilic microorganisms can be exploited (Castillo-Carvajal et al., 2014). Members of bacterial *Halomonas*, *Marinobacter*, *Rhodococcus*, *Dietzia*, *Gordonia*, *Pseudomonas*, and *Bacillus*, and archaeal *Halococcus*, *Haloferax*, and *Halobacterium* genera are common inhabitants of high salinity environments and potential HC biodegraders in these habitats (Belyaev et al., 2004). Moreover, KCl and $MgSO_4$ were used to stimulate halotolerant HC-oxidizing strains in oil-polluted soils and ponds (Al-Mailem et al., 2013). A salt-tolerant *Enterobacter cloacae* mutant with increased K^+ accumulation and exopolysaccharide level was obtained and used for bioaugmentation of oil- and salt-contaminated soil (Megharaj et al., 2011).

High environmental risks from the contamination of oxygen-depleted niches, such as shallow aquifers, deep sediments, and flooded soil, are caused by possible penetration of organic toxicants to drinking water and surface water bodies. For a long time biodegradation of HCs was presumed to be strictly aerobic process, and bioremediation of anoxic environments was out of the research focus. Nowadays, participation of anaerobic bacteria in the mineralization of organic xenobiotics was demonstrated through evidences. Kleinsteuber et al. (2012) showed in laboratory microcosms that biostimulation with electron acceptors (sodium nitrate, sodium sulfate, or iron chloride) enhanced anaerobic biodegradation of HCs. They applied activated carbon pellets with electron acceptors and nutrient sources to stimulate anaerobic degraders in shallow aquifers. Bioventing is another option for promoting in situ biodegradation of petroleum HCs in subsurface environments. This process provides enhanced oxygenation in a vadose zone and it is highly effective when paired with biosparging in a saturated zone (Ivshina et al., 2015).

CONCLUSIONS

Exploitation of hydrocarbon-oxidizing microorganisms' potential is inseparably associated with their sustainable use. The HOBs have a range of adaptations to survive and maintain high enzymatic activities in varying habitats, and therefore, they are inherently stable biological agents. In this context, special interest represents Gram-positive nonpathogenic bacteria with large and extremely flexible genomes, representatives of a vast and extensively explored class Actinobacteria, predominant in oil-contaminated biotopes and able not only to survive but actively function under harsh conditions. The selection of adequate (active, robust, ecologically, and biologically safe) strains further guarantees effects of bioindication and bioremediation using HC biodegraders. Nevertheless, to date, the ecophysiological and species diversity of HOB worldwide as well as their ecological tactics in highly diverse oil-contaminated environments are not sufficiently studied. Scant attention is paid to microorganisms' adaptations to pollutant-induced stresses, switches from one carbon sources to others, and horizontal catabolic gene transfer within microbial populations. The biological implications of multiple hydroxylation systems coexisting in a single strain and the ability to attack nonutilizable hydrocarbon substrates remain unclear. Moreover, the vast majority of microorganisms are uncultivable and we cannot yet culture them in laboratories for detailed study. Today, considerable efforts are being taken to isolate novel microorganisms and deposit them in collections as novel useful gene sources.

Biological Resource Centres (BRCs) and microbial collections registered with the World Data Centre for Microorganisms (WDCM, http://wdcm.nig.ac.jp) form powerful tools in the search for ideal ecobiotechnology strains (Chapters 10, 13, 14, and 15 of this book). Cultures deposited there are thoroughly characterized in terms of taxonomy and pathogenic/phytopathogenic properties, unlike numerous microbial isolates stored at working collections and often identified only to the genus level. There is also important that BRCs provide transport, deposit, and distribution of strains in accordance with international requirements (OECD, 2001) and guarantee their safety. Moreover, the comprehensive information on strain properties (isolation and growth conditions, substrate specificity, tolerance, etc.) is normally available in the BRC databases. In this context, collections specialized in maintaining the hydrocarbon-oxidizing microorganisms suitable for introducing into oil-contaminated environments have thus importance for bioremediation (Declerck et al., 2015; Ivshina, 2012; Ivshina and Kuyukina, 2013).

ACKNOWLEDGMENTS

This work has been partly supported by the extended national task on bioresource collections housed at the institutes of FASO Russia, the Russian Scientific Foundation (Project no. 14-14-00643) and by the Integrated Program of the Ural Branch of the Russian Academy of Sciences (Project no. 15-12-4-10).

REFERENCES

Abbasian, F., Lockington, R., Mallavarapu, M., Naidu, R., 2015. A comprehensive review of aliphatic hydrocarbon biodegradation by bacteria. Appl. Biochem. Biotechnol. 176, 670–699.

Abbasnezhad, H., Gray, M., Foght, J.M., 2011. Influence of adhesion on aerobic biodegradation and bioremediation of liquid hydrocarbons. Appl. Microbiol. Biotechnol. 92, 653–675.

Al-Mailem, D.M., Eliyas, M., Radwan, S.S., 2013. Bioremediation of oily hypersaline soil and water via potassium and magnesium amendment. Can. J. Microbiol. 59, 837–844.

Alvarez, H.M., 2003. Relationship between β-oxidation and the hydrocarbon-degrading profile and actinomycetes bacteria. Int. Biodeterior. Biodegr. 52, 35–42.

Alvarez, H.M., Silva, R.A., 2013. Metabolic diversity and flexibility for hydrocarbon biodegradation by Rhodococcus. In: Amoroso, M.J., Benimeli, C.S., Cuozzo, S.A. (Eds.), Actinobacteria. Application in Bioremediation and Production of Industrial Enzymes. CRC Press, London, New York, pp. 241–273.

Alvarez, H.M., Souto, M.F., Viale, A., Pucci, O.H., 2001. Biosynthesis of fatty acids and triacylglycerols by 2,6,10,14-tetramethyl pentadecane-grown cells of *Nocardia globerula* 432. FEMS Microbiol. 200, 195–200.

Applegate, B.M., Kehrmeyer, S.R., Sayler, G.S., 1998. A chromosomally based *tod-luxCDABE* whole-cell reporter for benzene, toluene, ethybenzene, and xylene (BTEX) sensing. Appl. Environ. Microbiol. 64 (7), 2730–2735.

Atlas, R.M., 1981. Microbial degradation of petroleum hydrocarbons: an environmental perspective. Microbiol. Rev. 45 (1), 180–209.

Atlas, R.M., Bartha, R., 1992. Hydrocarbon biodegradation and oil spill bioremediation. Marshall, K.C. (Ed.), Advances in Microbial Ecology, vol. 12, Plenum Press, New York, pp. 287–338.

Banik, R.M., Mayank, A.R., Prakash, R., Upadhyay, S.N., 2008. Microbial biosensor based on whole cell of *Pseudomonas* sp. for online measurement of *p*-nitrophenol. Sensor. Actuat. B Chem. 131, 295–300.

Barabás, G., Vargha, G., Szabó, I.M., Penyige, A., Damjanovich, S., Szöllösi, J., Matk, J., Hirano, T., Mátyus, A., Szabó, I., 2001. *n*-Alkane uptake and utilisation by *Streptomyces* strains. Antonie van Leeuwenhoek 79, 269–276.

Bechor, O., Smulski, D.R., Van Dyk, T.K., LaRossa, R.A., Belkin, S., 2002. Recombinant microorganisms as environmental biosensors: pollutants detection by *Escherichia coli* bearing *fabA'::lux* fusions. J. Biotechnol. 94 (1), 125–132.

Bej, A.K., Saul, D., Aislabie, J., 2000. Cold-tolerant alkane-degrading *Rhodococcus* species from Antarctica. Polar Biol. 23, 100–105.

Belkin, S., 2006. Genetically engineered microorganisms for pollution monitoring. Twardowska, I., Allen, H.E., Häggblom, M.M. (Eds.), Soil and Water Pollution Monitoring, Protection and Remediation, vol. 4, Springer, The Netherlands, pp. 147–160.

Belyaev, S.S., Borzenkov, I.A., Nazina, T.N., Rozanova, E.P., Glumov, I.F., Ibatullin, R.R., Ivanov, M.V., 2004. Use of microorganisms in the biotechnology for the enhancement of oil recovery. Microbiology 73 (5), 590–598.

Binazadeh, M., Karimi, I.A., Li, Z., 2009. Fast biodegradation of long chain *n*-alkanes and crude oil at high concentrations with *Rhodococcus* sp. Moj-3449. Enzyme Microb. Technol. 45, 195–202.

Bødtked, G., Hvidsten, I.V., Barth, T., Torsvik, T., 2009. Hydrocarbon degradation by *Dietzia* sp. A14101 isolated from an oil reservoir model column. Antonie van Leeuwenhoek 96, 459–469.

Bubinas, A., Giedraitytė, G., Kalėdienė, L., Nivinskiene, O., Butkiene, R., 2008. Degradation of naphthalene by thermophilic bacteria *via* a pathway, through protocatechuic acid. Cent. Eur. J. Biol. 3 (1), 61–68.

Cameron, P., 2012. Liability for catastrophic risk in the oil gas industry. Int. Energ. Law Rev. 6, 207–219.

Cappelletti, M., Fedi, S., Frascari, D., Ohtake, H., Turner, R.J., Zannoni, D., 2011. Analyses of both the *alkB* gene transcriptional start site and *alkB* promoter-inducing properties of *Rhodococcus* sp. strain BCP1 grown on *n*-alkanes. Appl. Environ. Microbiol. 77, 1619–1627.

Cappello, S., Calogero, R., Santisi, S., Genovese, M., Denaro, R., Genovese, L., Denaro, R., Genovese, L., Giuliano, L., Mancini, G., Yakimov, M.M., 2015. Bioremediation of oil polluted marine sediments: a bio-engineering treatment. Int. Microbiol. 18 (2), 127–134.

Castillo-Carvajal, L.C., Sanz-Martín, J.L., Barragán-Huerta, B.E., 2014. Biodegradation of organic pollutants in saline wastewater by halophilic microorganisms: a review. Environ. Sci. Pollut. Res. 21, 9578–9588.

Catalogue of Strains of Regional Specialized Collection of Alkanotrophic Microorganisms, 2016. Available from: http://www.iegmcol.ru/strains/index.html

CBD, 2001. Cartagena Protocol on Biosafety to the Convention on Biological Diversity. Secretariat of the Convention on Biological Diversity, Montreal, 30 pp.

CBD, 2011. Nagoya Protocol on Access to Genetic Resources and the Fair and Equitable Sharing of Benefits Arising from their Utilization to the Convention on Biological Diversity. Secretariat of the Convention on Biological Diversity, Montreal, 25 pp.

Christofi, N., Ivshina, I.B., 2002. Microbial surfactants and their use in field studies of soil remediation. J. Appl. Microbiol. 93 (6), 915–926.

Christofi, N., Ivshina, I.B., 2005. Microbial surfactants and their use in soil remediation. In: Fingerman, M., Nagabhushanam, R. (Eds.), Bioremediation of Aquatic and Terrestrial Ecosystems. Science Publishers, New Orleans, Louisiana, USA, pp. 311–327.

Coles, C.A., Patel, T.R., Akinnola, A.P., Helleurm, R.J., 2009. Influence of bulking agents, fertilizers and bacteria on the removal of diesel from a Newfoundland soil. Soil Sediment Contam. 18, 383–396.

Coon, M.J., 2005. Omega oxygenases: nonheme-iron enzymes and P450 cytochromes. Biochem. Biophys. Res. Commun. 338, 378–385.

Cuello, O.H., Caorlin, M.J., Reviglio, V.E., Carvajal, L., Juarez, C.P., de Guerra, E.P., Luna, J.D., 2002. *Rhodococcus globerulus* keratitis after laser *in situ* keratomileusis. J. Cataract Refract. Surg. 28, 2235–2237.

Dash, H.R., Mangwani, N., Chakraborty, J., Kumari, S., Das, S., 2013. Marine bacteria: potential candidates for enhanced bioremediation. Appl. Microbiol. Biotechnol. 97, 561–571.

Dawson, J.J., Iroegbu, C.O., Maciel, H., Paton, G.I., 2008. Application of luminescent biosensors for monitoring the degradation and toxicity of BTEX compounds in soils. J. Appl. Microbiol. 104 (1), 141–151.

De Carvalho, C.C.C.R., da Fonseca, M.M.R., 2005. Degradation of hydrocarbons and alcohols at different temperatures and salinities by *Rhodococcus erythropolis* DCL14. FEMS Microbiol. Ecol. 51, 389–399.

De Carvalho, C.C.C.R., Wick, L.Y., Heipieper, H.J., 2009. Cell wall adaptations of planktonic and biofilm *Rhodococcus erythropolis* cells to growth on C_5 to C_{16} *n*-alkane hydrocarbons. Appl. Microbiol. Biotechnol. 82, 311–320.

De Vasconcellos, S.P., Crespim, E., da Cruz, G.F., Senatore, D.B., Simioni, K.C.M., dos Santos Neto, E.V., Marsaioli, A.J., de Oliveira, V.M., 2009. Isolation, biodegradation ability and molecular detection of hydrocarbon degrading bacteria in petroleum samples from a Brazilian offshore basin. Org. Geochem. 40 (5), 574–588.

De Vasconcellos, S.P., Angolini, C.F.F., García, I.N.S., Dellagnezze, B.M., da Silva, C.C., Marsaioli, A.J., dos Santos Neto, E.V., de Oliveira, V.M., 2010. Reprint of: screening for hydrocarbon biodegraders in a metagenomic clone library derived from Brazilian petroleum reservoirs. Org. Geochem. 41 (9), 1067–1073.

Declerck, S., Willems, A., van der Heijden, M.G.A., Varese, G.C., Turkovskaya, O., Evtushenko, L., Ivshina, I., Desmeth, P., 2015. PERN: an EU–Russia initiative for rhizosphere microbial resources. Trends Biotechnol. 33 (7), 377–380.

Dellagnezze, B.M., de Sousa, G.V., Martins, L.L., Domingos, D.F., Limache, E.E.G., de Vasconcellos, S.P., da Cruz, G.F., de Oliveira, V.M., 2014. Bioremediation potential of microorganisms derived from petroleum reservoirs. Marine Pollut. Bull. 89 (1–2), 191–200.

Diplock, E.E., Mardlin, D.P., Killham, K.S., Paton, G.I., 2009. Predicting bioremediation of hydrocarbons: laboratory to field scale. Environ. Pollut. 157, 1831–1840.

Duarte, H.O., Droguett, E.L., Araújo, M., Teixeira, S.F., 2013. Quantitative ecological risk assessment of industrial accidents: the case of oil ship transportation in the coastal tropical area of northeastern Brazil. Hum. Ecol. Risk Assess. 19 (6), 1457–1476.

Emelyanova, E.V., Reshetilov, A.N., 2002. *Rhodococcus erythropolis* as the receptor of cell-based sensor for 2,4-dinitrophenol detection: effect of 'co-oxidation'. Process Biochem. 37, 683–692.

Etoumi, A., 2007. Microbial treatment of waxy crude oils for mitigation of wax precipitation. J. Petrol. Sci. Eng. 55, 111–121.

Feng, L., Wang, W., Cheng, J., Ren, Y., Zhao, G., Gao, C., Tang, Y., Han, W., Peng, X., Wang, L., 2007. Genome and proteome of long-chain alkane degrading *Geobacillus thermodenitrificans* NG80-2 isolated from a deep-subsurface oil reservoir. Proc. Natl. Acad. Sci. USA 104, 5602–5607.

Funhoff, E.G., Bauer, U., García-Rubio, I., Witholt, B., van Beilen, J.B., 2006. CYP153A6, a soluble P450 oxygenase catalizing terminal-alkane hydroxylation. J. Bacteriol. 188 (14), 5220–5227.

Galvão, T.C., Mohn, W.W., de Lorenzo, V., 2005. Exploring the microbial biodegradation and biotransformation gene pool. Trends Biotechnol. 23 (10), 497–506.

Goethals, D.T., Vereecke, D., Jaziri, M., van Montagu, M., Holsters, M., 2001. Leafy gall formation by *Rhodococcus fascians*. Annu. Rev. Phytopathol. 39, 27–52.

Gupta, S., Pathak, B., Fulekar, M.H., 2015. Molecular approaches for biodegradation of polycyclic aromatic hydrocarbon compounds: a review. Rev. Environ. Sci. Biotechnol. 14, 241–269.

Habe, H., Omori, T., 2003. Genetics of polycyclic aromatic hydrocarbon metabolism in diverse aerobic bacteria. Biosci. Biotechnol. Biochem. 67 (2), 225–243.

Hamamura, N., Yeager, C.M., Arp, D.J., 2001. Two distinct monooxygenases for alkane oxidation in *Nocardioides* sp. strain CF8. Appl. Environ. Microbiol. 67, 4992–4998.

Harms, H., Wells, M.C., van der Meer, J.R., 2006. Whole-cell living biosensors-are they ready for environmental application? Appl. Microbiol. Biotechnol. 70 (3), 273–280.

Heider, J., Schühle, K., 2013. Anaerobic biodegradation of hydrocarbons including methane. In: Rosenberg, E., DeLong, E.F., Lory, S., Stackebrandt, E. (Eds.), The Prokaryotes. Prokaryotic Physiology and Biochemistry. Spriger Verlag, Berlin Heidelberg, pp. 605–634.

Herter, S., Mikolasch, A., Schauer, F., 2012. Identification of phenylalkane derivatives when *Mycobacterium neoaurum* and *Rhodococcus erythropolis* were cultured in the presence of various phenylalkanes. Appl. Microbiol. Biotechnol. 93, 342–355.

Hua, F., Wang, H., 2012. Uptake modes of octadecane by *Pseudomonas* sp. DG17 and synthesis of biosurfactant. J. Appl. Microbiol. 112, 25–37.

Hua, F., Wang, H.Q., Zhao, Y.C., 2013. Trans-membrane transport of *n*-octadecane by *Pseudomonas* sp DG17. J. Microbiol. 51 (6), 791–799.

Ivshina, I.B., 2012. Current situation and challenges of specialized microbial resource centres in Russia. Microbiology 81 (5), 509–516.

Ivshina, I.B., Kuyukina, M.S., 2013. Turning Russian specialized microbial culture collections into resource centers for biotechnology. Trends Biotechnol. 31 (11), 609–611.

Ivshina, I.B., Oborin, A.A., Nesterenko, O.A., Kasumova, S.A., 1981. *Rhodococcus* bacteria in ground water from oil fields in the Perm Cisural region. Microbiology 50 (4), 531–538.

Ivshina, I.B., Berdichevskaya, M.V., Zvereva, L.V., Rybalka, L.V., Elovikova, E.A., 1995. Phenotypic characterization of alkanoprophic rhodococci from various ecosystems. Microbiology 64 (4), 507–513.

Ivshina, I.B., Kuyukina, M.S., Philp, J.C., Christofi, N., 1998. Oil desorption from mineral and organic materials using biosurfactant complexes produced by *Rhodococcus* species. World J. Microbiol. Biotechnol. 14, 711–717.

Ivshina, I.B., Kuyukina, M.S., Ritchkova, M.I., Philp, J.C., Cunningham, C.J., Christofi, N., 2001. Oleophilic biofertilizer based on a *Rhodococcus* surfactant complex for the bioremediation of crude oil-contaminated soil. AEHS Contam. Soil Sediment Water 4, 20–24.

Ivshina, I.B., Kuyukina, M.S., Krivoruchko, A.V., Plekhov, O.A., Naimark, O.B., Podorozhko, E.A., Lozinsky, V.I., 2013. Biosurfactant-enhanced immobilization of hydrocarbon-oxidizing *Rhodococcus ruber* on sawdust. Appl. Microbiol. Biotechnol. 97 (12), 5315–5327.

Ivshina, I.B., Kuyukina, M.S., Krivoruchko, A.V., Barbe, V., Fischer, C., 2014. Draft genome sequence of propane and butane oxidizing actinobacterium *Rhodococcus ruber* IEGM 231. Genome Announc. 2 (6.).

Ivshina, I.B., Kuyukina, M.S., Krivoruchko, A.V., Elkin, A.A., Makarov, S.O., Cunningham, C.J., Peshkur, T.A., Atlas, R.M., Philp, J.C., 2015. Oil spill problems and sustainable response strategies through new technologies. Environ. Sci. Process. Impacts 17, 1201–1219.

Iwabuchi, N., 2014. Selective stimulation of aromatic compound degradation by the indigenous marine bacterium Cycloclasticus for bioremediation of oil spills in the marine environment. In: Nojiri, H., Tsuda, M., Fukuda, M., Kamagata, Y. (Eds.), Biodegradative Bacteria: How Bacteria Degrade, Survive, Adapt, and Evolve. Springer, Japan, pp. 313–333.

Jegan, J., Vijayaraghavan, K., Senthilkumar, R., Velan, M., 2010. Naphthalene degradation kinetics of *Micrococcus* sp., isolated from activated sludge. Clean Soil Air Water 38 (9), 837–842.

Ji, Y., Mao, G., Wang, Y., Bartlam, M., 2013. Structural insights into diversity and *n*-alkane biodegradation mechanisms of alkane hydroxylases. Front. Microbiol. 4 (58.).

Jin, S., Makris, T.M., Bryson, T.A., Sligar, S.G., Dawson, J.H., 2003. Epoxidation of olefins by hydroperoxo-ferric cytochrome P450. J. Am. Chem. Soc. 125 (12), 3406–3407.

Kämpfer, P., Dott, W., Martin, K., Glaeser, S.P., 2014. *Rhodococcus defluvii* sp. nov., isolated from wastewater of a bioreactor and formal proposal to reclassify [*Corynebacterium hoagii*] and *Rhodococcus equi* as *Rhodococcus hoagii* comb. nov. Int. J. Syst. Evol. Microbiol. 64 (3), 755–761.

Kauppi, S., Sinkkonen, A., Romantschuk, M., 2011. Enhancing bioremediation of diesel-fuel-contaminated soil in a boreal climate: comparison of biostimulation and bioaugmentation. Int. Biodeter. Biodegr. 65, 359–368.

Kedlaya, I., Ing, M.B., Wong, S.S., 2001. *Rhodococcus equi* infections in immunocompetent hosts: case report and review. Clin. Infect. Dis. 32, 39–47.

Kleindienst, S., Paul, J.H., Joye, S.B., 2015. Using dispersants after oil spills: impacts on the composition and activity of microbial communities. Nat. Rev. Microbiol. 13, 388–396.

Kleinsteuber, S., Schleinitz, K.M., Vogt, C., 2012. Key players and team play: anaerobic microbial communities in hydrocarbon-contaminated aquifers. Appl. Microbiol. Biotechnol. 94, 851–873.

Koronelli, T.V., 1996. Principles and methods for raising the efficiency of biological degradation of hydrocarbons in the environment: review. Appl. Biochem. Microbiol. 32 (6), 519–526.

Kotani, T., Yamamoto, T., Yurimoto, H., Sakai, Y., Kato, N., 2003. Propane monooxygenase and NAD^+-dependent secondary alcohol dehydrogenase in propane metabolism by *Gordonia* sp. strain TY-5. J. Bacteriol. 185, 7120–7128.

Kotani, T., Kawashima, Y., Yurimoto, H., Kato, N., Sakai, Y., 2006. Gene structure and regulation of alkane monooxygenases in propane-utilizing *Mycobacterium* sp. TY-6 and *Pseudonocardia* sp. TY-7. J. Biosci. Bioeng. 102, 184–192.

Kotani, T., Yurimoto, H., Kato, N., Sakai, Y., 2007. Novel acetone metabolism in a propane-utilizing bacterium, *Gordonia* sp. strain TY-5. J. Bacteriol. 189, 886–893.

Kriszt, B., Táncsics, A., Cserháti, M., Tóth, Á., Nagy, I., Horváth, B., Nagy, I., Kukolya, J., 2012. *De novo* genome project for the aromatic degrader *Rhodococcus pyridinivorans* strain AK37. J. Bacteriol. 194 (5), 1247–1248.

Kulakov, L.A., Allen, C.C.R., Lipscomb, D.A., Larkin, M.J., 2000. Cloning and characterization of a novel *cis*-naphthalene dihydrodiol dehydrogenase gene (*narB*) from *Rhodococcus* sp. NCIMB12038. FEMS Microbiol. Lett. 182, 327–331.

Kumar, M., Leon, V., de Sisto Materano, A., Ilzins, O.A., 2007. A halotolerant and thermotolerant *Bacillus* sp. degrades hydrocarbons and produces tensio-active emulsifying agent. World J. Microbiol. Biotechnol. 23, 211–220.

Kuyukina, M.S., Ivshina, I.B., 2010. *Rhodococcus* biosurfactants: biosynthesis, properties and potential application. Alvares, H.M. (Ed.), Microbiology Monographs. Biology of Rhodococcus, vol. 16, Springer-Verlag, Berlin Heidelberg, pp. 292–313.

Kuyukina, M.S., Ivshina, I.B., Rychkova, M.I., Chumakov, O.B., 2000. Effect of cell lipid composition on the formation of nonspecific antibiotic resistance in alkanotrophic rhodococci. Microbiology 69, 51–57.

Kuyukina, M.S., Ivshina, I.B., Ritchkova, M.I., Philp, J.C., Cunningham, J.C., Christofi, N., 2003. Bioremediation of crude oil-contaminated soil using slurry-phase biological treatment and land farming techniques. Soil Sediment Contam. 12, 85–99.

Kuyukina, M.S., Ivshina, I.B., Baeva, T.A., Kochina, O.A., Gein, S.V., Chereshnev, V.A., 2015. Trehalolipid biosurfactants from nonpathogenic *Rhodococcus* actinobacteria with diverse immunomodulatory activities. New Biotechnol. 32 (3), 559–568.

Lang, S., Philp, J.C., 1998. Surface-active lipids in rhodococci. Antonie van Leeuwenhoek 74, 59–70.

Larkin, M.J., Kulakov, L.A., Allen, C.C.R., 2006. Biodegradation by members of the genus *Rhodococcus*: biochemistry, physiology, and genetic adaptation. Adv. Appl. Microbiol. 59, 1–29.

Larkin, M.J., Kulakov, L.A., Allen, C.C.R., 2010. Genomes and plasmids in Rhodococcus. Alvares, H.M. (Ed.), Microbiology Monographs. Biology of Rhodococcus, vol. 16, Springer-Verlag, Berlin Heidelberg, pp. 73–90.

Leahy, J.G., Batchelor, P.J., Morcomb, S.M., 2003. Evolution of the soluble diiron monooxygenases. FEMS Microbiol. Rev. 27, 449–479.

Li, Y., Wang, H., Hua, F., Su, M., Zhao, Y., 2014. Trans-membrane transport of fluoranthene by *Rhodococcus* sp. BAP-1 and optimization of uptake process. Bioresour. Technol. 155, 213–219.

Lin, H., Liu, J.-Y., Wang, H.-B., Ahmed, A.A.Q., Wu, Z.-L., 2011. Biocatalysis as an alternative fort the production of chiral epoxides: a comparative review. J. Mol. Catal. B Enzym. 72 (3), 77–89.

Margesin, R., 2000. Potential of cold-adapted microorganisms for bioremediation of oil-polluted Alpine soils. Int. Biodeter. Biodegr. 46, 3–10.

Martínková, L., Uhnáková, B., Pátek, M., Nešvera, J., Křen, V., 2009. Biodegradation potential of the genus *Rhodococcus*. Environ. Int. 35, 162–177.

McGenity, T.J., Folwell, B.D., McKew, B.A., Sanni, G.O., 2012. Marine crude-oil biodegradation: a central role for interspecies interactions. Aquat. Biosyst. 8, 10.

McLeod, M.P., Warren, R.L., Hsiao, W.W., Araki, N., Myhre, M., Fernandes, C., Miyazawa, D., Wong, W., Lillquist, A.L., Wang, D., Dosanjh, M., Hara, H., Petrescu, A., Morin, R.D., Yang, G., Stott, J.M., Schein, J.E., Shin, H., Smailus, D., Siddiqui, A.S., Marra, M.A., Jones, S.J., Holt, R., Brinkman, F.S., Miyauchi, K., Fukuda, M., Davies, J.E., Mohn, W.W., Eltis, L.D., 2006. The complete genome of *Rhodococcus* sp. RHA1 provides insights into a catabolic powerhouse. Proc. Natl. Acad. Sci. 103 (42), 15582–15587.

Megharaj, M., Ramakrishnan, B., Venkateswarlu, K., Sethunathan, N., Naidu, R., 2011. Bioremediation approaches for organic pollutants: a critical perspective. Environ. Int. 37, 1362–1375.

Na, K., Nagayasu, K., Kuroda, A., Takiguchi, N., Ikeda, T., Ohtake, H., Kato, J., 2005. Development of a genetic transformation system for benzene-tolerant *Rhodococcus opacus* strains. J. Biosci. Bioeng. 99 (4), 408–414.

Ni, Y., Wang, P., Song, H., Lin, X., Kokot, S., 2014. Electrochemical detection of benzo(a)pyrene and related DNA damage using DNA/hemin/nafion-graphene biosensor. Anal. Chim. Acta 821, 34–40.

OECD, 2001. Biological Resource Centers: Underpinning the Future of Life Sciences and Biotechnology. OECD Directorate for Science, Technology and Industry, Paris, 68 pp.

Paitan, Y., Biran, I., Shechter, N., Biran, D., Rishpon, J., Ron, E.Z., 2004. Monitoring aromatic hydrocarbons by whole cell electrochemical biosensors. Anal. Biochem. 335, 175–183.

Paliwal, V., Puranik, S., Purohit, H.J., 2012. Integrated perspective for effective bioremediation. Appl. Biochem. Biotechnol. 166, 903–924.

Panicker, G., Mojib, N., Aislabie, J., Bej, A.K., 2010. Detection, expression and quantitation of the biodegradative genes in Antarctic microorganisms using PCR. Antonie van Leeuwenhoek 97, 275–287.

Park, M., Tsai, S.-L., Chen, W., 2013. Microbial biosensors: engineered microorganisms as the sensing machinery. Sensors 13, 5777–5795.

Pérez-Pantoja, D., González, B., Pieper, D.H., 2010. Aerobic degradation of aromatic hydrocarbons. In: Timmis, K.N. (Ed.), Handbook of Hydrocarbon and Lipid Microbiology. Springer-Verlag, Berlin Heidelberg, pp. 800–837.

Peter, J., Hutter, W., Stöllnberger, W., Hampel, W., 1996. Detection of chlorinated and brominated hydrocarbons by an ion sensitive whole cell biosensor. Biosens. Bioelectron. 11 (12), 1215–1219.

Philp, J.C., Kuyukina, M.S., Ivshina, I.B., Dunbar, S.A., Christofi, N., Lang, S., Wray, V., 2002. Alkanotrophic *Rhodococcus ruber* as a biosurfactant producer. Appl. Microbiol. Biotechnol. 59 (2–3), 318–324.

Plotnikova, E.G., Altyntseva, O.V., Kosheleva, I.A., Puntus, I.F., Filonov, A.E., Gavrish, E.Y.u., Demakov, V.V., Boronin, A.M., 2001. Bacterial degraders of polycyclic aromatic hydrocarbons isolated from salt-contaminated soils and bottom sediments in salt mining areas. Microbiology 70, 51–58.

Popovtzer, R., Neufeld, T., Biran, D., Ron, E.Z., Rishpon, J., Shacham-Diamand, Y., 2005. Nanobiochip for toxicity detection in water. Nano Lett. 5 (6), 1023–1027.

Ramos, J.L., Marqués, S., van Dillewijn, P., Espinosa-Urgel, M., Segura, A., Duque, E., Krell, T., Ramos-González, M.-I., Bursakov, S., Roca, A., Solano, J., Femádez, M., Niqui, J.L., Pizamo-Tobias, P., Wittich, R.-M., 2011. Laboratory research aimed at closing the gaps in microbial bioremediation. Trends Biotechnol. 29 (12), 641–647.

Redmond, M.C., Valentine, D.L., Sessions, A.L., 2010. Identification of novel methane-, ethane-, and propane-oxidizing bacteria at marine hydrocarbon seeps by stable isotope probing. Appl. Environ. Microbiol. 76 (19), 6412–6422.

Reshetilov, A.N., Iliasov, P.V., Reshetilova, T.A., 2010. The microbial cell based biosensors. In: Somerset, V.S. (Ed.), Intelligent and Biosensors. Intech, pp. 289–321.

Riedel, K., Hensel, J., Rothe, S., Neumann, B., Scheller, F., 1993. Microbial sensors for determination of aromatics and their chloroderivatives. Part II: determination of chlorinated phenols using a Rhodococcus-containing biosensor. Appl. Microbiol. Biotechnol. 38, 556–559.

Rojo, F., 2009. Degradation of alkanes by bacteria. Environ. Microbiol. 11 (10), 2477–2490.

Ron, E.Z., Rosenberg, E., 2001. Natural roles of biosurfactants. Environ. Microbiol. 3, 229–236.

Ron, E.Z., Rosenberg, E., 2014. Enhanced bioremediation of oil spills in the sea. Curr. Opin. Biotechnol. 27, 191–194.

Sakai, Y., Takahashi, H., Wakasa, Y., Kotani, T., Yurimoto, H., Miyachi, N., Paul, P.V.V., Kato, N., 2004. Role of α-methylacyl coenzyme A racemase in the degradation of methyl-branched alkanes by *Mycobacterium* sp. strain P101. J. Bacteriol. 186, 7214–7220.

Sathishkumar, M., Binupriya, A., Baik, S.H., Yun, S.E., 2008. Biodegradation of crude oil by individual bacterial strains and a mixed bacterial consortium isolated from hydrocarbon contaminated areas. Clean Soil Air Water 36, 92–96.

Sayavedra-Soto, L.A., Chang, W.-N., Lin, T.-K., Ho, C.-L., Liu, H.-S., 2006. Alkane utilization by *Rhodococcus* strain NTU-1 alone and in its nature association with *Bacillus fusiformis* L-1 and *Ochrobactrum* sp. Biotechnol. Prog. 22, 1368–1373.

Sei, K., Sugimoto, Y., Mori, K., Kohno, T., 2003. Monitoring of alkane-degrading bacteria in a sea water microcosm during crude oil degradation by polymerase chain reaction based on alkane-catabolic genes. Environ. Microbiol. 5, 517–522.

Serebrennikova, M.K., Kuyukina, M.S., Krivoruchko, A.V., Ivshina, I.B., 2014. Adaptation of co-immobilized *Rhodococcus* cells to oil hydrocarbons in a column bioreactor. Appl. Biochem. Microbiol. 50, 265–272.

Sevilla, E., Yuste, L., Rojo, F., 2015. Marine hydrocarbonoclastic bacteria as whole-cell biosensors for *n*-alkanes. Microb. Biotechnol. 8 (4), 693–706.

Shennan, J.L., 2006. Utilisation of C_2-C_4 gaseous hydrocarbons and isoprene by microorganisms. J. Chem. Technol. Biotechnol. 81, 237–256.

Shukla, S.K., Mangwani, N., Rao, T.S., Das, S., 2014. Biofilm-mediated bioremediation of polycyclic aromatic hydrocarbons. In: Das, S. (Ed.), Microbial Biodegradation and Bioremediation. Elsevier, Oxford, pp. 203–232.

Sierra-García, I.N., Alvarez, J.C., de Vasconcellos, S.P., Pereira de Souza, A., dos Santos Neto, E.V., de Oliveira, V.M., 2014. New hydrocarbon degradation pathways in the microbial metagenome from Brazilian petroleum reservoirs. PLoS One 9 (2), e90087.

Smith, C.B., Tolar, B.B., Hollibaugh, J.T., King, G.M., 2013. Alkane hydroxylase gene (*alkB*) phylotype composition and diversity in northern Gulf of Mexico bacterioplankton. Front. Microbiol. 4 (370.).

Sorkhoh, N.A., Ghannoum, M.A., Ibrahim, A.S., Stretton, R.J., Radwan, S.S., 1990. Crude oil and hydrocarbon-degrading strains of *Rhodococcus rhodochrous* isolated from soil and marine environments in Kuwait. Environ. Pollut. 65, 1–17.

Sticher, P., Jaspers, M.C., Stemmler, K., Harms, H., Zehnder, A.J., van der Meer, J.R., 1997. Development and characterization of a whole-cell bioluminescent sensor for bioavailable middle-chain alkanes in contaminated groundwater samples. Appl. Environ. Microbiol. 63 (10), 4053–4060.

Stiner, L., Halverson, L.J., 2002. Development and characterization of a green fluorescent protein-based bacterial biosensor for bioavailable toluene and related compounds. Appl. Environ. Microbiol. 68 (4), 1962–1971.

Su, L., Jia, W., Houb, C., Lei, Y., 2011. Microbial biosensors: a review. Biosens. Bioelectron. 26, 1788–1799.

Suenaga, H., Ohnuki, T., Miyazaki, K., 2007. Functional screening of a metagenomic library for genes involved in microbial degradation of aromatic compounds. Environ. Microbiol. 9, 2289–2297.

Sutcliffe, I.C., 1998. Cell envelope composition and organization in the genus *Rhodococcus*. Antonie van Leeuwenhoek 74, 46–58.

Tamura, T., Hamachi, I., 2014. Recent progress in design of protein-based fluorescent biosensors and their cellular applications. ACS Chem. Biol. 9 (12), 2708–2717.

Tecon, R., 2015. Single-cell bacterial bioreporter assays to measure hydrocarbons. In: McGenity, T.J., Timmis, K.N., Fernandez, B.N. (Eds.), Hydrocarbon and Lipid Microbiology Protocols: Single-Cell and Single-Molecule Methods. Springer protocols handbooks. Springer-Verlag, Berlin Heidelberg, pp. 1–12.

Tecon, R., van der Meer, J.R., 2008. Bacterial biosensors for measuring availability of environmental pollutants. Sensors 8, 4062–4080.

Tizzard, A.C., Lloyd-Jones, G., 2007. Bacterial oxygenases: in vivo enzyme biosensors for organic pollutants. Biosens. Bioelectron. 22, 2400–2407.

Ufarté, L., Laville, É., Duquesne, S., Potocki-Veronese, G., 2015. Metagenomics for the discovery of pollutant degrading enzymes. Biotechnol. Adv. 33, 1845–1854.

Urlacher, V.B., Eiben, S., 2006. Cytochrome P450 monooxygenases: perspectives for synthetic application. Trends Biotechnol. 24 (7), 324–330.

Van Beilen, J.B., Funhoff, E.G., 2007. Alkane hydroxylases involved in microbial alkane degradation. Appl. Microbiol. Biotechnol. 74 (1), 13–21.

Van Beilen, J.B., Smits, T.H.M., Whyte, L.G., Schorcht, S., Röthlisberger, M., Plaggemeier, T., Engesser, K.-H., Witholt, B., 2002. Alkane hydroxylase homologues in gram-positive strain. Environ. Microbiol. 4, 676–682.

Van Beilen, J.B., Li, Z., Duetz, W.A., Smits, T.H.M., Witholt, B., 2003. Diversity of alkane hydroxylase systems in the environment. Oil Gas Sci. Technol. 58, 427–440.

Van Beilen, J.B., Funhoff, E.G., van Loon, A., Just, A., Kaysser, L., Bouza, M., Holtackers, R., Röthlisberger, M., Li, Z., Witholt, B., 2006. Cytochrome P450 alkane hydroxylases of the CYP153 family are common in alkane-degrading eubacteria lacking integral membrane alkane hydroxylases. Appl. Environ. Microbiol. 72 (1), 59–65.

Vilchez-Vargas, R., Geffers, R., Suárez-Diez, M., Conte, I., Waliczek, A., Kaser, V.-S., Kralova, M., Junca, H., Pieper, D., 2013. Analysis of the microbial gene landscape and transcriptome for aromatic pollutants and alkane degradation using a novel internally calibrated microarray system. Environ. Microbiol. 15 (4), 1016–1039.

Von der Weid, I., Marques, J.M., Cunha, C.D., Lippi, R.K., dos Santos, S.C.C., Rosado, A.S., Lins, U., Seldin, L., 2007. Identification and biodegradation potential of a novel strain of *Dietzia cinnamea* isolated from a petroleum-contaminated tropical soil. Syst. Appl. Microbiol. 30 (4), 331–339.

Wang, X.-B., Chi, C.G., Nie, Y., Tang, Y.-Q., Tan, Y., Wu, G., Wu, X.-L., 2011. Degradation of petroleum hydrocarbons (C_6-C_{40}) and crude oil by a novel *Dietzia* strain. Bioresour. Technol. 102 (17), 7755–7761.

Warhurst, A.W., Fewson, C.A., 1994. Biotransformations catalyzed by the genus *Rhodococcus*. Crit. Rev. Biotechnol. 14, 29–73.

Wasmund, K., Burns, K.A., Kurtböke, I., Bourne, D.G., 2009. Novel alkane hydroxylase gene (*alkB*) diversity in sediments associated with hydrocarbon seeps in the Timor Sea, Australia. Appl. Environ. Microbiol. 75 (23), 7391–7398.

Watrinson, R.J., Morgan, P., 1990. Physiology of aliphatic hydrocarbon-degrading microorganisms. Biodegradation 1, 79–92.

Wentzel, A., Ellingsen, T.E., Kotlar, H.K., Zotchev, S.B., Throne-Holst, M., 2007. Bacterial metabolism of long-chain *n*-alkanes. Appl. Microbiol. Biotechnol. 76, 1209–1221.

Wick, L., Munain, A.D., Springael, D., Harms, H., 2002. Responses of *Mycobacterium* sp. LB501T to the low bioavailability of solid anthracene. Appl. Microbiol. Biotechnol. 58, 378–385.

Yam, K.C., van der Geize, R., Eltis, L.D., 2010. Catabolism of aromatic compounds and steroids by Rhodococcus. Alvares, H.M. (Ed.), Microbiology Monographs. Biology of Rhodococcus, vol. 16, Springer-Verlag, Berlin Heidelberg, pp. 133–169.

Yergeau, E., Sanschagrin, S., Beaumier, D., Greer, C.W., 2012. Metagenomic analysis of the bioremediation of diesel-contaminated Canadian high arctic soils. PLoS One 7 (1), e30058.

Zhang, D., He, Y., Wang, Y., Wang, H., Wu, L., Aries, E., Huang, W.E., 2012. Whole-cell bacterial bioreporter for actively searching and sensing of alkanes and oil spills. Microb. Biotechnol. 5 (1), 87–97.

FURTHER READING

Kester, A.S., Foster, J.W., 1963. Diterminal oxidation of long-chain alkanes by bacteria. J. Bacteriol. 85 (4), 859–869.

Musat, T., 2015. The anaerobic degradation of gaseous, nonmethane alkane-from in situ processes to microorganisms. Comput. Struct. Biotechnol. J. 13, 222–228.

Chapter 7

An Overview of the Industrial Aspects of Antibiotic Discovery

Evan Martens* and Arnold L. Demain**

**Cempra, Inc., Chapel Hill, NC, United States; **RISE Institute, Drew University, Madison, NJ, United States*

INTRODUCTION

Secondary metabolites are organic compounds produced by animals, plants, and microorganisms that are not absolutely required for the survival of the producing organism. These compounds, often referred to as natural products, have helped combat pathogenic microorganisms. One of the most important types of microbial secondary metabolites are antibiotics, used to treat infectious diseases caused by bacteria and fungi. In 1900, infectious disease was the leading cause of human death. More recently, it was ranked second worldwide and number three in developed nations (Kraus, 2008; Nathan, 2004; Yoneyama and Katsumata, 2006). Since the discovery of penicillin by Alexander Fleming in 1928, antibiotics have revolutionized the field of medicine by saving millions of lives, eradicating human disease, reducing pain and suffering, improving the quality of life, and nearly doubling the human life span (Demain, 2006). About half of the drugs that we currently use were in fact discovered during the "golden age" of antibiotics from 1950 to 1960 (Davies, 2006). Some examples include the tetracyclines (1953), cephalosporins (1953), aminoglycosides (1957), and vancomycin (1958). Approximately 15,000 antibiotics are known and 160 of them are on the market (Ariyanchira, 2010). The composition of the antiinfective market in 2010 was cephalosporins at 23%, fluoroquinolones at 17%, penicillins at 16%, macrolides at 14%, carbapenems at 6%, and tetracyclines at 4%. Other classes of antimicrobial agents were at 11% and vaccines constituted 9% (Baltz, 2007).

Different Sources of Antibiotics

The vast majority of antibiotics can be classified as either (1) natural products of microorganisms, (2) those that are semisynthetically produced from natural products, or (3) chemically synthesized based on a known structure of the natural

Microbial Resources. http://dx.doi.org/10.1016/B978-0-12-804765-1.00007-2

product. They can be described as low molecular weight organic compounds, which are usually active at low concentrations against pathogenic microorganisms. Filamentous bacteria (actinomycetes) are Gram-positive organisms, many of which are found in the soil, that produce nearly 75% of the antibiotics (Bronson and Barrett, 2001). A single genus, *Streptomyces*, is responsible for producing 75% of these compounds. For example, strains of *Streptomyces hygroscopicus* are capable of making about 200 antibiotics. The nonfilamentous bacteria produce a much smaller amount, approximately 12% of known antibiotics, while the antibiotic yield from the filamentous fungi is 20%.

In the search for new antibiotics, many of the new products are made by chemists via modification of natural antibiotics. This process is known as "semisynthesis." As early as 1974, over 20,000 penicillins, 4000 cephalosporins, 2500 tetracyclines, 1000 rifamycins, 500 kanamycins, and 500 chloramphenicols had been prepared by semisynthesis. More than 350 agents have reached the world market as antimicrobials. They include (1) natural products, (2) semisynthetic antibiotics, and (3) synthetic chemicals (Tahlan and Jensen, 2013). The commercial antibiotics include the cephalosporins (45%), penicillins (15%), quinolones (11%), tetracyclines (6%), macrolides (5%); the remainder include the aminoglycosides, ansamycins, glycopeptides, lipopeptides, and polyenes. The strictly synthetics include the sulfa drugs, azoles, oxazolidinones (linezolid), quinolones, and fluoroquinolones.

Classes of Antibiotics

Classes of antibiotics include the beta-lactams (penicillins, cephalosporins), tetracyclines, macrolides, fluoroquinolones, aminoglycosides, monobactams, carbapenems, and glycopeptides. Each antibiotic has a unique mechanism of action (such as inhibition of protein synthesis, cell wall synthesis, nucleic acid synthesis, etc.), and are either bacteriostatic or bactericidal.

β-Lactams

Approximately 40 β-lactam compounds have been used in medicine, including penicillins, cephalosporins, monobactams, clavams, carbapenems, and monocyclic β-lactams (Banik, 2013). Over 50% of worldwide prescriptions are for β-lactams. The titer of penicillin G is over 100 g/L. Chemical synthesis of novel β-lactams has been reviewed (Paradkar, 2013). Interestingly, some β-lactams have activities, such as cholesterol-lowering and antitumor.

Resistance to the β-lactam antibiotics is largely attributable to β-lactamase enzymes (about 450 in existence) that are produced by certain bacteria. However, even after 60 years, β-lactams are still widely used, due in large part to the discovery of β-lactamase inhibitors, that is, new carbapenems, such as doripenem, clavulanic acid, imipenem, meropenem, ertapenem, all of which have been approved for clinical use. Augmentin is a combination of the semisynthetic amoxicillin and clavulanic acid. It is used orally against bacterial infections, such as sinusitis, pneumonia, bronchitis, and otitis media. Global sales of Augmentin were over $2.1 billion in 2013 (SciBX, 2013).

An interesting new drug is a combination of the β-lactams ceftolozane and tazobactam (Chopra and Roberts, 2001). It is an intravenous drug combination for nosocomial Gram-negative infections. Ceftolozane is a cephalosporin with activity against MDR-*Pseudomonas aeruginosa* and most *Enterobacteriaceae*. Its combination with tazobactam broadens its activity spectrum to include extended spectrum β-lactamase (ESBL)-producing pathogens. ESBLs confer resistance to most β-lactams including penicillins, cephalosporins, and the monobactam aztreonam.

Tetracyclines

These were the first broad-spectrum antibiotics. The first tetracycline, that is, chlortetracycline, was discovered at the Cyanamid Company in 1948. They have played a significant role in combating intracellular pathogens, such as chlamydiae, mycoplasmas, rickettsiae, and protozoan parasites (Nicolaou et al., 2009). In addition, they are also effective at treating Lyme disease, the ulcer-causing bacterium *Helicobacter pylori*, and potential bioterrorism-related diseases, such as anthrax, plague, and tularemia. With the emergence of resistance due to antibiotic efflux, the second-generation, semisynthetic tetracyclines (minocycline and doxycycline) were developed. These drugs helped limit the extent of antibiotic efflux, due to the fact that they were more lipophilic and able to be taken up to a greater degree by the resistant strains. However, resistance based on ribosomal protection was soon to follow, which prompted the approval of a third-generation tetracycline, tigecycline, in 2005. Another tetracycline, oxytetracycline, is produced by *Streptomyces rimosus* at over 100 g/L. Ten tetracyclines have been utilized as antibiotics (Deshpande et al., 2008). Total tonnage of tetracyclines amounted to 5000 metric tons in 2009.

Macrolides

Actinomycetes, such as *Streptomyces*, *Micromonospora*, *Saccharopolyspora*, and *Actinoplanes* are also responsible for the production of the polyketide macrolides. Erythromycin was the first macrolide approved, in the 1950s, and was then followed by oleandomycin, pikromycin, amphotericin B, midecamycin, josamycin, and carbomycin. Tylosin became useful in livestock and veterinary medicine. It was determined that these early macrolides had several problems, which included acid instability, poor bioavailability, and rapid elimination. This prompted the discovery of the very important semisynthetic, second-generation macrolides, that is, clarithromycin (6-0-methyl erythromycin A; Biaxin) and azithromycin (Zithromax). These compounds had a lower rate of elimination from the body, were dosed only once or twice per day, and also had enhanced acid stability. In an effort to combat resistance, the third-generation "ketolides," which were semisynthetic derivatives of erythromycin, became important in the late 1980s. One of the ketolides is telithromycin (Ketek), which was approved in Europe in 2001 and in the United States of America in 2004. However, telithromycin was found to be associated with serious adverse events, such as exacerbation of myasthenia gravis, retinal detachment, Achilles tendon rupture, and idiosyncratic hepatotoxicity, which subsequently led to restrictions on its use in 2007.

Macrolides bind to the 50s subunit of the bacterial ribosome and inhibit protein synthesis. They are used to treat respiratory tract pathogens, including *Streptococcus pneumoniae*, *Haemophilus influenzae*, *Moraxella catarrhalis*, as well as atypical organisms, such as *Legionella pneumophila* and *Mycoplasma pneumoniae*. In addition, macrolides are effective against gonorrhea, chlamydia, Lyme disease, peptic ulcers, and skin and soft tissue bacteria (staphylococci, *Propionibacterium acnes*, and *Streptococcus pyogenes*).

Quinolones and Fluoroquinolones

The quinolone and fluoroquinolone groups, which represent a synthetic class of antibiotics, inhibit DNA gyrase and have broad-spectrum activity. The first quinolone discovered was chloroquine, used primarily for the treatment and prevention of malaria. It was later followed by nalidixic acid, cinoxacin, rosoxacin, pipemidic acid, piromidic acid, and oxolinic acid. Further research and experimentation showed that adding a fluorine at the C6 position led to an enhanced activity of the quinolones, prompting the emergence of the fluoroquinolones. Some examples of notable fluoroquinolones include ciprofloxacin, ofloxacin, enoxacin, perfloxacin, moxifloxacin and levofloxacin. Despite their effectiveness, these drugs are associated with a variety of adverse effects, such as retinal detachment, Achilles tendon rupture, prolongation of the QT interval, and peripheral neuropathy. Since they are broad-spectrum agents, they also tend to wipe out the normal intestinal flora and can cause *Clostridium difficile* colitis/diarrhea (Xiao-Juan et al., 2013).

Oxazolidinones

These antibiotics have a new mode of action, that is, interference with the initiation of translation. They inhibit protein synthesis in growing bacteria by binding to the 50S ribosomal subunit and inhibiting formation of the 70S ribosomal initiation complex. The totally synthetic oxazolidinone antibacterials were discovered by DuPont in the 1980s. Linezolid (Zyvox of Pfizer/Pharmacia) inhibits vancomycin-resistant bacteria. DuPont did not commercialize these antibiotics but Upjohn did. Linezolid was approved in 2000 for use against methicillin-resistant *Staphylococcus aureus* (MRSA). The drug possesses Gram-positive activity against *S. aureus, Streptococcus spp., and Enterococcus spp.*, and lacks cross-resistance to every clinically significant resistance mechanism tested. Linezolid is active not only against MRSA but also against vancomycin-resistant enterococci (VRE), penicillin-resistant pneumococci, cephalosporin-resistant bacteria, multidrug-resistant *Mycobacterium tuberculosis* (MDR-TB), *Mycobacterium avium*, and some anaerobes. It is bacteriostatic, orally active, and has some activity against Gram-negative organisms as well. Clinical resistance has recently occurred by a single base mutation resulting in generation of a methyltransferase (Humphries, 2013).

About two billion people are infected with *Mycobacterium tuberculosis* each year and 10% of them develop active tuberculosis (TB). The TB organism usually lives latently in cells, such as macrophages and then becomes active in the lung. The usual antibiotic regime includes isoniazid, pyrazinamide, ethambutol, and rifamycin as a four compound combination. For 50 years, there had been no new anti-TB drug but in 2012, the FDA approved Janssen Therapeutics' new bedaquiline (Situro) which inhibits ATP synthesis in the cell membrane, thus preventing energy generation and replication. AstraZeneca is conducting clinical trials of AZD5847, an oxazolidinone which can kill *Mycobacteium tuberculosis* inside macrophages. Twelve vaccines are also in clinical trials against TB. MDR-TB is on the rise with 50,000 new cases each year (Genilloud and Vicente, 2012).

Bacterial DNA Gyrase Inhibitors

These include novobiocin, chlorobiocin, and coumermycin A1 discovered back in the 1950s. A new member, kibdelomycin, was more recently discovered using an antisense approach to screen a collection of natural product extracts (Baltz, 2014). It is produced by *Kibdelosporangium* sp and has strong activity against *S. aureus*, MRSA, *S. pneumoniae*, *Enterococcus faecalis*, and *H. influenzae*. It represents the first novel class of potent inhibitors of bacterial type II topoisomerase and DNA gyrase activities with Gram-positive antibacterial activity.

Peptides

Peptides represent another important class of antibiotics, with over 1000 known compounds (Felnagle et al., 2008; Glaser, 2013). Fifty-five peptides have uses, such as antibiotics, for chronic constipation, irritable bowel syndrome with constipation, as a pulmonary surfactant, for anemia due to chronic kidney disease in adults on dialysis, as a proteasome inhibitor for multiple myeloma, for short bowel syndrome, type 2 diabetes and for Cushing's disease (Goldstein et al., 2010). The annual market is $15 billion and many are made by chemical synthesis.

The antimicrobial peptide group includes vancomycin, teicoplanin, streptogramins, the bacteriocins and other compounds. The glycopeptide vancomycin, the so-called "drug of last resort" for years, has been used to treat infections caused by antibiotic-resistant bacteria. However, with the emergence of VRE, other antibiotics, such as the lipoglycopeptide teicoplanin, daptomycin, linezolid, synercid, and televancin were developed. Telavancin, which was FDA-approved in 2009, is a vancomycin analog and contains a deacylaminoethyl moiety. It is more active versus *C. difficile* than vancomycin (Eisenstein et al., 2010). Teicoplanin is approved for vancomycin-resistant *C. difficile*. The advantages of telavancin over vancomycin and teicoplanin include better in vivo efficacy, better pharmacokinetics, greater safety, easier administration, and better activity against resistant staphylococci, streptococci, and enterococci. The cyclic lipopeptides include daptomycin and ramoplanin. Daptomycin is produced by Streptomyces roseosporus, and acts against Gram-positive bacteria, including VRE, MRSA, and penicillin-resistant *S. pneumoniae*. It is used for complicated skin and soft tissue infections, *S. aureus* bacteremia and endocarditis. It kills by binding irreversibly to the cell membrane, and disrupting plasma membrane function without penetrating into the cytoplasm. When it inserts its tail into the membrane, it causes efflux of K and membrane depolarization, thus destroying the ion concentration gradient. Thus, the target bacterium cannot make ATP or take in nutrients. Daptomycin was discovered at the Eli Lilly and Company laboratories in the 1970s but was thought to be toxic. It was then licensed to Cubist Pharmaceuticals Inc. (now part of Merck & Co.) and FDA-approved for Gram-positive skin infections in 2003 and for S. aureus bacteremia and endocarditis in 2006 (Mandal et al., 2014). The use of recombinant probiotics with antimicrobial peptides as a dual strategy to improve the immune response in immune-compromised patients has been reviewed by S.M. Mandal et al. (2014).

Acyl depsipeptides (ADEPs) can kill persisters which are rare, slow-growing bacteria, generally identical to the rest of the population (Conlon et al., 2013). They occur at 1 in 10,000 to a million cells of a rapidly growing population. ADEPs work well especially when combined with rifampicin, linezolid, or ciprofloxacin. ADEP action against Gram-positive bacteria involves activation of protein clpP, which is a subunit of the protease Clp. Clp is found in all bacteria and functions in protein-quality control and in the regulated degradation of specific proteins.

Additional glycopeptides include dalbavancin, used for the treatment of MRSA, as well as oritavancin. Via transcriptional engineering, production of teicoplanin by *Actinoplanes teichomyceticus* has been improved (Wang et al., 2014). Other improvements in teicoplanin production have been reported by L. Horbal et al. (2014). Teicoplanin is two to four times as active as the as vancomycin against methicillin-resistant and -sensitive *S. aureus* and is much less toxic.

Romidepsin is an interesting depsipeptide, which inhibits histone deacetylase and is produced by the Gram-negative bacterium *Chromobacterium violaceum* (VanderMolen et al., 2011). The organism was isolated from soil by Ueda et al. (1994) of the Fujisawa Corp. (now Astellas). It was further investigated by the US National Cancer Institute (NCI). The Fujisawa Co. carried out clinical trials in 2002 after the NCI confirmed the anticancer activity. It was licensed to Gloucester Pharmaceuticals in the United States of America in 2004, which was later acquired by the Celgene Corp. Romidepsin is active against cutaneous T-cell lymphoma (CTCL) and was approved by the FDA in 2009. In the United States of America, there are 1500 new cases and 500 deaths per year from CTCL.

The bacteriocins are produced by a number of bacteria (Maffioli et al., 2012). Comprising part of the bacteriocins are the lantibiotics, that is, peptides produced ribosomally and modified posttranslationally and the best known is nisin. Lantibiotics are gene-encoded, small (19–39 amino acid residues), and contain one or more lanthionine or methyl-lanthionine residues. They act against Gram-positive bacteria including MRSA, enterococci, and *C. difficile* but are inactive against Gram-negative bacteria. The latter is probably due to the outer membrane of Gram-negatives preventing the lantibiotic from entering the cell and reaching the outer side of the cytoplasmic membrane. In addition to nisins A and Z, the bacteriocins consist of subtilin, nukacin ISK-1, several lacticins, mersacidin, actagardine, cinnamycin, SapB, sublancin, gallidermin, plantaricin W, mutacin, streptins, bovicin, nukacin, and epidermins. A few are in clinical trials. Over 50 lantibiotics are produced by Gram-positive bacteria, such as *Staphylococcus* and *Streptomyces*. Despite the fact that Type A lantibiotics are effective in vivo with a low toxicity risk and potent, broad-spectrum activity (Asaduzzaman and Sonomoto, 2009), they have not been used frequently for treatment (Smith and Hillman, 2008). Their limited use is likely attributed to an expensive and time-consuming manufacturing process, along with high cost.

Fosfomycin

Fosfomycin was introduced by Merck in 1969. Its structure is (1*R*, 2*S*)-1,2-epoxy-propylphosphonic acid. It inhibits the initial enzyme of cell wall formation, that is, uridine diphosphate-*N*-acetylglucosamine enolpyruvyl transferase (MurA). It has successfully cured infections of the lower urinary tract, is active against methicillin- and vancomycin-resistant *S. aureus* infections causing phagocytosis, and has immunomodulatory activity. It has been the first choice for treatment of cystitis and is the only-FDA-approved drug for acute cystitis during pregnancy. It is also used for diabetic foot infections and is active against ciprofloxacin-resistant *Escherichia coli*, carbapenem-resistant *Enterobacteriaceae* (CRE), VRE (Pogue et al., 2013), and colistin-resistant *Enterobacteriacae* (Chen et al., 2012).

Actinomycins

About 30 species of *Streptomyces* and *Micromonospora* produce actinomycins. Actinomycin D has been used against Wilms Tumor and childhood rhabdomyosarcoma.

A marine-derived *Streptomyces avermitilis* strain MS449 produces actinomycins D and X_2 with potent activity against *Mycobacterium tuberculosis* (Emri et al., 2013). Tuberculosis causes over 2,000,000 deaths per year and is the leading cause of death among people with HIV/AIDS.

Antifungal Agents

Two hundred species of fungi are pathogenic to mammals. Most such infections are self-limiting but they can be deadly to immune-compromised patients, for example, systemic fungal infections have been responsible for death of 50% of leukemia patients. Fungal infections doubled in frequency from the 1980s to the 1990s with bloodstream infections increasing five-fold with an observed mortality of 55%. There was an increasing incidence of candidiasis, cryptococcosis, and aspergillosis, especially in AIDS patients. Aspergillosis failure rates exceeded 60%. Fungal infections occurred often after transplant operations, usually involving species of *Candida* and *Aspergillus*. Pulmonary aspergillosis has been the main factor involved in the death of recipients of bone marrow transplants and *Pneumocystis carinii* has been the number one cause of death in patients with AIDS from Europe and North America. Fungal infections have been responsible for 40% of nosocomial infection deaths, for example, *Candida* infections have killed between 3,000 and 11,000 people per year in the United States of America due to nosocomial candidemia.

Antifungal agents include the azoles, allylamines, echinocandins, polyene macrolides, amphotericin B, and beauvericin. Azoles are the largest group and include fluconazole, itraconazole, voriconazole, posaconazole, racuconazole, and albuconazole. The azoles act by competitively inhibiting lanosterol 14-alpha-demethylase, the key enzyme in formation of steroids. Inhibition leads to ergosterol depletion, accumulation of lanosterol and other 14-methyl sterols, causing inhibition of fungal growth. The allylamines prevent growth from developing, thus blocking growth on skin; the leader is Lamisil. Polyenes are used topically.

The echinocandins (candins) are cyclic lipopeptides and more than 20 natural echinocandins have been isolated. They are cyclic lipo-hexapeptides synthesized on nonribosomal peptide synthase complexes by different ascomycota fungi (Emri et al., 2013). Although weakly soluble and with a rather narrow spectrum of activity, echinocandins have been used against *Candida* and *Aspergillus* infections due to their unique mode of action. They were the first antifungals that inhibited cell wall biosynthesis. They inhibit (1,3)-beta-glucan synthase and thus the biosynthesis of the 1,3-glucan layer of the *Candida albicans* cell wall; they are relatively nontoxic. They include the semisynthetic FDA-approved anidulafungin, caspofungin, and micafungin. Anidulafungin acts against *Candida* biofilms (Rosato et al., 2013). Most of these biofilms occur on implanted devices, such as catheters. Anidulafungin has been shown to be more active than amphotericin B, fluconazole, and intraconzole for this application. Caspofungin acetate of Merck & Co. (pneumocandin, Cancidas), produced by *Zalerion arboricola*, which inhibits cell wall formation via inhibition of beta-1,3-glucan synthase, was approved in 2000. It is administered as an aerosol for prophylaxis against *Pneumocystis carinii*. It is also active against *Candida*, *Aspergillus*, and *Histoplasma*. More than 20 natural echinocandins have been isolated. Other echinocandin derivatives are Astellas' micafungin (FK-463) and

Versicor/Pfizer's anidulafungin (V-echinocandid, LY-303366). Other echinocandins include aculeacins, mulundocandins, pneumocandins, sporiofungins, catechol-sulfate echinocandins, and cryptocandin.

A lipopeptide-producing *Bacillus amyloliquefaciens* inhibits clinical isolates of *C. albicans* (Song et al., 2013). The cyclic lipopeptide contains a heptapeptide and a 3-hydroxy fatty acid with 15 carbon atoms. *C. albicans* is the most virulent agent infecting immune-compromised patients with HIV infections. Most of the available drugs against *C. albicans* are toxic. Previous cyclic lipopeptides include surfactin, iturin and fengycin. The new compound was shown to kill drug-resistant *C. albicans* as well as the pathogen *Candida tropicalis.*

Fungi are important in nosocomial blood stream infections which kill 40% of patients. Upon screening 20,000 microbial extracts for ability to synergize a low dose of ketoconazole (KTC), 12 were found to be broad-spectrum in activity. Seven of these showed some activity against human hepatoma cells, the most effective of which was beauvericin (BEA). In mice infected with *Candida parapsilosis,* the combination of BEA at 0.5 mg/kg and KTC at 0.5 mg/kg prolonged survival and reduced fungal counts in animal kidneys, lungs, and brains. These effects were not observed even at very high doses of KTC alone (50 mg/kg). KTC inhibits lanosterol 14-demethylase, blocking sterol synthesis in fungi and mammals, and is used against *Candida* and mold infections. However, it is toxic, and known to cause hepatitis. *C. parapsilosis,* used as a test strain in this study, is the second most commonly isolated fungus in clinical laboratories, following *C. albicans*. The antifungal effect of the combination was much better than for either alone in vitro and in an immune-compromised mouse model. BEA in the combination broadened the KTC spectrum on drug-resistant strains and reduced its side effects. Although BEA has little antifungal activity, the combination was fungicidal, as compared to the fungistatic activity of KTC. The combination showed no negative effect on human liver HepG2 cells.

Antifungal polyene macrolides are made by soil actinomycetes, and are also active against parasites, enveloped viruses, and prion diseases. However, they have toxicity toward mammalian cells and exhibit poor distribution in tissues. They consist of a large family of polyketides, including nystatin, amphotericins, candicidin, pimaricin, and rimocidin. They are polyhydroxylated macrocyclic lactones of 20–40 carbon atoms with three to eight conjugated double bonds. They also have a deoxyamino sugar residue. They act by interaction with ergosterol to form transmembrane channels which cause leakage of cellular K^+ and Mg^{2+}, subsequently leading to fungal death.

Antiviral Agents

Dengue virus kills 390 million people yearly and there is no readily available vaccine to use (Normile, 2013). It is spread by *Aedes aegypti* mosquitos and causes excruciating joint pain and can lead to death. The first antiviral agent used in medicine was idoxuridine (3-iodo-2'-deoxyuridine) back in the mid-20th century. Since then, about 50 drugs have been approved, about half against HIV

(De Clercq, 2010). Worldwide, 33 million people have HIV, 2.7 million were newly infected, and 2 million died. In the United States of America, 1 million people had HIV and 50,000 new cases developed. The virus mutates easily and becomes resistant to medication. Although there is no cure available to date, efforts to develop a drug or a vaccine have been discussed by L. Hock (2013).

Antivirals discovered over the years include idoxuridine, interferon, nigericin, acyclovir, azidothymidine, actinohivin, cyanovirin-N, fuzeon (enfuvirtide), juniferdin, sanglifehrins, tunicamycin, and Isentress (raltergravir). Sanglifehrins are polyketides made by Isentress, discovered at Merck & Co., and FDA-approved for HIV in 2007. It was the first inhibitor of integrase, an enzyme which inserts HIV DNA into human DNA. Merck supplies its antiretroviral drug to African countries at no cost. Nigericin inhibits the vaccinia pox virus, Semliki Forest virus, polio virus, and influenza virus replication. Sanglifehrins are polyketides made by *Streptomyces* sp. A92-308110. They were originally studied as immunosuppressants but were later found to inhibit the hepatitis C virus (Gregory et al., 2011). Also active against the hepatitis C virus is tunicamycin.

Additional Products of Industrial Significance

Important groups of products include (1) herbicides (bialaphos, glufosinate), (2) antiparasitic agents including coccidiostats and avermectin, (3) antimalarial agents, and (4) agricultural fungicides. Important coccidiostats are polyethers, such as monensin, lasalocid, and salinomycin. Monensin is also used as a growth promoter in ruminants. Avermectin has proved to be an important drug in the treatment of onchocerciasis (river blindness) both in the United States of America and in Africa. The disease occurrence amounts to 17.7 million annual cases with 99% in Africa. Great progress against river blindness has taken place in Africa due to the free distribution of avermectin by Merck and Co., the developer of the drug.

Antimalarial agents include chloroquine, primaquine, doxycycline, clindamycin, fosmidomycin, azithromycin, and artemisinin. Malaria affected nearly 225 million people in 2012 and killed 655,000 of these (Halford, 2012; Paddon et al., 2013). The major cause of death of young children in underdeveloped countries is malaria. The disease is caused by five species of *Plasmodium*. *Plasmodium falciparum* is the most well-known, causing 650,000 deaths per year, mainly children within sub-Saharan Africa. It is responsible for over 90% of global deaths from malaria. *Plasmodium vivax* has attracted recent attention because it causes 70%–90% of malarial cases in Asia and South America, 50%–60% of the cases in Southeast Asia and the Western Pacific, and 1%–10% in Africa. *Plasmodium vivax* is carried by at least 71 mosquito species.

Primaquine has traditionally been used against malaria but today its use is restricted due to side effects and weak activity against blood stages of the parasite. The isoprenoid artemisinin is a potent antimalarial and is made naturally by the sweet wormwood *Artemisia annua* (George et al., 2015). It is used by itself or part of antimalarial combination therapies (ACTs). Using *Saccharomyces cerevisiae* genetically engineered in *E. coli,* artemisinic acid was produced at a yield of 25 g/L and then converted into the semisynthetic artemisinin. Artemisinin has been approved by WHO and is being produced commercially by Sanofi.

A new drug class has been found that acts against malaria (McNamara et al., 2013), which contains imidazopyrazines, which act against the intracellular development of many *Plasmodium* species at each stage of infection in the invertebrate host. The imidazopyrazines interact with the ATP-binding pocket of the lipid kinase phosphatidyl-inositol 4-hydroxy kinase, which alters the intracellular distribution of phosphatidyl-inositol-4-phosphate. Recent developments in malarial medicines has been reviewed by TN Wells et al. (2015). In 2014, a new 8-aminoquinolone called tafenoquine completed clinical trials and a single dose was found to have the same effect as 14 days of primaquine. Another potentially new drug is the spiroindolone KAE609, that is, cipargamin, which passed through Phase I clinical trials and has been tested on human patients in Thailand. It kills the malarial parasite even faster than artesunate, is seven times more potent than artesunate and 40 times more potent than 4-aminoquinolones. It might be a single dose cure and cheaper than current drugs. Novartis will submit it for approval in 2017. Another recent development is the discovery of a quinoline diamine called DDD 107498 by scientists at the University of Dundee in Scotland, which blocks protein translation in *Plasmodium falciparum* and kills the parasite (Baragaña et al., 2015; Lavelle, 2015). The group screened a library of over 4,700 small molecules and chemically improved the best one by 120-fold. This drug might cost only $1 per dose. DDD 107498 is undergoing safety trials and is about to enter clinical trials.

Agricultural fungicides include methylacrylates based on the structures of the antifungal oudemansins and strobilurins produced by basidiomycetous fungi of the genera *Oudemansiella* and *Strobilurus* respectively. Leading products include azoxystrobin and fenpiclonil. The latter is a semisynthetic version of the *Pseudomonas* antibiotic pyrrolnitrin. Other important agricultural fungicides include the polyoxins which are peptide-nucleosides inhibiting chitin synthetase.

Antibiotic Resistance Development

Antibiotic-resistant bacteria cause 28,000 deaths annually in the United States of America and 2 million Americans are infected with such bacteria each year, according to the CDC (Ernst, 2015). Although the mechanism is not understood, antibiotics in feed cause weight gain in livestock. This leads to the use of tremendous amounts of antibiotics in animal growth. Denmark eliminated the nontherapeutic use of antibiotics in livestock. This decreased the number of antibiotic-resistant strains in both the animals and in the Danish population. Thus, resistance to antibiotics has decreased in that country. It was estimated that worldwide antibiotic use in food animal production would increase by 67% from 2010 to 2030, from about 63,000 tons to about 106,000 tons (Price et al., 2015; Van Boeckel et al., 2015). Increased use of antibiotics in agriculture can quickly increase antibiotic resistance among human pathogens that enter livestock. This has already occurred with MRSA. In the United States of America, antibiotic use in animals represents about 80% of total antibiotic consumption. Such use of antibiotics in pigs, poultry, and cattle creates ideal conditions for emergence and spread of antimicrobial-resistant bacteria in animals and in humans. This is on the rise since countries, such as Brazil, Russia, India, China, and South Africa have increased livestock production and decreased rice and wheat production for their populations of rising income. Thus, the transition to animal protein-based diets and the increase in antibiotic resistance is closely linked. Also, use of antibiotics in fish and shrimp farming is increasing, especially in Southeast Asia. A strong fighter against the use for animal growth of important antibiotics used by humans is Stuart Levy of Tufts University in Boston (Azvolinsky, 2015). Levy showed in 1976 that feeding low doses of antibiotics led to the replacement in the animal's intestine of normal microbiota to antibiotic-resistant bacteria in humans who lived and worked on the farm and then into animal handlers. In 1993, Levy published "The Antibiotic Paradox: How Miracle Drugs are destroying the Miracle."

In March of 2015, the US White House issued a plan to cut microbial infections by half within 5 years (Erickson, 2015). The plan attempted to stop the unnecessary application of antibiotics for growth promotion of farm animals. Unfortunately, it does allow use of antibiotics to help animals survive unsanitary, crowded, and stressful confinement conditions. The White House also hopes to fund the discovery of new antibiotics to kill drug-resistant bacteria and the development of diagnostic tests to rapidly detect these pathogens. The plan is called "the National Action Plan for Combating Antibiotic Resistant Bacteria" and is focused on *C. difficile*, carbapenem-resistant *Enterobacteriaceae* and MRSA, hoping to cut 50% of such infections by 2020. It aims for the elimination of the use of medically-important antibiotics for growth promotion in animals. About 80% of US-produced antibiotics are used in agriculture, supposedly mainly for disease prevention.

Scale of Antibiotic Production and Market Sales

Semisynthesis, which involves the chemical modification of naturally produced compounds, has yielded many new and useful antibiotics. To illustrate this point, since 1974, over 20,000 semisynthetic penicillins, 4,000 cephalosporins, 2,500 tetracyclines, 1,000 rifamycins, 500 kanamycins, and 500 chloramphenicols have been prepared. Semisynthesis represents a promising approach for the development of new antibiotics, including those that will combat the growing problem of antimicrobial resistance.

Nearly 100,000 tons of antibiotics are produced worldwide, with penicillins G and V leading the way at 60,000 tons. Of these, 25,000 tons represent the products used for direct medical use. The remaining 35,000 tons per year is converted to the penicillin precursor, 6-aminopenicillanic acid (6-APA) for semisynthesis of ampicillin, amoxicillin and other penicillins) and to 7-ADCA for production of semisynthetic cephalosporins (Soetaert and Vandamme, 2010). Semisynthetic β-lactams are used widely and have the following production levels in tons per year: amoxicillin at 15,000, ampicillin at 5,000, cephalexin at 4,000, and cefadroxil at 1000 tons. Approximately 2,500 tons of cephalosporins and 5,500 tons of tetracyclines are produced per year, while glycopeptides, such as vancomycin and teicoplanin amount to a total of 9,000 tons. Production of the macrolide erythromycin is over 4,000 tons per year. About 1000 tons are used as erythromycin A whereas the rest is chemically converted to 1500 tons of the leading macrolide, azithromycin A, 1500 tons of clarithromycin, and 400 tons of roxithromycin.

In 2005, the global sales of oral antibiotics alone was $25 billion (Christoffersen, 2006). The total antibacterial market reached $42 billion in 2009. 45% was due to sales of cephalosporins and fluoroquinolones. Nearly half of the $6.4 billion market for all quinolones was dominated by the fluoroquinolones, such as levofloxacin. The composition of the antimicrobial market is cephalosporins at 23%, fluoroquinolones at 17%, penicillins at 16%, macrolides at 14%, vaccines at 9%, carbapenems at 6%, tetracyclines at 4%, and other classes at 11%. The vaccine market is at $3.9 billion.

The 2011 market for antifungal agents was $6.1 billion, including allylamines, azoles, polyene macrolides, and others (Carlson, 2012). Azoles inhibit the key enzyme for fungal growth, and are the largest group of the earlier mentioned antifungals.

Decrease in Antibiotic Discovery and Challenges Faced by the Pharmaceutical Industry

Despite the emergence of new pathogens and antibiotic resistance, there has been a decline in the rate of antibiotic discovery. Since 1980, over 30 new diseases have been identified, including HIV/AIDS, Ebola virus, Hanta virus, *Cryptosporidium*, Legionnaires' disease, Lyme disease, and *E. coli* 0157:H7. The occurrence of antibiotic-resistant bacteria, such as *Pseudomonas aeruginosa*, MRSA, VRE, *Neisseria gonorrhoeae*, multidrug resistant TB (MDR-TB), and *Klebsiella pneumoniae* carbapenemase-producing bacteria (KPC) further emphasize the need for new and effective antimicrobials. However, discovery of new drugs in general is not simple anymore. It takes 11.5–16 years to discover one drug from 11,000 lead compounds (Walsh and Wencewicz, 2014). Although the golden age of antibiotic discovery lasted from 1940 to 1960, the next 50 years dealt mainly with medicinal chemistry in which the structures of the original drugs were chemically modified.

Several factors have contributed to the decrease in new drug discovery by the pharmaceutical industry over the past 20 years. These include high costs for clinical trials and a lengthy period of time for completion of preclinical and clinical development. In addition, governments have provided companies with a short window for sales before patent expiration (when products become generics). Pharmaceutical organizations are also faced with numerous regulatory hurdles. To complicate these matters, the discovery of antibiotics that will be effective against resistant bacteria has been difficult. Lastly, the sales of antibiotics are limited due to their relatively short treatment period compared to drugs taken on a chronic basis for conditions, such as hypercholesterolemia, diabetes, and hypertension.

The decrease in drug discovery efforts has had its greatest effect on antibiotics. New antibiotics are urgently needed. The development of antibiotic-resistant pathogens is a major public health crisis, leading to significant morbidity and mortality. There are also naturally resistant bacteria in existence, such as *Pseudomonas aeruginosa*, *Stenotrophomonas maltophilia*, *Enterococcus faecium*, *Burkholderia cepacia*, and *Acinetobacter baumannii* which cause serious infections. About 2 million people in the United States of America are infected with antibiotic-resistant bacteria, resulting in 23,000 deaths per year. Furthermore, some current drugs on the market, notably the fluoroquinolone class, are associated with toxicity and adverse events.

From 1990 to 2005, several large pharmaceutical companies focused on combinatorial chemistry. Their attempts at developing synthetic molecules failed in large part because of a lack of diverse and pharmacologically active compounds. Subsequently, this had led to a decrease in new drug approvals, FDA and orphan drug applications, and new chemical entities. As a result, much of the drug discovery efforts are now dependent on small pharmaceutical companies and the biotechnology industry. Despite the decline in antibiotic discovery, it is important to emphasize that some new and useful drugs have emerged. These include the candins, the semisynthetic glycylcyclines and streptogramins, the further development of the unutilized antibiotics, daptomycin and lipiarmycin, and the application of underutilized products, such as the glycopeptide, teicoplanin.

The downturn in antibiotic research by the major pharmaceutical companies has seen some reversal recently. In 2013, Roche returned to the antibiotic arena, partnering with Polyphor to conduct Phase II clinical trials on POL 7080, a macrocyclic peptide-mimetic against *Pseudomonas aeruginosa* infections. AstraZeneca, one of the few companies not giving up on antibiotics, partnered with FOB Synthesis on a combination of a beta-lactamase inhibitor with carbapenems. Sanofi, which had dropped antibiotic studies in 2004, has partnered with Fraunhofer-Gesellschaft on discovery of new antiinfectives. Cooperation of pharmaceutical companies with smaller organizations is being aided by the GAIN Act in the United States of America (2012), which provides five additional years of market exclusivity for new antibiotics.

Potential Solutions for Discovery of New and Effective Antibiotics

Natural products represent a promising source of new and effective antibiotics. Therefore, it is essential that research efforts be devoted to the continued discovery of secondary metabolites. Combinatorial biosynthesis has been a useful approach for finding new products. It involves creating a hybrid molecule through genetic modification of a producing organism and horizontal gene transfer (Demain, 2002; Felnagle et al., 2008; Hopwood et al., 1985; McAlpine, 1998; Wong and Khosla, 2012). This process has yielded new erythromycins, spiramycins, tetracenomycins, anthracyclines, and nonribosomal peptides. High-throughput screening of natural product libraries is also needed, along with the use of natural product scaffolds for combinatorial chemistry (Breinbauer et al., 2002; Kingston and Newman, 2002; Waldemann and Breinbauer, 2002). It has been demonstrated that high-throughput screening of natural product collections achieves a higher hit rate than synthetic compounds within combinatorial libraries (Breinbauer et al., 2002). Other opportunities for drug discovery include genomics, proteomics, and metabolomics (Demain and Zhang, 2005). Further exploration of biodiversity, genome mining, and systems biology are additional fields made possible due to technological advancements in drug discovery.

A major advance in antibiotic discovery has come from the laboratories of Kim Lewis and Slava Epstein at Northeastern University in Boston. It was known that only a small number of microorganisms from nature, for example, 1%, can be isolated in artificial media in Petri dishes. Uncultured bacteria account for 99% of all species in nature. The Epstein group first used a diffusion chamber incubated in situ in which diffusion provided the microorganisms with their naturally occurring growth factors. Many more cultures were isolated from nature by this method. They then devised an isolation chip (called the IChip) composed of several hundred miniature diffusion chambers (Nichols et al., 2010). The IChip recovered many more isolates from nature than even the diffusion chamber. Many of the species isolated have never been isolated before by standard techniques. Isolation from seawater was only 4% of inoculated cells which formed colonies in Petri dishes, as compared to 20% with the diffusion chamber and 34% with the IChip. The figures from soil were 7%, 31%, and 33% respectively. The organisms growing in the IChip were more novel than those growing by standard methods. The IChip has 192 tiny wells. In using the IChip, a natural environment, for example, mud, is diluted so that approximately one bacterial cell is in each channel. The chamber is then sandwiched between two semipermeable membranes and buried in the mud for 1–2 weeks. Once a cell grows in the chamber and forms colonies, it is likely to grow in Petri dishes.

By use of the IChip, Ling et al. (2015) isolated 10,000 isolates from nature and 25 potentially new antibiotics were identified from these isolates. One of the microorganisms, *Eleftheria terrae* (a new species of β-proteobacteria), produced a new antibiotic. Through genome sequencing, it was found to be a member of a new genus related to aquabacteria which are Gram-negative organisms, and not known to produce antibiotics. The antibiotic was named teixobactin and it is an unusual depsipeptide containing enduracididine, methyl phenylalanine, and four D-amino acids. Its biosynthetic gene cluster consists of two large nonribosomal peptide synthetase-coding genes. The antibiotic inhibits cell wall synthesis via binding to a highly conserved motif of lipid II (precursor of peptidoglycans) and lipid III (precursor of cell wall teichoic acid). Its target, lipid II, is a polyprenyl-coupled cell envelope precursor which is readily accessible on the outside of Gram-positive bacteria. The specific target for teixobactin is the pyrophosphate-sugar moiety of lipid II. No mutants of *S. aureus* or *Mycobacterium tuberculosis* are resistant to teixobactin. It has excellent activity against Gram-positive pathogens including strains that are drug-resistant. such as MRSA. Potency versus most Gram-positive species was below 1 μg/mL. It is exceptionally active versus *C. difficile* and *Bacillus anthracis*. Its killing activity of *Staphylococcus aureus* was greater than that of vancomycin. It was ineffective versus most Gram-negative bacteria except an *E. coli* strain with a defective outer membrane permeability barrier. No toxicity to mammalian cells was observed. Teixobactin is an inhibitor of peptidoglycan synthesis with no effect on label incorporation into DNA. Resistance to teixobactin could not be developed. The antibiotic is active in the presence of serum, is stable, and has good microsomal stability. It is active in mice versus MRSA and *S.pneumoniae*. The producer of teixobactin is a Gram-negative bacterium and its outer membrane prevents teixobactin from reentring into the producing organism. Teixobactin is distinct from glycopeptides, lantibiotics, and defensins. It shows no cross-resistance with vancomycin, which also is a lipid II binder. In addition, there were no negative effects observed in toxicological in vitro investigation, that is, hemolysis, cytotoxicity, etc.

Virulence is being considered as a target for new antibiotic discovery (Escaich, 2010). Virulence factors are essential in vivo for the invading organism to colonize, invade host tissue, adapt to various environments in the host, subvert host functions, and overcome the defenses of the host. Virulence inhibitors combat quorum sensing, adhesion and colonization including pili formation, secretion systems, sortases, iron metabolism, and resistance to host innate immunity.

A more thorough examination of the different geographical and climatic locations around the world will help increase natural product discovery and ensure microbial diversity. Interestingly, the vast majority of microorganisms (95%–99.9%) have not been cultured in the laboratory. Therefore, there is a significant number of organisms in nature that could be analyzed for their secondary metabolite production. To put this into perspective, a sample of 30 g of soil may contain up to 500,000 rare bacterial species and about 20,000 common bacterial species. It is thought that anywhere from 0.25 to 9.9 million fungal species may exist, while nearly 65,000–70,000 species have already been identified (Burrill, 2012). Continued efforts must be devoted to exploring both aquatic and terrestrial ecosystems around the globe. The government must also play a role in order to ensure that new and effective antibiotics are developed. Currently, the time from patent approval to when products become generic is 22 years. This period is much too short, when considering how much time it takes a pharmaceutical industry to complete the discovery efforts (2–10 years), preclinical (4 years), and clinical development (about 6 years for Phase I–III trials). In addition, there is 1 year for FDA review and approval as well as another year for postmarketing testing (Ernst and Young, 2000; Burrill, 2012; Watkins, 2002). Although the US government has set up a translational center for new drug discovery, they must also expand the time before products become generics.

REFERENCES

Ariyanchira, S., 2010. Overcoming antibiotic-resistant bacteria. Gen. Eng. Biotechnol. News 30 (15), 18.

Asaduzzaman, S.M., Sonomoto, K., 2009. Lantibiotics: diverse activities and unique modes of action. J. Biosci. Bioeng. 107, 475–487.

Azvolinsky, A., 2015. Resistance Fighter. Scientist 29 (6), 1–6.

Baltz, R.H., 2007. Antimicrobials from actinomycetes: back to the future. Microbe 2, 125–131.

Baltz, R.H., 2014. Combinatorial biosynthesis of cyclic lipopeptides antibiotics: a model for synthetic biology to accelerate the evolution of secondary metabolic biosynthetic pathways for nonribosomal peptides. ACS Synth. Biol. 3 (10), 748–758.

Banik, B.K., 2013. Synthesis and biological studies of novel β-lactams. In: Brahmachari, G. (Ed.), Chemistry and Pharmacology of Naturally Occurring Bioactive Compounds. CRC Press, Boca Raton, pp. 31–72.

Baragaña, B., Hallyburton, I., Lee, M.C., Norcross, N.R., Grimaldi, R., Otto, T.D., Proto, W.R., Blagborough, A.M., Meister, S., Wirjanata, G., Ruecker, A., 2015. A novel multiple-stage antimalarial agent that inhibits protein synthesis. Nature 522 (7556), 315–320.

Breinbauer, R., Vetter, J.R., Waldmann, H., 2002. From protein domains to drug candidates—natural products as guiding principles in the design and synthesis of compound libraries. Angew. Chem. Int. Ed. 41, 2879–2890.

Bronson, J.J., Barrett, J.F., 2001. Quinolone, everninomycin, glycylcycline, carbapenem, lipopeptide and cephem antibiotics in clinical development. Curr. Med. Chem. 8, 1775–1793.

Burrill, G.S., 2012. Personalized medicine or blockbusterology. BioPharm 15 (4), 46–50.

Carlson, B., 2012. As fungal infections expand, so does market. Gen. Eng. Biotech. News 32 (4), 11.

Chen, C., Song, F., Wang, Q., Abdel-Mageed, W.M., Guo, H., Fu, C., Hou, W., Dai, H., Liu, X., Yang, N., Xie, F., 2012. A marine-derived *Streptomyces* sp. MS449 produces high yield of actinomycin X2 and actinomycin D with potent anti-tuberculosis activity. Appl. Microbiol. Biotechnol. 95 (4), 919–927.

Chopra, I., Roberts, M., 2001. Tetracycline antibiotics: mode of action, applications, molecular biology, and epidemiology of bacterial resistance. Microbiol. Mol. Biol. Rev. 65 (2), 232–260.

Christoffersen, R.E., 2006. Antibiotics—an investment worth making? Nat. Biotechnol. 24, 1512–1514.

Conlon, B.P., Nakayasu, E.S., Fleck, L.E., LaFleur, M.D., Isabella, V.M., Coleman, K., Leonard, S.N., Smith, R.D., Adkins, J.N., Lewis, K., 2013. Activated ClpP kills persisters and eradicates a chronic biofilm infection. Nature 503 (7476), 365–370.

Davies, J., 2006. Where have all the antibiotics gone? Can. J. Infect. Dis. Med. Microbiol. 17 (5), 287–290.

De Clercq, E., 2010. Highlights in the discovery of antiviral drugs: a personal retrospective. J. Med. Chem. 53 (4), 1438–1450.
Demain, A.L., 2002. Prescription for an ailing pharmaceutical industry. Nat. Biotechnol. 20, 331.
Demain, A.L., 2006. From natural products discovery to commercialization: a success story. J. Ind. Microbiol. Biotechnol. 33, 486–495.
Demain, A.L., Zhang, L., 2005. Natural products and drug discovery. In: Zhang, L., Demain, A.L. (Eds.), Natural Products: Drug Discovery and Therapeutic Medicine. Humana Press, Totowa, pp. 3–29.
Deshpande, A., Pant, C., Jain, A., Fraser, T.G., Rolston, D.D., 2008. Do fluoroquinolones predispose patients to *Clostridium difficile* associated disease? A review of the evidence. Curr. Med. Res. Opin. 24 (2), 329–333.
Eisenstein, B.I., Oleson, Jr., F.B., Baltz, R.H., 2010. Daptomycin: from the mountain to the clinic, with essential help from Francis Tally, MD. Clin. Infect. Dis. 50 (Suppl 1), S10–S15.
Emri, T., Majoros, L., Tóth, V., Pócsi, I., 2013. Echinocandins: production and applications. Appl. Microbiol. Biotechnol. 97 (8), 3267–3284.
Erickson, B.E., 2015. White House targets antibiotic-resistant bacteria. Chem. Eng. News 93 (10), 6.
Ernst, S., 2015. Resistant. Am. Lab. 47 (3), 4.
Ernst, Young, 2000. The Ernst & Young Fifteenth Annual Report on the Biotechnology Industry. Ernst & Young LLP, Palo Alto, CA.
Escaich, S., 2010. Novel agents to inhibit microbial virulence and pathogenicity. Expert Opin. Ther. Patents. 20 (10), 1401–1418.
Felnagle, E.A., Jackson, E.E., Chan, Y.A., Podevels, A.M., Berti, A.D., McMahon, M.D., Thomas, M.G., 2008. Nonribosomal peptide synthetases involved in the production of medically relevant natural products. Mol. Pharm. 5 (2), 191–211.
Genilloud, O., Vicente, F., 2012. Chapter 11: novel approaches to exploit natural products from microbial resources. In: Genilloud, O., Vicente, F. (Eds.), Drug Discovery from Natural Products. Royal Society of Chemistry, Cambridge, UK, pp. 221–248.
George, K.W., Alonso-Gutierrez, J., Keasling, J.D., Lee, T.S., 2015. Isoprenoid drugs, biofuels, and chemicals-artemisinin, farnesene, and beyond. Adv. Biochem. Eng. Biotechnol. 148, 355–389.
Glaser, V., 2013. Scaling up peptide drugs. Gen. Eng. Biotechnol. News 33 (7), 54–55.
Goldstein, E.J., Citron, D.M., Tyrrell, K.L., Warren, Y.A., 2010. Bactericidal activity of televancin, vancomycin, and metronidazole against *Clostridium difficile*. Anaerobe 16 (3), 220–222.
Gregory, M.A., Bobardt, M., Obeid, S., Chatterji, U., Coates, N.J., Foster, T., Gallay, P., Leyssen, P., Moss, S.J., Neyts, J., Nur-e-Alam, M., 2011. Preclinical characterization of naturally occurring polyketide cyclophilin inhibitors from the sanglifehrin family. Antimicrob. Agents Chemother. 55 (5), 1975–1981.
Halford, B., 2012. Easier path to leading antimalarial drug. Chem. Eng. News 90 (52), 21.
Hock, L., 2013. A Distant Hope. R&D magazine 55 (6), 18–21.
Hopwood, D., Malpartida, F., Kieser, H.M., Ikeda, H., Duncan, J., Fujii, I., Rudd, B.A.M., Floss, H.G.I., mura, S., 1985. Production of hybrid antibiotics by genetic engineering. Nature 314, 642–644.
Horbal, L., Kobylyanskyy, A., Truman, A.W., Zaburranyi, N., Ostash, B., Luzhetskyy, A., Marinelli, F., Fedorenko, V., 2014. The pathway-specific regulatory genes, tei15* and tei16*, are the master switches of teicoplanin production in *Actinoplanes teichomyceticus*. Appl. Microbiol. Biotechnol. 98 (22), 9295–9309.
Humphries, C., 2013. Latency: a sleeping giant. Nature 502, S14–S15.
Kingston, D.G., Newman, D.J., 2002. Mother Nature's combinatorial libraries; their influence on the synthesis of drugs. Curr. Opin. Drug. Disc. Devel. 5 (2), 304–316.

Kraus, C.N., 2008. Low hanging fruit in infectious disease drug development. Curr. Opin. Microbiol. 11 (5), 434–438.

Lavelle, J., 2015. Novel antimalarial agent has new mode of action. Chem. Eng. News 93 (25), 6.

Ling, L.L., Schneider, T., Peoples, A.J., Spoering, A.L., Engels, I., Conlon, B.P., Mueller, A., Schäberle, T.F., Hughes, D.E., Epstein, S., Jones, M., 2015. A new antibiotic kills pathogens without detectable resistance. Nature 517 (7535), 455–459.

Maffioli, S.I., Monciardini, P., Sosio, M., Donadio, S., 2012. Chapter 7: new lantibiotics from natural and engineered strains. In: Genilloud, O., Vicente, O. (Eds.), Drug Discovery from Natural Products. Royal Society of Chemistry, Cambridge, UK, pp. 116–139.

Mandal, S.M., Silva, O.N., Franco, O.L., 2014. Recombinant probiotics with antimicrobial peptides: a dual strategy to improve immune response in immunocompromised patients. Drug Discov. Today 19 (8), 1045–1050.

McAlpine, J., 1998. Unnatural natural products by genetic manipulation. In: Sapienza, D.M., Savage, L.M. (Eds.), Natural Products II: New Technologies to Increase Efficiency and Speed. Internat Bus Comm, Southborough, MA, pp. 251–278.

McNamara, C.W., Lee, M.C., Lim, C.S., Lim, S.H., Roland, J., Nagle, A., Simon, O., Yeung, B.K., Chatterjee, A.K., McCormack, S.L., Manary, M.J., 2013. Targeting Plasmodium PI(4)K to eliminate malaria. Nature 504 (7479), 248–253.

Nathan, C., 2004. Antibiotics at the crossroads. Nature 431, 899–902.

Nichols, D., Cahoon, N., Trakhtenberg, E.M., Pham, L., Mehta, A., Belanger, A., Kanigan, T., Lewis, K., Epstein, S.S., 2010. Use of ichip for high-throughput in situ cultivation of "uncultivable" microbial species. Appl. Environ. Microbiol. 76 (8), 2445–2450.

Nicolaou, K.C., Chen, J.S., Edmonds, D.J., Estrada, A.A., 2009. Recent advances in the chemistry and biology of naturally occurring antibiotics. Angew. Chem. Int. Ed. Engl. 48 (4), 660–719.

Normile, D., 2013. Tropical medicine. Surprising new dengue virus throws a spanner in disease control efforts. Science 342 (6157), 415.

Paddon, C.J., Westfall, P.J., Pitera, D.J., Benjamin, K., Fisher, K., McPhee, D., Leavell, M.D., Tai, A., Main, A., Eng, D., Polichuk, D.R., 2013. High-level semi-synthetic production of the potent antimalarial artemisinin. Nature 496 (7446), 528–532, 2013.

Paradkar, A., 2013. Clavulanic acid production by *Streptomyces clavuligerus*: biogenesis, regulation and strain improvement. J. Antibiot. 66 (7), 411–420.

Pogue, J.M., Marchaim, D., Abreu-Lanfranco, O., Sunkara, B., Mynatt, R.P., Zhao, J.J., Bheemreddy, S., Hayakawa, K., Martin, E.T., Dhar, S., Kaye, K.S., 2013. Fosfomycin activity versus carbapenem-resistant Enterobacteriaceae and vancomycin-resistant Enterococcus, Detroit, 2008-10. J. Antibiot. 66 (10), 625–627.

Price, L.B., Koch, B.J., Hungate, B.A., 2015. Ominous projections for global antibiotic use in food-animal production. Proc. Natl. Acad. Sci. USA 112 (18), 5554–5555.

Rosato, A., Piarulli, M., Schiavone, B.P., Catalano, A., Carocci, A., Carrieri, A., et al., 2013. In vitro effectiveness of anidulafungin against Candida sp. biofilms. J. Antibiot. 66 (12), 701–704.

SciBX, 2013. Special Issue on Antibiotic Resistance, from Cubist Pharmaceuticals. Addressing the unmet need for anti-infective development (Rising to the challenge of multidrug-resistant bacteria, Cubist is building a robust anti-infective development portfolio).

Smith, L., Hillman, J.D., 2008. Therapeutic potential of type A (I) lantibiotics, a group of cationic peptide antibiotics. Curr. Opin. Microbiol. 11, 401–408.

Soetaert, W., Vandamme, E.J. (Eds.), 2010. Industrial Biotechnology: Sustainable Growth and Economic Success. first ed. Wiley-VCH, Weinheim.

Song, B., Rong, Y.J., Zhao, M.X., Chi, Z.M., 2013. Antifungal activity of the lipopeptides produced by *Bacillus amyloliquefaciens* anti-CA against *Candida albicans* isolated from clinic. Appl. Microbiol. Biotechnol. 97 (16), 7141–7150.

Tahlan, K., Jensen, S.E., 2013. Origins of the β-lactam rings in natural products. J. Antibiot. 66 (7), 401–410.

Ueda, H., Nakajima, H., Hori, Y., Fujita, T., Nishimura, M., Goto, T., Okuhara, M., 1994. FR901228, a novel antitumor bicyclic depsipeptide produced by *Chromobacterium violaceum* No. 968. I. Taxonomy, fermentation, isolation, physico-chemical and biological properties, and antitumor activity. J. Antibiot. 47 (3), 301–310.

Van Boeckel, T.P., Brower, C., Gilbert, M., Grenfell, B.T., Levin, S.A., Robinson, T.P., Teillant, A., Laxminarayan, R., 2015. Global trends in antimicrobial use in food animals. Proc. Natl. Acad. Sci. USA 112 (18), 5649–5654.

VanderMolen, K.M., McCulloch, W., Pearce, C.J., Oberlies, N.H., 2011. Romidepsin (Istodax, NSC 630176, FR901228, FK228, depsipeptide): a natural product recently approved for cutaneous T-cell lymphoma. J. Antibiot. 64 (8), 525–531.

Waldemann, H., Breinbauer, R., 2002. Nature provides the answer. Screening 3 (6), 46–48.

Walsh, C.T., Wencewicz, T.A., 2014. Prospects for new antibiotics: a molecule-centered perspective. J. Antibiot. 67 (1), 7–22.

Wang, H., Yang, L., Wu, K., Li, G., 2014. Rational selection and engineering of exogenous principal sigma factor (σ(HrdB)) to increase teicoplanin production in an industrial strain of *Actinoplanes teichomyceticus*. Microb. Cell Fact. 13, 10.

Watkins, K.J., 2002. Fighting the clock. Chem. Eng. News 80 (4), 27–34.

Wells, T.N., Hooft van Huijsduijnen, R., Van Voorhis, W.C., 2015. Malaria medicines: a glass half full? Nat. Rev. Drug Discov. 14 (6), 424–442.

Wong, F.T., Khosla, C., 2012. Combinatorial biosynthesis of polyketides-a perspective. Curr. Opin. Chem. Biol. 16 (1-2), 117–123.

Xiao-Juan, X., Wen-Xiang, H., Jia-Jun, L., 2013. Study of linezolid and its resistance mechanism. Chi. J. Antibiot. 38 (6), 411–414, 429.

Yoneyama, H., Katsumata, R., 2006. Antibiotic resistance in bacteria and its future for novel antibiotic development. Biosci. Biotechnol. Biochem. 70, 1060–1075.

Chapter 8

Accessing Marine Microbial Diversity for Drug Discovery

Lynette Bueno Pérez and William Fenical
Center for Marine Biotechnology and Biomedicine, Scripps Institution of Oceanography, University of California, La Jolla, CA, United States

INTRODUCTION

During the years 1981–2014, unmodified or modified natural products, in addition to molecules inspired by the structures of natural products, were responsible for 51% of all new drug approvals. When biological and vaccine drugs are excluded, natural products small molecules or derivatives then contribute 65% of all approved drugs during this time period (Newman and Cragg, 2016). The discovery of penicillin in 1928 by Alexander Fleming from the fungus, *Penicillium chrysogenum* (previously *P. notatum*), revealed the potential of microorganisms as valuable sources for drug discovery (Bennett and Chung, 2001). Fleming's findings resulted in decades of intense investigations of terrestrial microorganisms that concluded with the discovery of more than 120 antibiotics during the "Great Antibiotic Era." (Aminov, 2010). However, in the mid-1990s, the pharmaceutical industries lost interest in microbial natural products-based drug discovery resulting in a marked decreased in the number of approved antibiotics. The lack of new antibiotics and overuse of antibiotics in human treatment and in animal feeds, resulted in a steep increase in resistant bacterial infections, including those from methicillin-resistant *Staphylococcus aureus* (MRSA), multidrug-resistant (MDR) *Pseudomonas aeruginosa*, and many others (Boucher et al., 2009; Boucher, 2010). As a result, in 2016 there is an urgent need for new antibacterials to treat these menacing new infectious pathogens. In this regard, new approaches and new environmental sources are of the utmost importance for the discovery of new antibiotic scaffolds. During the *Great Antibiotic Era*, the microbial complexity of the world's oceans were never explored (Aminov, 2010). It had been continually mentioned that marine bacteria were difficult to culture. Perhaps most important, however, was the prevailing notion that microorganisms in the sea were not unique and largely

Microbial Resources. http://dx.doi.org/10.1016/B978-0-12-804765-1.00008-4

reflected the same taxa as found on land (Aminov, 2010). Of course, it is now abundantly clear that the isolation of unique marine bacteria requires innovative methods that reflect the environment in which they live and that the nutrients and methods applied to soil samples may not be appropriate (Choi et al., 2015). More contemporary studies clearly show that marine environments are a massive source for taxonomically-diverse microorganisms (Béjà et al., 2002; Giovannoni et al., 1990; Stach et al., 2003).

Although marine microbial drug discovery is still in its infancy, and the oceans remain an environment still avoided by most pharmaceutical industries, the ocean has proven itself as a highly significant drug resource. Seven marine-derived drugs are on the market and at least 25 others are currently in clinical trials (Mayer, 2014). It is clear that the amazing abundance and diversity of marine bacteria represent a promising resource for drug discovery.

The revolutionary classification of life by Carl Woese using ribosomal RNA (rRNA) sequencing data into three life domains: Bacteria, Archaea, and Eukarya changed how precision bacterial taxonomy was undertaken (Woese et al., 1990). The relationships of bacteria could, for the first time, be highly fine-tuned avoiding the often subjective classification by biochemical methods. Innovations in metagenomics have allowed the bacterial diversity of environmental samples to be defined leading to the recognition that massive taxa of marine bacteria have yet to be cultivated (Kennedy et al., 2010). In addition, the subsequent development of low-cost genomic sequencing and gene annotation has led to a revolution in the understanding of the diversity of secondary metabolite biosynthetic pathways present in bacteria, particularly in the drug-rich actinobacteria.

In this mini-review, we briefly discuss marine bacteria, analyzing the diversity of marine habitats, sampling methods, isolation, and cultivation for drug discovery. The review will not be comprehensive, but rather we have selected aspects of microbial discovery that we have learned over more than 25 years of trial and error. Most of the examples are from our work that illustrate several useful findings and conclusions. In addition, we present a few case studies to illustrate successful drug development with molecules in clinical trials or with potential to become clinically approved entities.

THE MARINE MICROBIAL ENVIRONMENT

The oceans, covering 71% of the Earth's surface, contains a diversity of distinct microbial habitats, with the predominance of certain bacterial Phyla inhabiting different ocean zones in what is known as vertical stratification (Table 8.1) (Munn, 2011; Qian et al., 2011). Environmental conditions at the surface waters (the sea-atmosphere layer) are characterized by high concentrations of dissolved oxygen, organic matter, high temperatures (25–30°C), moderate salinity levels, and phototrophic (sunlight availability for photosynthesis) conditions favoring the growth of a diverse number of microbial communities (Dubilier et al., 2008;

TABLE 8.1 General Characteristics of the Ocean's Vertical Stratification Zones

Ocean zones	Depth (m)	Environmental traits	Major bacterial phyla
Neuston	0	↑ [O_2], organic matter, temperature,	Cyanobacteria
		microbial density, and diversity,	
		↓salinity	
Epipelagic	0–200	↑ [O_2], organic matter, temperature,	Cyanobacteria
		microbial density, and diversity,	
		↓salinity	
Mesopelagic	200–1000	↓[O_2], organic matter, temperature,	Proteobacteria,
		microbial density, and diversity,	Actinobacteria
		↑salinity, sulfur, methane, pressures	
Bathypelagic	1000–4000	↓[O_2], organic matter, temperature,	Firmicutes
		microbial density, and diversity,	
		↑salinity, sulfur, methane, pressures	

Qian et al., 2011). As you increase in ocean depth, there is a decrease in temperature, pressure, dissolved oxygen, organic matter, and sunlight, and an increase in salinity, sulfur, and methane concentrations due to microbial degradation of organic matter fallen into the sea, at times favoring chemosynthetic (organisms that derive their energy from the oxidation of sulfur or methane) microbial habitats (Dubilier et al., 2008; Qian et al., 2011). Due to the temperatures and pressures of the deep sea, there is a decrease in microbial population from 4×10^5 cells/mL in surface waters to less than 10^5 cells/mL in the deep ocean waters, accounting for a marked decreased in microbial diversity.

Hydrothermal vents, in the Atlantic and Pacific mid-ocean ridges, present an extreme and highly specialized environment harboring diverse microbial life (Munn, 2011; Thornburg et al., 2010). Hydrothermal vents are formed when cold ocean waters permeate into the cracks of the heated ocean's crust, heating deep ocean water, and exchanging elements and transition metals between the heated rocks and the hydrothermal fluids. As these hydrothermal fluids reach the magma, temperatures reaching up to 400°C, produce an explosion of rising water (known as smoker chimneys) releasing methane gas, metal sulfides, iron, manganese oxides, and silicates, that then deposits in the ocean's floor (Munn, 2011; Thornburg et al., 2010). The dynamic environment of the hydrothermal vents with varying temperatures and chemical compositions hosts diverse habitats for microorganisms along the surrounding heated water, sediment, and chimney's walls. Chemosynthetic and hyperthermophilic archaea and bacteria growing at temperatures up to 121°C dominate these habitats (Munn, 2011; Thornburg et al., 2010). The bacterial Phyla, Actinobacteria, Bacteroidetes, and Firmicutes, are found in hydrothermal vent areas. However, Proteobacteria, dominate these habitats comprising one-third of the microbial composition (Thornburg et al., 2010).

SAMPLING MARINE MICROENVIRONMENTS

Marine Water and Sediments

Many tools are available to collect water samples and sediments that range from tethered Niskin bottles, cylinders, or grabbers lowered down to the deep sea from ships on the water surface to submarines (Anderson, 2003). Tethered plastic bottles, cylinders, or samplers simply hand-thrown into the sea, are commonly used to collect water samples and sediments (Anderson, 2003). In our work, we have repeatedly used small sampling devices, such as grabs and coring devices from small boats. Connecting these samplers to electric fishing reels and using high strength Dacron fishing line (300–500 lb breaking strength) has allowed us to sample up to 2500 m. In an interesting experiment, an SIO colleague assisted us to design and construct an autonomous sampling device. In theory, this device could be deployed without tether to the surface, it would descend to the ocean bottom, take a core sample, drop chain weights, and being buoyant return to the surface. This worked well to a point until the sampler became stuck in extra-soft ocean sediments and was not retrieved. We then realized that the massive inconsistencies of bottom sediments could not be predicted making this approach less usable. Examples of a sediment grab, coring device, and our experimental autonomous sampler are illustrated later (Fig. 8.1). We have found these devices to provide reliable sediment collections with high replicate numbers at minimal cost.

A more sophisticated, but costly way to collect sediments from the ocean floor is by the use of submersible vehicles. Three types of submersibles are used: the human occupied submarine, remotely operated, and autonomous unmanned vehicles (Alvin FAQs, 2006; Underwater vehicles, 2015). The human occupied submersibles can accommodate a small group of scientists, usually up to three people (Alvin FAQs, 2006; Underwater vehicles, 2015). For the remotely operated underwater vehicles (ROVs), tethered robots controlled by scientists on the surface navigate along the ocean floor collecting samples. The autonomous submersibles are programmed robots sent out to the ocean's bottom, without the need for constant remote control by scientists (Underwater vehicles, 2015).

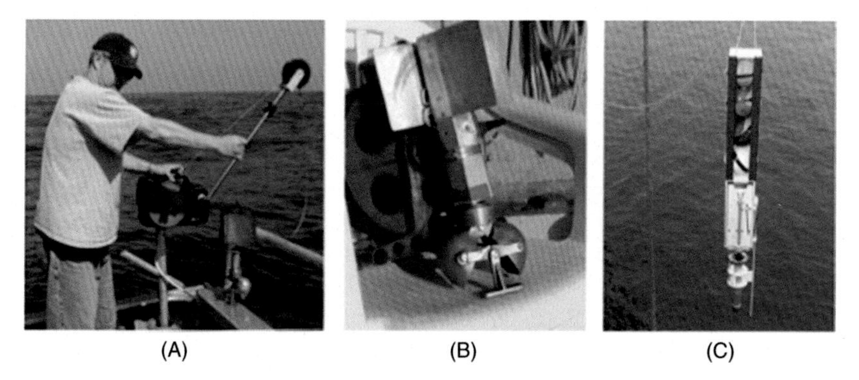

FIGURE 8.1 **Various simple sampling devices to access marine sediments.** (A) Deep ocean fishing reel adapted to deploy the small sediment grab shown in B. (B) A spring-loaded sediment grab. (C) An autonomous sediment sampler designed and constructed at SIO by Kevin Hardy.

These submersible vehicles have automated arms that reach out to collect sediments and are usually equipped with cameras and videos to record the deep ocean microbial life (Alvin FAQs, 2006; Underwater vehicles, 2015). The direct observations recorded by scientists, as well as in videos, presents an advantage of the use of submarines over the simple tethered coring devices and grabbers. The biggest disadvantages of these underwater vehicles are the high cost of operation, about $45,000 per day (Underwater vehicles, 2015).

SELECT HABITAT MARINE BACTERIA

It is unmistakable that marine bacteria are adapted to almost every niche in the ocean, both deep and shallow. Components driving these adaptations are temperature, salinity, pressure, and the presence of significant nutrients to support growth. The reader is referred to classic review papers that provide an enlarged discussion of this concept (D'Amico et al., 2006; Fuhrman and Azam, 1983; Giovannoni and Rappé, 2000; Pratt and Waddell, 1959; Somero, 1992). The bacteria that thrive in seawater are, in general, specific groups of considerable fundamental interest, but unlikely to be a significant resource for drug discovery. Of major significance is the SAR 11 Complex, exemplified by the isolation of *Pelagibacter ubique*, a small alphaproteobacterium that is the major microbe in seawater (Rappé et al., 2002). Because of its abundance in both saline and fresh waters, it may be considered the most abundant bacterium on planet Earth. *P. ubique* is characterized by its extremely small genome size (1,308,759 bp), the smallest of all free-living organisms (Giovannoni et al., 2005). The genome of *P. ubique* contains the bare essentials for life, consequently these bacteria seem to lack secondary metabolite genes.

Symbiotic bacteria are ubiquitous in the oceans, from shallow waters to deep oceans and hydrothermal vents. Some common symbiont hosts of marine bacterial symbionts are worms, mussels, clams, and shrimps, but symbionts have also been found to live in association with whale carcasses or sunken woods (Dubilier et al., 2008). Many hosts obtain a nutritional benefit from their symbiotic bacteria as is the case of the gutless *Riftia pachyptila* tubeworm. *R. pachyptila* endosymbionts oxidize sulfur to synthesize organic compounds that are used as energy source for the host (Dubilier et al., 2008). On occasions, multiple symbionts are found associated with the same host. For example, the *Bathymodiolus* mussels found in the deep sea and hydrothermal vents, harbor sulfur and methane oxidizing endosymbionts, providing an advantage for the host with its ability to adapt to changes in the environment (Dubilier et al., 2008).

Sponges (Phylum Porifera) predominate on tropical reefs and are also present in the deep sea and polar waters. Sponges are a prolific source of endo- and epibionts (extracellular symbionts) (Hentschel et al., 2006). The most abundant bacterial symbionts belong to the phyla Proteobacteria (29%), followed by Acidobacteria (23%), and Chloroflexi (22%). Actinobacteria and Cyanobacteria, commonly known as major producers of bioactive compounds, were also found associated with sponges in 12 and 4% of the cases, respectively (Hentschel et al., 2006). Access to surface and endosymbionts is one of the rapidly expanding growth areas in marine microbial drug discovery. Access to surface bacteria presents little problems, but of course, it is difficult to predict success. We are still in a formative time in which the specific requirements to culture marine bacteria are being explored and defined. Surfaces are easily examined and they can simply be swabbed and transferred to agar plates for cultivation.

Over the past 30 years, the concept of invertebrate endosymbiont production of secondary metabolites has been one of the main themes in marine natural products chemistry. In this brief review, it would be impossible to adequately cover the massive literature on this subject and to critique and evaluate the hundreds of papers devoted to this subject. Instead, we will present a brief overview of the major advances in this field, generally being made within the past decade. Looking at the past literature, it has been generally accepted, based upon the presence of intratissue bacteria and upon structure comparisons (similar structures being bacterial metabolites) that symbionts are responsible for the massive diversity of secondary metabolites present in marine sponges, ascidians, and other invertebrates. This generalization needs to be carefully evaluated in view of the massive numbers of metabolites isolated from marine invertebrates and the relatively few focused studies generating solid evidence.

A variety of methods have been employed to approach this problem, including direct bacterial isolation from the host and chemical analysis, cultivation of the symbiont, and more recently genetic methods involving gene cluster cloning and cluster analysis illustrating the specific genes likely to be responsible for secondary metabolite construction. The fact that bacterial endosymbionts are difficult (impossible, some say) to cultivate has severely hampered the final solution to this significant question.

Over the past decade, considerable progress has been made to evaluate endosymbiont secondary metabolite synthesis. The excellent work of the Schmidt and Long groups, has clearly illustrated that the cyanobacterial endosymbiont *Prochloron didemni*, found in large abundance in ascidians of the family Didemnidae, produce the patellamides and related cyclic peptides characteristic of these ascidians (Long et al., 2005; Schmidt and Donia, 2010; Schmidt et al., 2005). The Schmidt group was able to describe the ribosomal gene cluster responsible for patellamide production, clone the cluster in *E. coli* and observe patellamides production in culture (Schmidt and Donia, 2010; Schmidt et al., 2005), while Long and coworkers used shotgun cloning and heterologous expression to illustrate production by *Prochloron* ascidians (Long et al., 2005). Subsequent studies by the Schmidt group have expanded understanding of *Prochloron* biosynthesis in beautiful ways to illustrate this as a general attribute of this family of ascidians and their symbionts (Schmidt et al., 2012). Cyanobacterial symbionts are observed in other invertebrates and they appear to also contribute to the overall biosynthesis of secondary metabolites. Cyanobacterial symbionts have been observed in a variety of sponges and some appear to play a central role in the production of secondary metabolites in select sponges of the genus *Dysidea* (Simmons et al., 2008).

A group of invertebrates that standout as clear examples of symbiont production of secondary metabolites is the taxon of sponges of the order Lithistida. This is a group of largely tropical sponges found in both shallow and deeper oceans to 1000 m (Pomponi et al., 2001). Of the lithistids, the Indopacific sponge *Theonella swinhoei* clearly stands out as a unique, highly chemically-prolific source of secondary metabolites (Bewley et al., 1996). Over a 35 year period of investigation, numerous polypeptides (e.g., the amazing polytheonamides) and polyketides (oonamide A, theopederin) and even some nitrogenous terpenes have been reported from this source (Bewley et al., 1996; Blunt et al., 2015). What stood out was that sponge samples from different locales had different chemical components. Although it is quite possible that this sponge is taxonomically complex, thus providing an explanation of its chemical variability, the novelty and high bioactivity of this sponge led to assumptions that bacterial symbionts were responsible for chemical variability. Early studies were instrumental in confirming the presence of symbionts, leading to studies providing significant evidence that the uncultured symbiont, labeled *Candidatus Entotheonella palauensis*, was the producer of the secondary metabolites (Bewley et al., 1996; Schmidt et al., 2000). Unfortunately, all attempts to cultivate this symbiont have failed, thus knowledge of the systematics of the bacterium and its biosynthetic capacity had not been available. However, in the last decade, new molecular methods evolved that have allowed significant progress in understanding the taxonomic and biosynthetic foundation of secondary metabolite synthesis by these sponge symbionts. Using single cell genomic sequencing of *Entotheonella*, the Piel group in Zurich have now been able to pin-point the specific gene clusters responsible for the synthesis of virtually all isolated and some unknown secondary metabolites (Wilson et al., 2014). In addition, phylogenetic analysis of the ribosomal 16S gene showed that this bacterium is entirely distinct and may form its own Phylum suggested as the *Tectomicrobia*. At present, this unique taxon is limited with high confidence to the lithistid sponges. However, based upon sequence probes of a large number of taxonomically diverse sponges, the presence of similar gene sequences (95.5%–99.9%) seems to indicate that this unique symbiont is much more widely distributed (Wilson et al., 2014).

CULTURING MARINE BACTERIA FOR DRUG DISCOVERY

Less than 1% of the microorganisms present in the natural environment have been cultured in the laboratory (Joint et al., 2010; Zhang, 2005). While this statistic may seem like an impediment to success, it is clear that these statistics are old and outdated. Using newer methods, such as infinite dilution single cell cultivation (Fröhlich and König, 2000) low nutrient culture media, long incubation periods, and innovative selection tools, have allowed a much greater percentage of marine bacteria to be isolated (Bollmann et al., 2007; Choi et al., 2015). Once isolated, these bacteria are likely to be cultivated in sufficient scale to support chemical studies. However, complex microorganism interactions during laboratory isolation procedures, such as differential growth rates and media condition requirements, complicate isolations and contribute to the low "culturability" of marine bacteria in the laboratory (Joint et al., 2010; Zhang, 2005). Several reports have documented the benefits of using ultra-low nutrient culture media and even media only composed of 4% agar for initial isolations (Choi et al., 2015). Perhaps even more important is that marine bacteria adapt slowly to growth on agar and may require long incubation times before colonies can be observed. In our work, we have found that a predominance of the bacterial colonies observed after long incubation times, often greater than 4 weeks, were novel taxa not yet cultivated. Some of these unique genera were also found to produce simple alkaloidal antibiotics (Choi et al., 2015).

The marine actinomycetes, one of the prime focus groups for drug discovery, has not proven to be especially difficult to isolate. Recognizing that bacteria have different requirements for the production of secondary metabolites, it appears necessary however to cultivate using a diversity of conditions, perhaps different temperatures, different media components (e.g., addition of inorganic salts and trace elements), at varying incubation times, and in both static solid agar media or shake flasks to determine the optimal parameters for the particular strains (Joint et al., 2010; Zhang, 2005). Also, the coculture of bacteria has been used as a strategy to improve the production of bioactive compounds (Zhang, 2005).

Bacteria in the laboratory are commonly grown in solid stationary agar media or in shake flasks. Some bacteria may grow better in agar while others may prefer the liquid shake flasks. Because shake flask cultivation can be scaled to larger culture volumes, and this approach can be easily undertaken with minimal resources, this method is likely to be preferred. The biggest advantage of culturing bacteria in flasks is that colonies can grow in suspension throughout all the liquid media providing for high microbial density growth space, whereas in agar microorganisms have very limited space to grow. Flasks can be incubated in rotatory shakers increasing aeration and the amount of dissolved oxygen in the culture. A drawback of culturing bacteria in flasks is that the increase in turbidity associated with high microbial population growth makes the assessment of possible contamination somewhat difficult.

Culture flasks usually provide milligram scale production of pure active compounds, which is sufficient for full characterization and activity testing of novel bioactive metabolites. However, several hundred grams of the active compound are likely to be needed for clinical trials and commercialization purposes.

The use of industrial fermenters is the second option employed by pharmaceutical industries to generate large-scale production of bioactive natural products. An example is the supply for clinical trials of the potent cytotoxic compound, salinosporamide A, aka Marizomib (see case study later), isolated from the obligate marine actinomycete *Salinispora tropica* (Fenical et al., 2009). Industrial fermenter tanks containing 500–1500 L of *S. tropica* led to the purification of 260 mg/L of salinosporamide A (Fenical et al., 2009). Although commercial fermenters are effective and conditions can be rigorously controlled, fermenters are expensive and require the attention of professionals for success. Academic laboratories are not likely to have the resources to use these tools.

EXAMPLES OF OBLIGATE MARINE MICROBIAL METABOLITES

Although deep ocean hydrothermal vent environments have been suggested to provide significant opportunities for drug discovery, few bioactive secondary metabolites have been isolated from hydrothermal vent bacteria (Andrianasolo et al., 2009, 2012; Homann et al., 2009). The ammonificins A–D (**1-4**, Fig. 8.2), purified from the bacterium *Thermovibrio ammonificans* living near the chimney walls of a hydrothermal vent on the East Pacific Rise (9°50′N) are excellent examples. Ammonificins C and D (**3-4**) were shown to induce apoptosis of W2 cells at 2 and 3 μM, respectively (Andrianasolo et al., 2009). Other deep ocean habitats have also been examined. One notable example is the isolation

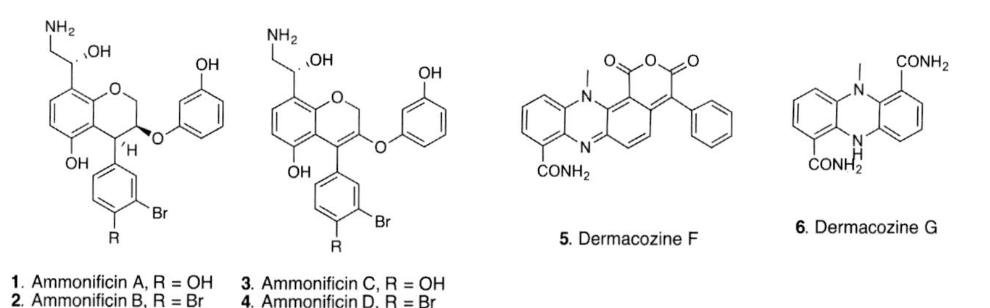

FIGURE 8.2 **Select structures of metabolites produced by hydrothermal vent and deep ocean sediment bacteria.**

of the unicellular actinomycete *Dermacoccus abyssi,* strains MT1.1 and MT1.2, isolated from deep ocean sediments in the Mariana Trench (−10,898 m). Cultivation of these strains yielded a series on new metabolites, the dermacozines. Dermacozines F (**5**) and G (**6**, Fig. 8.2), derived from anthranilic acid metabolism were cytotoxic toward the leukemia cell line K562 at low micromolar levels (Abdel-Mageed et al., 2010; Wagner et al., 2014).

By far, the most extensive marine microbial studies have involved bacteria isolated from marine sediments, both shallow and moderately deep (<4000 m). Marine sediments are similar to terrestrial soil in that they can be the ultimate sink for settling organic matter and nutrients. Sediments can thus be rich in bacteria and other microorganisms, but can also be patchy with areas of very low in nutrients in some samples. Hundreds of sediment-derived bacteria, emphasizing the actinomycetes, have been studied yielding many unique and unprecedented new chemical structures. These studies are in yearly review by the New Zealand Group in *Natural Products Reports* (Blunt et al., 2015). It would be difficult to highlight the many unique metabolites isolated, but those molecules that are characterized by high levels of halogenation seem most relevant. A group of molecules, precedented by the historic discovery of pentabromopseudilin (**7**, Fig. 8.3) in 1966, clearly illustrate marine bacterial biosynthesis involving bromination (Burkholder et al., 1966). Unique, halogenated metabolites include the marinopyrroles (**8–9**, Fig. 8.3) (Hughes et al., 2008), chlorizidine (**10**, Fig. 8.3) (Alvarez-Mico et al., 2013) isolated from marine actinomycetes, and the diversity of bromopyrroles (**11**, Fig. 8.3), and bromobipyrroles (**12**, Fig. 8.3) from marine *Pseudoalteromonas* strains (Agarwal et al., 2014). Of significant interest are the chlorinated merochlorins A-D (**13–16**, Fig. 8.3) produced by the marine streptomycete strain CNH-189 (Kaysser et al., 2012). These compounds share an amazing biosynthesis of the carbon skeletons of which are produced by chlorination and rearrangement.

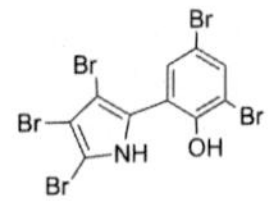

7. Pentabromopseudilin
8. Marinopyrrole A (X = H) 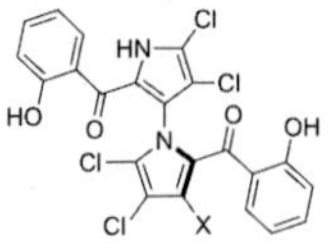
9. Marinopyrrole B (X = Br)
10. Chlorizidine 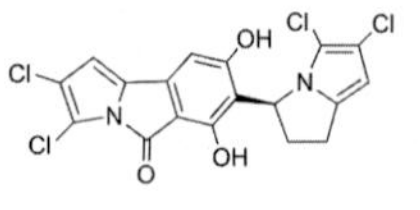
11. Tetrabromopyrrole 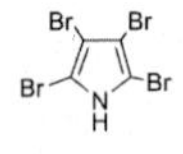
12. Hexabromobipyrrole 
13. Merochlorin A 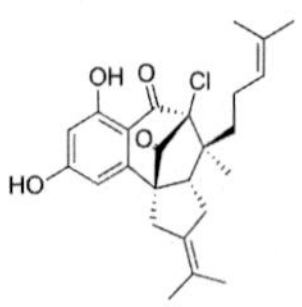
14. Merochlorin B 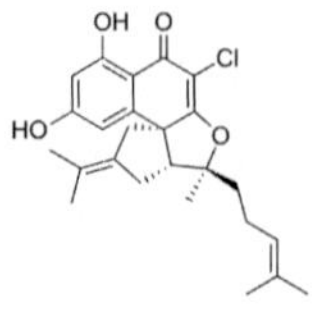
15. Merochlorin C 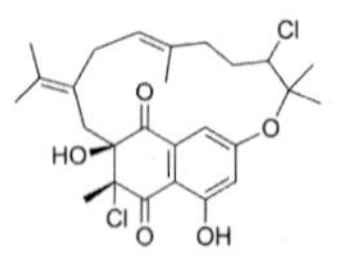
16. Merochlorin D 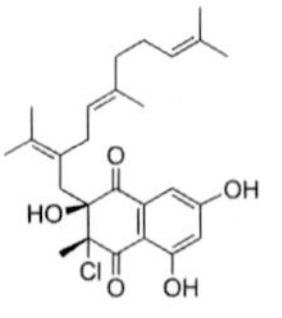

FIGURE 8.3 **Select halogenated metabolites illustrating halogenation in marine bacteria.**

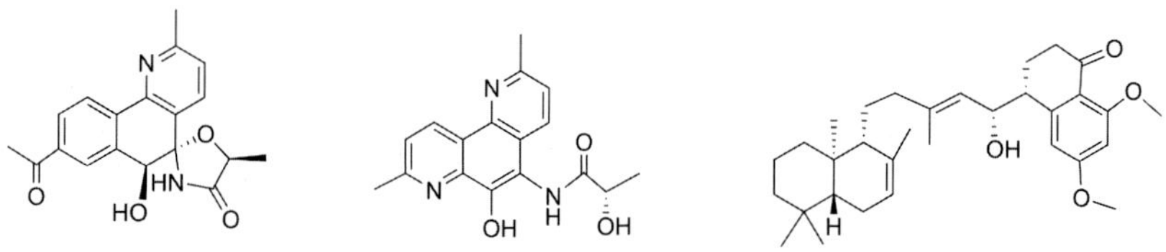

17. Actinobenzoquinoline A 18. Actinophenanthroline A 19. Actinoranone

FIGURE 8.4 **Select metabolites from marine-derived *Streptomyces* strains that require seawater for growth.**

One of the most common genera isolated from marine sediments are actinomycetes of the common soil-derived genus *Streptomyces*. *Streptomyces* strains isolated from marine sediments are taxonomically complex with some strains closely related to those from soil, while others appear taxonomically-unique and have been shown to have an obligate requirement for seawater (Prieto-Davó et al., 2008). The proposal that actinomycetes are rinsed into the sea as spores, survive for significant periods, and later germinate in nutrient media is likely to be valid (Prieto-Davó et al., 2008). However, the streptomycetes that require seawater appear to be adapted to life under saline conditions. *Streptomyces* strains that require seawater can produce some unusual structures, such as the alkaloids actinobenzoquinolines and actinophenanthrolines (**17–18,** Fig. 8.4) from our marine *Streptomyces* strain CNH-189, and actinoranone (**19,** Fig. 8.4) from the seawater-requiring streptomycete strain CNQ-027 (Feling et al., 2003; Nam et al., 2013). It is noteworthy that these unique compounds can be related to the requirement of the strain for seawater. Perhaps this criterion should be considered to indicate strains that are adapted to the marine environment.

CASE STUDIES OF MARINE DRUG DISCOVERY

In this section, we have picked three classes of marine microbial metabolites based upon their demonstrated or anticipated importance in drug discovery. Their isolation and salient pharmacological features of these compounds are included.

SALINOSPORAMIDE A

Salinosporamide A (**20**, Fig. 8.5) was initially isolated in 2002 from a sediment obligate marine actinomycete identified as *S. tropica* (strain CNB-392) (Feling et al., 2003). Initially, *Salinispora* spp. were assigned as new *Micromonospora* species, but in 2005, further 16S rRNA gene sequence analysis of a larger number of strains revealed these strains as belonging to a new genus (Fenical et al., 2009; Jensen et al., 2015; Lam et al., 2010). Cytotoxicity-guided purification of the extract from a 20 L cultivation of *S. tropica*, strain CNB-392, led to the purification of 150 mg of **20** (Feling et al., 2003). The unique

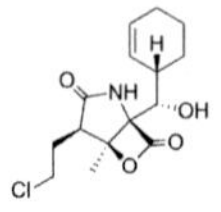

20. Salinosporamide A
21. Abyssomicin B 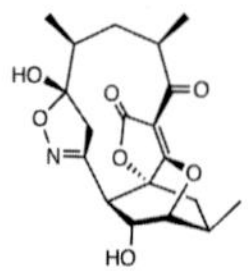
22. Abyssomicin C, CO = β 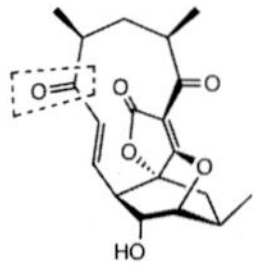
24. Atrop-abyssomicin C, CO = α
23. Abyssomicin D 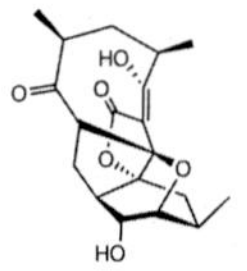
25. Abyssomicin J 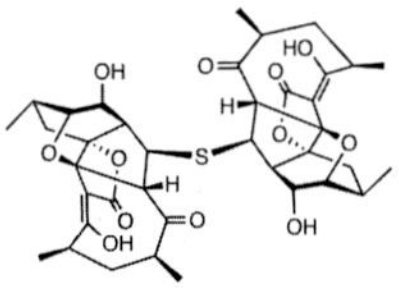
26. Abyssomicin 2 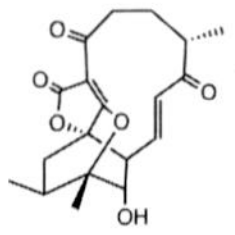
27. Thiocoraline, R = OH 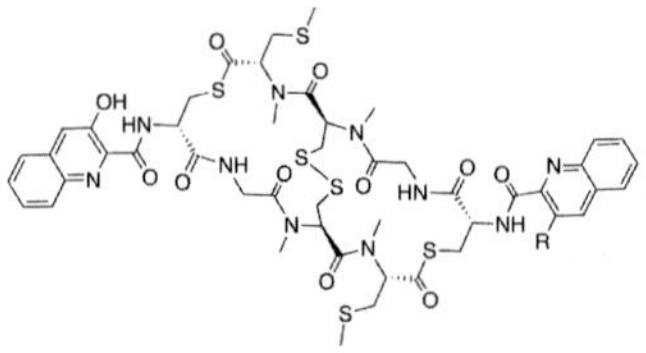
28. 22'-Deoxythiocoraline, R = H

FIGURE 8.5 **Metabolites described in the case studies.**

structure of salinosporamide A (**20**) presents a fused γ-lactam-*β*-lactone ring system containing a cyclohexenyl carbinol and chloroethyl functional groups (Feling et al., 2003; Fenical et al., 2009; Groll et al., 2006; Jensen et al., 2015; Lam et al., 2010). The close similarity of the bicyclic ring in salinosporamide A (**20**) to the well-known proteasome inhibitor omuralide immediately suggested that this compound may also target the 20S proteasome (Feling et al., 2003; Groll et al., 2006; Jensen et al., 2015). Salinosporamide A (**20**) showed potent cytotoxicity against HCT-116 human colon cells with an IC_{50} value of 11 ng/mL (Feling et al., 2003).

The mechanism of action of salinosporamide A (**20**) was confirmed as an inhibitor of the 20S proteasome, a validated target for cancer chemotherapy (Feling et al., 2003; Fenical et al., 2009; Groll et al., 2006; Jensen et al., 2015; Lam et al., 2010). The activity of **20** against purified rabbit and human 20S proteasome showed potent inhibition of the three proteolytic subunits, chymotrypsin-like (CT-L), trypsin-like (T-L), and caspase-like (C-L), with IC_{50} values in the low pM to mid nM range (Feling et al., 2003; Fenical et al., 2009; Groll et al., 2006; Jensen et al., 2015; Lam et al., 2010). The crystal structure of the complex of **20** with the yeast 20S proteasome confirmed that salinosporamide A (**20**) perfectly occupies all three catalytic sites (CT-L, T-L, and C-L) of the proteasome (Groll et al., 2006). In vivo studies of salinosporamide A (**20**) in mice confirmed the inhibition of proteasome activity, however the irreversible aspects of binding were recognized as unique (Feling et al., 2003; Fenical et al., 2009; Groll et al., 2006; Jensen et al., 2015; Lam et al., 2010). Binding of **20** occurs by interaction of the *β*-lactone ring of salinosporamide A (**20**) and the hydroxyl group of a threonine residue in the proteasome active site, followed by the formation of a cyclic ether ring and chlorine elimination of **20** (Groll et al., 2006). A downstream effect observed from the proteasome inactivation is the inhibition of NF-*κ*B activation (Feling et al., 2003; Fenical et al., 2009;

Groll et al., 2006; Jensen et al., 2015; Lam et al., 2010). Salinosporamide A (**20**), administered IV twice a week (0.15 mg/Kg) or orally (0.25 mg/Kg) showed efficacy in a mouse multiple myeloma xenograft model (Fenical et al., 2009). Later, salinosporamide A (**20**) was also shown to be effective in mouse colon cancer xenograft models (Jensen et al., 2015).

Due to the potent in vitro and in vivo activity of salinosporamide A (**20**), coupled with a known and validated proteasome target led Nereus Pharmaceuticals to license the compound from UC-San Diego and initiate preclinical studies ultimately leading to its accelerated advancement into human clinical trials (Feling et al., 2003; Groll et al., 2006; Jensen et al., 2015; Lam et al., 2010; Schmidt et al., 2012). Two major difficulties encountered during the preclinical development of salinosporamide A (**20**) were the large-scale production under saline fermentation conditions and the relative instability of the β-lactone ring in aqueous solutions (Groll et al., 2006; Jensen et al., 2015; Lam et al., 2010; Schmidt et al., 2012). The discovery production yield of **20** using seawater cultivation was approximately 7.5 mg/L (Feling et al., 2003). During development, it became clear that the seawater-based fermentation medium was a problem for use in industrial fermenters due to its corrosive effects. Also, the fermentation culture required the inclusion of animal-derived constituents and natural seawater, which are not accepted in current Good Manufacturing Practices (cGMP) (Feling et al., 2003; Fenical et al., 2009; Jensen et al., 2015; Lam et al., 2010). Creative media formulation studies by the Nereus fermentation group, headed by Ray Lam, allowed the substitution of natural seawater with noncorrosive salts and other media components consistent with appropriate guidelines (Feling et al., 2003; Fenical et al., 2009; Jensen et al., 2015; Lam et al., 2010). In addition, the use of an adsorbent resin added to the fermentation broth allowed salinosporamide A (**20**) to be adsorbed as it was produced, thus isolating the sensitive beta-lactone ring from hydrolytic ring opening. An 18-fold increase in the production of salinosporamide A (**20**) was realized (Feling et al., 2003; Fenical et al., 2009; Jensen et al., 2015; Lam et al., 2010), resulting in a final yield of 350 mg/L, which provided sufficient material for clinical trials (Lam et al., 2010). Salinosporamide A (**20**) produced by this process is recovered in overall 50% yield with greater than 98% purity (Jensen et al., 2015; Lam et al., 2010).

Salinosporamide A (**20**), in its unmodified natural form, has completed Phase I trials in patients with lymphoma, malignant glioma, melanoma, multiple myeloma, nonsmall cell lung cancer, and pancreatic cancer (ClinicalTrials.gov, 2016). In addition, it is currently in Phase II clinical trials in patients with relapsed or refractory multiple myeloma (ClinicalTrials.gov, 2016). Interestingly, given the sensitivity of the lactone ring, salinosporamide A (**20**) is active when administered orally. Clinical trials have been conducted as a single agent and in multiple agent chemotherapy. It has been shown to be effective in patients with resistance to other proteasome inhibitors (e.g., bortezomib) (Jensen et al., 2015; Lam et al., 2010).

ABYSSOMICINS

In 2004, a new class of polyketide antibiotics, the abyssomicins, were identified from the deep-sea sediment-derived actinomycete, *Verrucosispora* strain AB 18-032, collected in the sea of Japan at a depth of 289 m (Bister et al., 2004). Bioassay-guided isolation, following the inhibitory activity of *Verrucosispora* extract against *p*-aminobenzoate (*p*ABA) biosynthesis, yielded the abyssomicins B–D (**21-23**, Fig. 8.5), which were purified and identified by NMR and single crystal X-ray crystallographic analysis. Subsequent chemical investigations with strain AB 18-032 revealed the presence of abyssomicins G and H in addition to abyssomycin C (**22**), and atrop-abyssomicin C (**24**, Fig. 8.5) (Keller et al., 2007). Abyssomicin C (**22**) was active against Gram-positive bacteria including MRSA and vancomycin multiresistant *S. aureus,* with MIC values of 4 and 13 μg/mL, respectively (Bister et al., 2004). Interestingly, atrop-abyssomicin C (**24**) displayed more potent antibiotic activity than abyssomicin C (**22**) (Keller et al., 2007). Furthermore, abyssomicin C (**22**) and atrop-abyssomicin C (**24**) showed antituberculosis activities against *Mycobacterium smegmatis* and *M. tuberculosis* (Wang et al., 2013). The enone Michael acceptor present in these latter abyssomicins, but absent in the inactive abyssomicin derivatives, is believed to be essential for activity (Bister et al., 2004; Keller et al., 2007). An apparent contradiction to this statement was the isolation of the thioether dimer abyssomicin J (**25**, Fig. 8.5) from *Verrucosispora* strain MS100128, which while active does not contain the enone functionality (Wang et al., 2013). Nevertheless, it is speculated that **25** is a prodrug that goes through a reverse Michael addition activated by cytochrome P450 enzymes, thus supporting the importance of the enone functionality for the activity of abyssomicin-class compounds (Wang et al., 2013). The target of the abyssomicins, the *p*-aminobenzoic acid biosynthesis pathway, is present in microorganisms but absent in humans, thus this mechanism provides promise for the next generation of antibiotics (Bister et al., 2004; Wang et al., 2013).

Other abyssomicins with antiretroviral activity have been isolated from the marine-derived actinomycete of the genus *Streptomyces* (León et al., 2015). Bioassay-guided purification of the extract, using an in vitro assay for the reactivation of latent HIV virus, resulted in the purification of abyssomycins 2-5. Abyssomicin 2 (**26**, Fig. 8.5) produced partial viral reactivation in cells obtained from HIV patients undergoing traditional antiretroviral therapy. Traditional antiretroviral agents are effective to prevent the progression of the disease but do not target latent viral cells. Consequently, abyssomicin 2 (**26**) with this unique activity has the potential for development as a cotreatment for HIV (León et al., 2015).

THIOCORALINES

Thiocoraline (**27**, Fig. 8.5) is a unique cyclic depsipeptide isolated from a *Micromonospora* strain (L-13-ACM2-092), from a marine coral collected from the Indian Ocean (Pérez Baz et al., 1997; Romero et al., 1997). Thiocoraline

(**27**) displayed strong antimicrobial activity against the Gram-positive bacteria, *S. aureus* and *B. subtilis* with identical MIC values of 0.05 μg/mL and showed an MIC value of 0.03 μg/mL against *Micrococcus luteus* (Romero et al., 1997). In addition, thiocoraline (**27**) showed potent in vitro cytotoxicity against the human alveolar epithelial cell line A549, the human melanoma cell line MEL28 and the murine leukemia cell line P388 with IC_{50} values of 0.002 μg/mL. The compound exhibited lower potency against human colon cancer HT29 cells with an IC_{50} value of 0.01 μg/mL (Romero et al., 1997). A potent cytotoxic thiocoraline analogue, 22'-deoxythiocoraline (**28**, Fig. 8.5), was produced by a *Verrucosispora* sp. (WMMA107 strain) obtained from the sponge *Chondrilla caribensis f. caribensis* (Wyche et al., 2011). 22'-Deoxythiocoraline (**28**) was cytotoxic against A549 cells with an EC_{50} value of 0.13 μM (Wyche et al., 2011).

The antibiotic mechanism of action of thiocoraline (**27**) was found to be the inhibition of RNA synthesis, and to a lesser extent, DNA synthesis (Romero et al., 1997). Also, the high potency of **27** against the human medullar thyroid carcinoma MTC-TT cell line (IC_{50} value of 7.6 nmol/L) was shown to be produced by cell cycle arrest at the G1 phase and activation of the Notch signaling pathway (Tesfazghi et al., 2013). Similarly, in vitro studies with thiocoraline (**27**) using human pancreatic carcinoma BON and human bronchopulmonary carcinoma H727 cells showed nanomolar cytotoxicities (Tesfazghi et al., 2013). Thiocoraline (**27**) treatment in BON and H727 cells also caused the activation of the Notch pathway but a G2/M cell cycle arrest was observed with these cell lines (Tesfazghi et al., 2013). in vivo studies with thiocoraline (**27**), administered in micelles to mice containing BON tumors, resulted in 63% reduction in tumor volume compared to the vehicle control. All mice survived the study with no signs of toxicity (Wyche et al., 2011). Given the consistent potency of thiocoraline (**27**), its novel mechanisms of action, and the analogs being discovered, this class of agents clearly presents opportunities for drug development. It is not clear where the development of this compound stands, but the development of natural products is a complex and lengthy process that frequently lasts for many years.

CONCLUDING REMARKS

In this brief synopsis, we have attempted to provide an overview of a massive field still in its infancy. The huge diversity we have seen in our bacterial isolation program, and now recognized worldwide, coupled with the growing knowledge on how to isolate and culture marine bacteria, defines the massive potential of this source in drug discovery. Unfortunately, the pharmaceutical industries, especially those in the US, are largely no longer investing in microbial drug discovery. Although there are indications of a resurgence of interest in natural product drugs, leads are licensed mainly after massive preclinical development has been undertaken by biotech or academic laboratories. This avoidance of marine drug resources has had, and will continue to have, a chilling impact on funding levels for natural products research. One can easily argue, and provide solid statistical evidence, that the oceans are a significant resource for drug discovery. It is thus discouraging that this unequivocal evidence is not being heard.

ACKNOWLEDGMENT

LBP thanks the NIH for fellowship support under Grant 5 R01 GM085770.

REFERENCES

Abdel-Mageed, W.M., Milne, B.F., Wagner, M., Schumacher, M., Sandor, P., Pathom-aree, W., et al., 2010. Dermacozines, a new phenazine family from deep-sea dermacocci isolated from a Mariana Trench sediment. Org. Biomol. Chem. 8, 2352–2362.

Agarwal, V., El Gamal, A.A., Yamanaka, K., Poth, D., Kersten, R.D., Schorn, M., et al., 2014. Biosynthesis of polybrominated aromatic organic compounds by marine bacteria. Nat. Chem. Biol. 10, 640–647.

Alvarez-Mico, X., Jensen, P.R., Fenical, W., Hughes, C.C., 2013. Chlorizidine, a cytotoxic 5H-pyrrolo[2,1-a]isoindol-5-one-containing alkaloid from a marine *Streptomyces* sp. Org. Lett. 15, 988–991.

Alvin FAQs, 2006. Woods Hole Oceanographic Institution. Available from: <http://www.whoi.edu/page.do?pid=10995>

Aminov, R.I., 2010. A brief history of the antibiotic era: lessons learned and challenges for the future. Front. Microbiol. 1, 1–7.

Anderson G., 2003. Tools of the oceanographer: sampling equipment. Available from: <http://www.marinebio.net/marinescience/01intro/tosamp.htm>

Andrianasolo, E.H., Haramaty, L., Rosario-Passapera, R., Bidle, K., White, E., Vetriani, C., et al., 2009. Ammonificins A and B, hydroxyethylamine chroman derivatives from a cultured marine hydrothermal vent bacterium, *Thermovibrio ammonificans*. J. Nat. Prod. 72, 1216–1219.

Andrianasolo, E.H., Haramaty, L., Rosario-Passapera, R., Vetriani, C., Falkowski, P., White, E., et al., 2012. Ammonificins C and D, hydroxyethylamine chroman derivatives from a cultured marine hydrothermal vent bacterium, *Thermovibrio ammonificans*. Marine Drugs 10, 2300–2311.

Béjà, O., Suzuki, M.T., Heidelberg, J.F., Nelson, W.C., Preston, C.M., Hamada, T., et al., 2002. Unsuspected diversity among marine aerobic anoxygenic phototrophs. Nature 415, 630–633.

Bennett, J.W., Chung, K.-T., 2001. Alexander Fleming and the discovery of penicillin. Adv. Appl. Microbiol. 49, 163–184.

Bewley, C.A., Holland, N.D., Faulkner, D.J., 1996. Two classes of metabolites from *Theonella swinhoei* are localized in distinct populations of bacterial symbionts. Experientia 52, 716–722.

Bister, B., Bischoff, D., Ströbele, M., Riedlinger, J., Reicke, A., Wolter, F., et al., 2004. Abyssomicin C-A polycyclic antibiotic from a marine *Verrucosispora* strain as an inhibitor of the *p*-aminobenzoic acid/tetrahydrofolate biosynthesis pathway. Angew. Chem. Int. Ed. 43, 2574–2576.

Blunt, J.W., Copp, B.R., Keyzers, R.A., Munro, M.H.G., Prinsep, M.R., 2015. Marine natural products. Nat. Prod. Rep. 32, 116–211.

Bollmann, A., Lewis, K., Epstein, S.S., 2007. Incubation of environmental samples in a diffusion chamber increases the diversity of recovered isolates. Appl. Environ. Microbiol. 73, 6386–6390.

Boucher, H.W., 2010. Challenges in anti-infective development in the era of bad bugs, no drugs: a regulatory perspective using the example of bloodstream infection as an indication. Clin. Infect. Dis. 30, S4–S9.

Boucher, H.W., Talbot, G.H., Bradley, J.S., Edwards, Jr., J.E., Gilbert, D., Rice, L.B., et al., 2009. Bad bugs, no drugs: no ESKAPE! An update from the Infectious Diseases Society of America. Clin. Infect. Dis. 48, 1–12.

Burkholder, P.R., Pfister, R.M., Leitz, F.H., 1966. Production of a pyrrole antibiotic by a marine bacterium. Appl. Microbiol. 14, 649–653.

Choi, E.J., Nam, S.-J., Paul, L., Beatty, D., Kauffman, C.A., Jensen, P.R., et al., 2015. Previously uncultured marine bacteria linked to novel alkaloid production. Chem. Biol. 22, 1270–1279.

ClinicalTrials.gov., 2016. U.S. National Library of Medicine, U.S. National Institutes of Health, U.S. Department of Health and Human Services. Available from: https://clinicaltrials.gov/ct2/results?term=salinosporamide+A&Search=Search

D'Amico, S., Collins, T., Marx, J.-C., Feller, G., Gerday, C., 2006. Psychrophilic microorganisms: challenges for life. EMBO Rep. 7, 385–389.

Dubilier, N., Bergin, C., Lott, C., 2008. Symbiotic diversity in marine animals: the art of harnessing chemosynthesis. Nat. Rev. Microbiol. 6, 725–740.

Feling, R.H., Buchanan, G.O., Mincer, T.J., Kauffman, C.A., Jensen, P.R., Fenical, W., 2003. Salinosporamide A: a highly cytotoxic proteasome inhibitor from a novel microbial source, a marine bacterium of the new genus *Salinospora*. Angew. Chem. Int. Ed. 42, 355–357.

Fenical, W., Jensen, P.R., Palladino, M.A., Lam, K.S., Lloyd, G.K., Potts, B.C., 2009. Discovery and development of the anticancer agent salinosporamide A (NPI-0052). Bioorg. Med. Chem. 17, 2175–2180.

Fröhlich, J., König, H., 2000. New techniques for isolation of single prokaryotic cells. FEMS Microbiol. Rev. 24, 567–572.

Fuhrman, J.A., Azam, F., 1983. Adaptations of bacteria to marine subsurface waters studied by temperature response. Mar. Ecol. Prog. Ser. 13, 95–98.

Giovannoni, S.J., Rappé, M.S., 2000. Evolution, diversity and molecular ecology of marine prokaryotes. In: Kirchman, D. (Ed.), Microbial Ecology of the Oceans. John Wiley & Sons, New York, pp. 47–84.

Giovannoni, S.J., Britschgi, T.B., Moyer, C.L., Field, K.G., 1990. Genetic diversity in Sargasso Sea bacterioplankton. Nature 345, 60–63.

Giovannoni, S.J., Tripp, H.J., Givan, S., Podar, M., Vergin, K.L., Baptista, D., et al., 2005. Genome streamlining in a cosmopolitan oceanic bacterium. Science 309, 1242–1245.

Groll, M., Huber, R., Potts, B.C.M., 2006. Crystal structures of salinosporamide A (NPI-0052) and B (NPI-0047) in complex with the 20S proteasome reveal important consequences of β-lactone ring opening and a mechanism for irreversible binding. J. Am. Chem. Soc. 128, 5136–5141.

Hentschel, U., Usher, K.M., Taylor, M.W., 2006. Marine sponges as microbial fermenters. FEMS Microbiol. Ecol. 55, 167–177.

Homann, V.V., Sandy, M., Tincu, J.A., Templeton, A.S., Tebo, B.M., Butler, A., 2009. Loihichelins A-F, a suite of amphiphilic siderophores produced by the marine bacterium *Halomonas* LOB-5. J. Nat. Prod. 72, 884–888.

Hughes, C.C., Prieto-Davo, A., Jensen, P.R., Fenical, W., 2008. The marinopyrroles, antibiotics of an unprecedented structure class from a marine *Streptomyces* sp. Org. Lett. 10, 629–631.

Jensen, P.R., Moore, B.S., Fenical, W., 2015. The marine actinomycete genus *Salinispora*: a model organism for secondary metabolite discovery. Nat. Prod. Rep. 32, 738–751.

Joint, I., Mühling, M., Querellou, J., 2010. Culturing marine bacteria—an essential prerequisite for biodiversity. Microb. Biotechnol. 3, 564–575.

Kaysser, L., Bernhardt, P., Nam, S.-J., Loesgen, S., Ruby, J.G., Skewes-Cox, P., et al., 2012. Merochlorins A–D, cyclic meroterpenoid antibiotics biosynthesized in divergent pathways with vanadium-dependent chloroperoxidases. J. Am. Chem. Soc. 134, 11988–11991.

Keller, S., Nicholson, G., Drahl, C., Sorensen, E., Fiedler, H.-P., Süssmuth, R.D., 2007. Abyssomicins G and H and atrop-abyssomicin C from the marine Verrucosispora strain AB-18-032. J. Antibiot. (Tokyo) 60, 391–394.

Kennedy, J., Flemer, B., Jackson, S.A., Lejon, D.P.H., Morrissey, J.P., O'Gara, F., et al., 2010. Marine metagenomics: new tools for the study and exploitation of marine microbial metabolism. Marine Drugs 8, 608–628.

Lam, K.S., Lloyd, G.K., Neuteboom, S.T.C., Palladino, M.A., Sethna, K.M., Spear, M.A., et al., 2010. From natural products to clinical trials: NPI-0052 (salinosporamide A), a marine actinomycete-derived anticancer agent. In: Buss, A.D., Butler, M.S. (Eds.), Natural Products Chemistry for Drug Discovery. Royal Society of Chemistry, Cambridge, UK, pp. 355–373.

León, B., Navarro, G., Dickey, B.J., Stepan, G., Tsai, A., Jones, G.S., et al., 2015. Abyssomicin 2 reactivates latent HIV-1 by a PKC-and HDAC-independent mechanism. Org. Lett. 17, 262–265.

Long, P.F., Dunlap, W.C., Battershill, C.N., Jaspars, M., 2005. Shotgun cloning and heterologous expression of the patellamide gene cluster as a strategy to achieving sustained metabolite production. Chembiochem. 6, 1760–1765.

Mayer MSM, 2014. The global marine pharmaceuticals pipeline. Available from: <http://marinepharmacology.midwestern.edu>

Munn, C., 2011. Marine Microbiology: Ecology & Applications, second ed. Garland Science Taylor & Francis Group, New York, NY; Abingdon, UK, (Chapter 1).

Nam, S.-J., Kauffman, C.A., Paul, L.A., Jensen, P.R., Fenical, W., 2013. Actinoranone, a cytotoxic meroterpenoid of unprecedented structure from a marine adapted *Streptomyces* sp. Org. Lett. 15, 5400–5403.

Newman, D.J., Cragg, G.M., 2016. Natural products as sources of new drugs from 1981 to 2014. J. Nat. Prod. 79, 629–661.

Pérez Baz, J., Cañedo, L.M., Fernández Puentes, J.L., Silva Elipe, M.V., 1997. Thiocoraline, a novel depsipeptide with antitumor activity produced by a marine *Micromonospora*. II. Physico-chemical properties and structure determination. J. Antibiot. 50, 738–741.

Pomponi, S.A., Kelly, M., Reed, J.K., Wright, A.E., 2001. Diversity and bathymetric distribution of lithistid sponges in the tropical western Atlantic region. Bull. Biol. Soc. Wash. 10, 344–353.

Pratt, D.B., Waddell, G., 1959. Adaptation of marine bacteria to growth in media lacking sodium chloride. Nature 183, 1208–1209.

Prieto-Davó, A., Fenical, W., Jensen, P.R., 2008. Comparative actinomycete diversity in marine sediments. Aquat. Microb. Ecol. 52, 1–11.

Qian, P.-Y., Wang, Y., Lee, O.O., Lau, S.C.K., Yang, J., Lafi, F.F., et al., 2011. Vertical stratification of microbial communities in the Red Sea revealed by 16S rDNA pyrosequencing. ISME J. 5, 507–518.

Rappé, M.S., Connon, S.A., Vergin, K.L., Giovannoni, S.J., 2002. Cultivation of the ubiquitous SAR11 marine bacterioplankton clade. Nature 418, 630–633.

Romero, F., Espliego, F., Pérez Baz, J., García de Quesada, T., Grávalos, D., De La Calle, F., et al., 1997. Thiocoraline, a new depsipeptide with antitumor activity produced by a marine *Micromonospora*. I. Taxonomy, fermentation, isolation, and biological activities. J. Antibiot. 50, 734–737.

Schmidt, E.W., Donia, M.S., 2010. Life in cellulose houses: symbiotic bacterial biosynthesis of ascidian drugs and drug leads. Curr. Opin. Biotechnol. 21, 827–833.

Schmidt, E.W., Obraztsova, A.Y., Davidson, S.K., Faulkner, D.J., Haygood, M.G., 2000. Identification of the antifungal peptide-containing symbiont of the marine sponge *Theonella swinhoei* as a novel delta-proteobacterium, "*Candidatus Entotheonella palauensis*". Marine Biol. 136, 969–977.

Schmidt, E.W., Nelson, J.T., Rasko, D.A., Sudek, S., Eisen, J.A., Haygood, M.G., Patellamide A, et al., 2005. and C biosynthesis by a microcin-like pathway in *Prochloron didemni*, the cyanobacterial symbiont of *Lissoclinum patella*. Proc. Natl. Acad. Sci. USA 102, 7315–7320.

Schmidt, E.W., Donia, M.S., McIntosh, J.A., Fricke, W.F., Ravel, J., 2012. Origin and variation of tunicate secondary metabolites. J. Nat. Prod. 75, 295–304.

Simmons, T.L., Coates, R.C., Clark, B.R., Engene, N., Gonzalez, D., Esquenazi, E., et al., 2008. Biosynthetic origin of natural products isolated from marine microorganism–invertebrate assemblages. Proc. Natl. Acad. Sci. USA 105, 4587–4594.

Somero, G.N., 1992. Adaptations to high hydrostatic pressure. Ann. Rev. Physiol. 54, 557–577.

Stach, J.E.M., Maldonado, L.A., Masson, D.G., Ward, A.C., Goodfellow, M., Bull, A.T., 2003. Statistical approaches for estimating actinobacterial diversity in marine sediments. Appl. Environ. Microbiol. 69, 6189–6200.

Tesfazghi, S., Eide, J., Dammalapati, A., Korlesky, C., Wyche, T.P., Bugni, T.S., et al., 2013. Thiocoraline alters neuroendocrine phenotype and activates the Notch pathway in MTC-TT cell line. Cancer Med. 2, 734–743.

Thornburg, C.C., Zabriskie, T.M., McPhail, K.L., 2010. Deep-sea hydrothermal vents: potential hot spots for natural products discovery? J. Nat. Prod. 73, 489–499.

Underwater vehicles, 2015. Woods Hole Oceanographic Institution. Available from: <https://www.whoi.edu/main/topic/underwater-vehicles>

Wagner, M., Abdel-Mageed, W.M., Ebel, R., Bull, A.T., Goodfellow, M., FiedlerHP, et al., 2014. Dermacozines H–J isolated from a deep-sea strain of *Dermacoccus abyssi* from Mariana Trench sediments. J. Nat. Prod. 77, 416–420.

Wang, Q., Song, F., Xiao, X., Huang, P., Li, L., Monte, A., et al., 2013. Abyssomicins from the South China sea deep-sea sediment *Verrucosispora* sp.: natural thioether michael addition adducts as antitubercular prodrugs. Angew. Chem. Int. Ed. 52, 1231–1234.

Wilson, M.C., Mori, T., Rückert, C., Uria, A.R., Helf, M.J., Takada, K., et al., 2014. An environmental bacterial taxon with a large and distinct metabolic repertoire. Nature 506, 58–62.

Woese, C.R., Kandler, O., Wheelis, M.L., 1990. Towards a natural system of organisms: proposal for the domains Archaea, Bacteria, and Eucarya. Proc. Natl. Acad. Sci. USA 87, 4576–4579.

Wyche, T.P., Hou, Y., Braun, D., Cohen, H.C., Xiong, M.P., Bugni, T.S., 2011. First natural analogs of the cytotoxic thiodepsipeptide thiocoraline A from a marine *Verrucosispora* sp. J. Org. Chem. 76, 6542–6547.

Zhang, L., 2005. Integrated approaches for discovering novel drugs from microbial natural products. In: Zhang, L., Demain, A.L. (Eds.), Natural Products: Drug Discovery and Therapeutic Medicine. Humana Press Inc, Totowa, NJ, pp. 33–55.

FURTHER READING

Nam, S.-J., Kauffman, C.A., Jensen, P.R., Moore, C.E., Rheingold, A.L., Fenical, W., 2015. Actinobenzoquinoline and actinophenanthrolines A−C, unprecedented alkaloids from a marine actinobacterium. Org. Lett. 17, 3240–3243.

Sogin, M.L., Morrison, H.G., Huber, J.A., Welch, D.M., Huse, S.M., Neal, P.R., et al., 2006. Microbial diversity in the deep sea and the underexplored "rare biosphere". Proc. Natl. Acad. Sci. USA 103, 12115–12120.

Wyche, T.P., Dammalapati, A., Cho, H., Harrison, A.D., Kwon, G.S., Chen, H., et al., 2014. Thiocoraline activates the Notch pathway in carcinoids and reduces tumor progression in vivo. Cancer Gene Ther. 21, 518–525.

Chapter 9

Cryptic Pathways and Implications for Novel Drug Discovery

Kozo Ochi
Hiroshima Institute of Technology, Saeki-ku, Hiroshima, Japan

INTRODUCTION

The bacterial taxa producing the largest number of chemically diverse secondary metabolites belong to the members of the order *Actinomycetales*, especially the streptomycetes. Other major sources of these metabolites include soil bacilli, myxococci, and pseudomonads. Fungi also produce an estimated 42% of antibiotic products. Recent advances in DNA sequencing technologies have enabled entire genomes to be sequenced rapidly and inexpensively. Although sequencing of the genomes of *Streptomyces coelicolor* A3(2), *S. avermitilis*, *S. griseus*, and *Saccharopolyspora erythraea* showed that each species contains genes that encode enzymes synthesizing 20 or more potential secondary metabolites, only some of these enzymes are produced during fermentation (Bentley et al., 2002; Ikeda et al., 2003; Ohnishi et al., 2008; Oliynyk et al., 2007). Similarly, the model fungus *Aspergillus indulans* is potentially able to produce 32 polyketides, 14 nonribosomal peptides and two indole alkaloids, but, to date, only about 50% of these secondary metabolites have been identified. Thus, the number of bacterial and fungal secondary metabolites identified so far may represent only the tip of the iceberg. Homologous and heterologous expression of these cryptic secondary metabolite-biosynthetic genes, which are often "silent" under ordinary laboratory fermentation conditions, may lead to the discovery of novel secondary metabolites. The success of this approach depends on finding methods to induce or enhance the expression of cryptic or poorly expressed pathways, yielding material for structural determination, and biological testing. In activating silent genes, it is particularly important to activate the transcription, rather than the biosynthesis process because these genes are "sleeping" at the transcriptional level (Fig. 9.1). This chapter addresses current progress in the activation of these pathways, describing methods for activating silent genes, and especially focusing on genetic

Microbial Resources. http://dx.doi.org/10.1016/B978-0-12-804765-1.00009-6

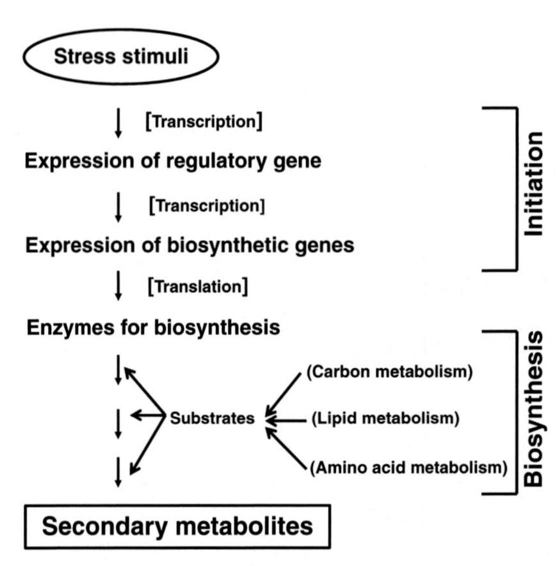

FIGURE 9.1 **Outline of pathways for the production of secondary metabolites, including initiation of transcription and biosynthesis.**

manipulation of transcription and translation, the utilization of elicitors and co-cultivation (Baltz, 2011; Challis, 2014; Chiang et al., 2011; Choi et al., 2015; Fedorenko et al., 2015; Müller and Wink, 2014; Reen et al., 2015; Zhu et al., 2014).

HOMOLOGOUS EXPRESSION OF SILENT GENES

OSMAC Approach

Several strategies have been formulated to identify novel natural products through the awakening of silent gene clusters. These strategies include the one strain many compounds (OSMAC) approach, in which new compounds can be induced under different fermentation conditions. Secondary metabolite profiles can be strongly influenced by varying cultivation conditions. Small changes in the composition of growth media can not only alter the amounts of specific compounds, but may result in the production of a completely different set of molecules. This approach is not only effective for the discovery of novel compounds, but can optimize antibiotic production. However, this approach can be laborious, and it is uncertain whether conditions can be found to stimulate the synthesis of all the interesting and useful compounds that organisms can potentially produce (Reen et al., 2015).

Ribosome and RNA Polymerase Engineering

The notion of "ribosome engineering" originally came from results with *Streptomyces lividans*. Although *S. lividans* normally does not produce antibiotics due to the dormancy of antibiotic biosynthesis genes, a strain with an altered ribosomal S12 protein that confers resistance to streptomycin produced abundant quantities of the blue-pigmented antibiotic actinorhodin. Furthermore, the bacterial alarmone ppGpp, produced on ribosomes in response to nutrient starvation, was found to bind to RNA polymerase (RNAP), eventually initiating the production of antibiotics. These results suggested that modifying RNAP by, for example, introducing a rifampicin resistance mutation may mimic the ppGpp-bound form of RNAP. The latter shows enhanced affinity to the promoter region of genes that regulate the synthesis of secondary metabolites, activating the expression of biosynthetic gene clusters. These observations led to the development of a ribosome engineering method targeting S12, RNAP, and other ribosomal proteins and translation factors, thus activating or enhancing the production of secondary metabolites. Ribosome engineering is applicable to silent gene activation and strain improvement, resulting in the identification of novel secondary metabolites, as well as to the enhancement of enzyme production and tolerance to toxic chemicals. Combinations of various drug-resistance mutations can further enhance bacterial productivity, for example, as demonstrated by introducing the octuple drug-resistance mutations into *S. coelicolor* 1147 (Fig. 9.2). The octuple mutant C8 produced huge amounts (1.63 g/L) of the antibiotic actinorhodin, 180-fold higher than the level produced by the wild type 1147.

Ribosome engineering is characterized by simplicity, consisting of the isolation of spontaneously developed drug-resistant mutants. This method does not require induced mutagenesis or any genomic information and provides a rational approach of enhancing bacterial capabilities for industrial applications. The mechanisms underlying the generation of these mutants have been studied extensively in *Streptomyces* and *Bacillus*. Streptomycin-resistant (Sm^r) *rpsL*-mutant ribosomes, which carry an amino acid substitution in S12 that confers a high-level resistance to streptomycin, are more stable than wild-type ribosomes, indicating that increased stability may enhance protein synthesis during the late growth phase of bacteria. Increased expression of the translation factor ribosome recycling factor also contributes to the enhanced protein synthesis of the *rpsL* K88E mutant in late growth phase. That is, both the greater stability of the 70S ribosomes and the elevated levels of ribosome recycling factor resulting from the K88E mutation are responsible for enhanced protein synthesis during late growth phase, with the latter being responsible for antibiotic overproduction.

The *rpoB* gene, which encodes RNAP β-subunit, and the *rpsL* gene, which encodes ribosomal protein S12, confer resistance to the antibiotics rifampicin and streptomycin, respectively. Certain mutations in these genes can activate silent or weakly expressed genes of bacteria, leading to the discovery of novel antibacterial agents as represented by discovery of the piperidamycins (Fig. 9.3).

The activation of silent genes by the *rpoB* mutations was attributed, at least in part, to the increased affinity of mutant RNAP for silent gene promoters. The introduction of the *rpoB* mutation S487L (corresponding to S433L

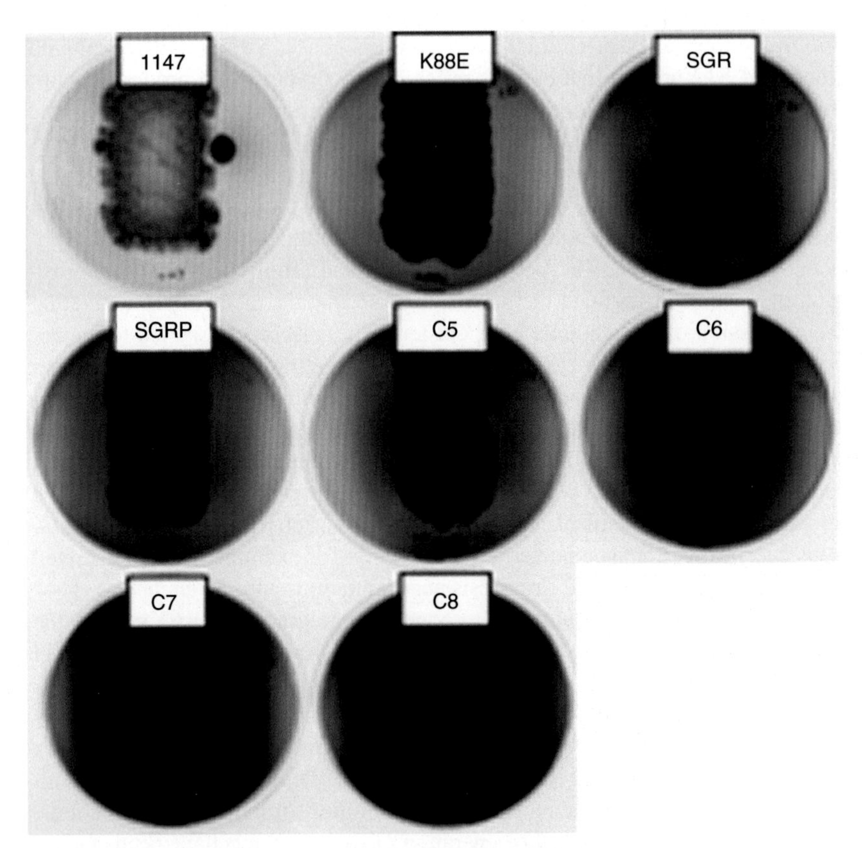

FIGURE 9.2 **Antibiotic production in drug-resistant mutants.** Strains were inoculated onto GYM plates, which were incubated for 8 days. The reverse sides of the plates are shown to illustrate the blue antibiotic actinorhodin. 1147, wild-type; C8, octuple mutant (*str gen rif par gnt fus tsp lin*) resistant to streptomycin, gentamicin, rifampicin, paromomycin, geneticin, fusidic acid, thiostrepton, and lincomycin.

Piperidamycin	R_1	R_2
A	H	H
D	H	CH_3
F	CH_3	CH_3

FIGURE 9.3 **The chemical structures of the antibacterial compounds isolated.**

of *S. coelicolor*) into a *B. subtilis* strain also yielded cells that overproduced an aminosugar antibiotic, 3,3′-neotrehalosadiamine (NTD), the production of which is dormant in the wild-type strain. Drug-resistance mutations in *Nocardiopsis* sp, such as those conferring resistance to streptomycin or rifampicin, result in broad alterations in metabolic phenotype and secondary metabolism. Similarly, *rsmG* mutations in *Streptomyces.* spp, which are responsible for a low level of resistance to streptomycin, activate not only antibiotic production but the expression of other secondary metabolite-biosynthetic genes. Loss of the m^7G modification in 16S rRNA results in resistance to streptomycin, providing a molecular basis for *rsmG* mutation-induced streptomycin resistance. *S. coelicolor* carrying *rsmG* mutations exhibit enhanced expression of the *metK* gene encoding *S*-adenosylmethionine (SAM) synthetase, eventually leading to the overproduction of antibiotics (Fig. 9.4). Furthermore, the addition of SAM to streptomycetes induces their overproduction of antibiotics, indicating that SAM is a common intracellular signal molecule for the onset of secondary metabolism in *Streptomyces*.

A recent study showed the broad applicability of the *rpoB* mutation method to the expression of cryptic secondary metabolite-biosynthetic gene clusters.

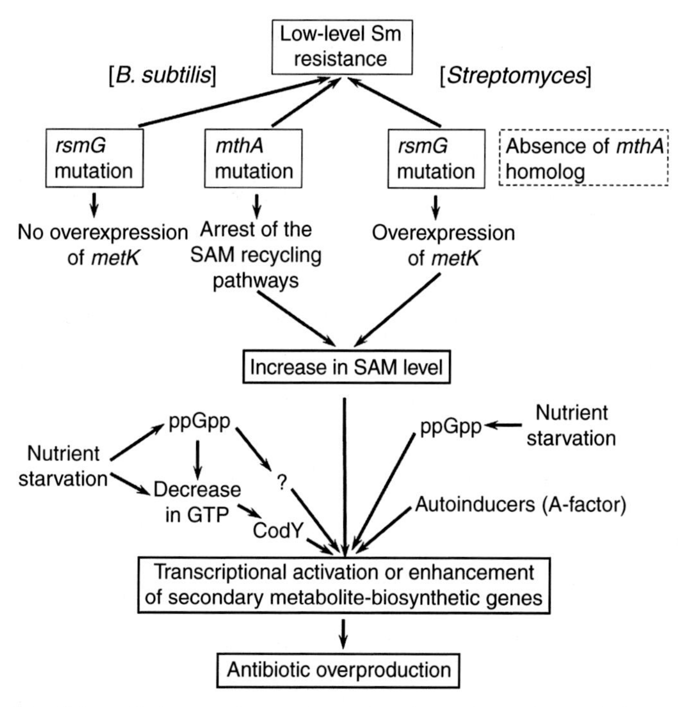

FIGURE 9.4 **Schematic showing the signal transduction pathways in *B. subtilis* and *Streptomyces* spp., from the *mthA* or *rsmG* mutation to the enhancement of antibiotic production or the activation of silent genes.**

Use of this method in *S. griseus* resulted in marked transcriptional activation of 18 genes belonging to 18 secondary metabolite-biosynthetic gene clusters, with some showing a 50–70-fold increase. Transcriptional analysis of 15 genes belonging to 15 cryptic secondary metabolite-biosynthetic gene clusters in *S. erythraea* showed that the H437Y and H437R mutations, which effectively enhanced erythromycin production, also were activated the cryptic genes of *S. erythraea*, with 6 of these 15 genes showing greater than or equal to threefold increase in transcription. Similarly, the S433L and H437Y mutations, which enhanced actinorhodin production, activated the cryptic genes of *S. coelicolor*. Amino acid alterations at the H437 position (corresponding to positions H482 and H526 in *B. subtilis* and *E. coli*, respectively) were often effective in activating cryptic pathways. Available evidence suggests that a number of specific effects of *rpoB* Rif^R mutations may be due to their specific impact on transcription dynamics, including promoter binding/isomerization and transcription elongation/termination, which do not directly correlate with each other.

The recent finding, that several actinomycetes possess two *rpoB* genes, has suggested a new strategy of activating silent gene expression. Two *rpoB* paralogs, *rpoB(S)* [rifampicin-sensitive wild-type *rpoB*] and *rpoB(R)* [rifampicin-resistant mutant-type *rpoB*], provide *Nonomuraea* sp. strain 39727 with two functionally distinct and developmentally regulated RNAPs. The product of *rpoB(R),* the expression of which increases after transition to stationary phase, is characterized by five amino acid substitutions (H426N, S431N, F445M, S474Y, and M581D) located within or close to the rifampicin resistance cluster.

The expression of *rpoB(R)* was found to markedly activate antibiotic biosynthesis, with the *rpoB(R)*-specific H426N mutation essential in activating secondary metabolism. Other *rif* cluster-associated *rpoB(R)*-specific missense mutations likely interact functionally with the H426N mutation, leading to *rpoB(R)* having marked effects. Mutant-type, or duplicated, *rpoB* often exists in nature, with *rpoB* gene polymorphisms detected in five of 75 inherently rifampicin-resistant actinomycetes isolated from nature. Most of these polymorphisms, however, were present in rare actinomycetes, not in *Streptomyces* spp. Moreover, a *Nonomuraea terrinata* strain with a single (*rpoB*(R)) *rpoB* showed much less growth (representing primary metabolism) and sporulation and antibiotic production (representing the developmental strategy) than strains with a duplicated (*rpoB*(S) + *rpoB*(R)), especially under stressful conditions. These findings suggested that *rpoB* duplication has physiological significance. This technology may have greater potential than a simple Rif^R selection method, currently used to improve the production of secondary metabolites. Indeed, the introduction of *rpoB*(R) was much more effective in boosting antibiotic production than the introduction, for example, of the H426Y mutation alone. It is also of interest to assess whether various *rpoB(R)* forms found in nature are more capable than *Nonomuraea rpoB(R)* of activating silent bacterial genes (Alifano et al., 2015; Hosaka et al., 2009; Ochi and Hosaka, 2013; Ochi et al., 2014; Tala et al., 2009; Tanaka et al., 2013).

Regulatory Gene Activation

New natural products encoded by silent gene clusters may be unlocked by the constitutive expression of genes encoding putative regulators present in the same gene cluster. Constitutive overexpression of a putative pathway-specific large ATP-binding LuxR (LAL)-type regulator was shown to successfully induce the expression of the silent type I modular PKS gene cluster in *Streptomyces ambofaciens*, enabling the identification of a unique structural class of polyketides with promising antitumor activity. These findings suggested that the constitutive expression of these pathway-specific activators represents a powerful approach for the discovery of novel bioactive natural products, because genes that encode proteins involved in the biosynthesis of secondary metabolites are often controlled by pathway-specific activators (Laureti et al., 2011).

Elicitors

Elicitors are signal compounds that stimulate the activation or synthesis of other compounds. Upon sensing input signals, producer strains exhibit a discrete response, which may alter metabolite profiles. In particular, rare earth elements (REEs) are involved in the overexpression of antibiotics and in the activation of silent or weakly expressed genes in bacteria. Addition of very low concentrations of REEs to cultures of various *Streptomyces* strains has been found to enhance antibiotic production 2–25-fold. For example, scandium (Sc) was found to increase the transcription of the pathway-specific positive regulatory gene *actII*-ORF4 in *S. coelicolor*. Moreover, addition to the medium of low concentrations of Sc or La resulted in a 2.5–to 12-fold increase in the expression of nine genes belonging to nine cryptic secondary metabolite-biosynthetic gene clusters. HPLC analysis of ethyl acetate-extractable metabolites showed that several compounds were present only in REE-treated cultures. Moreover, addition to the medium of sublethal concentrations (one-third of the MIC) of $ScCl_3$ or $LaCl_3$ markedly enhanced the transcription of silent genes of *S. griseus*, indicating that these REEs were stress stimuli in activating secondary metabolism. Because of its feasibility, this approach should facilitate the discovery of new biologically active compounds. An important advantage of using REEs is that this method does not require any gene engineering technology or genomic information on the strains examined.

Other elicitors for secondary metabolism include *N*-acetyl-glucosamine, which blocks *S. coelicolor* development and antibiotic production on nutrient-rich media through the global transcriptional regulator DasR, whereas both processes are stimulated on nutrient-poor media. Intriguingly, transcription of an otherwise cryptic biosynthetic gene cluster was stimulated by the addition of *N*-acetyl-glucosamine into the nutrient-poor (i.e., synthetic) medium. DasR links nutrient stress to antibiotic production by *Streptomyces*. A high concentration of *N*-acetylglucosamine, perhaps mimicking its accumulation after autolytic degradation of the vegetative mycelium, may be a major checkpoint for the onset of secondary metabolism. The response is transmitted to antibiotic pathway-specific activators through the transcriptional repressor DasR. As expected, the *dasR* mutant was found to awaken cryptic gene clusters encoding hypothetical antibiotics, indicating that this method is applicable to the activation of silent gene clusters.

Antibiotics are likely the best known bacterial secondary metabolites. Recent study clarified that these agents act as signaling molecules and mediate outcomes other than cell death. Sublethal concentrations of antibiotics were recently reported to generate diverse metabolic responses in *Streptomyces*. For example, sublethal concentrations of lincomycin, which inhibits protein synthesis by binding to ribosomes, induced the expression in *Streptomyces* strains of genes involved in secondary metabolism. *S. lividans* 1326 grown in the presence of lincomycin at 1/12 or 1/3 of its MIC produced several antibacterial compounds that were not detected in cells grown in lincomycin-free medium. In relation to these work, cross-culture of *S. coelicolor* with *S. erythraea* (erythromycin producer) or *B. subtilis* (bacilysin producer) activated the production of blue-colored antibiotic actinorhodin by *S. coelicolor* (Fig. 9.5), representing the effectiveness of coclutivation in inducing the production of cryptic metabolites (see subsequent sections).

Identification of elicitors that induce the expression of silent gene clusters under standard conditions is a challenge. A high throughput platform was recently developed to identify elicitors of silent gene clusters in bacteria to assess the expression of known cryptic gene clusters of interest. For example, studies in *Burkholderia thailandensis* showed that sublethal concentrations of trimethoprim served as a global activator of secondary metabolism by inducing at least five biosynthetic pathways (Imai et al., 2015; Ochi et al., 2014; Seyedsayamdost, 2014; Zhu et al., 2014).

Cocultivation

In nature, microorganisms coexist within complex communities. Cocultivation, also called mixed cultivation, has been shown effective in inducing the production of cryptic metabolites. *Bacillus subtilis* can induce secondary metabolite production in fungi, as shown by the formation of macrocarpon C. To date, only a few examples of gene regulatory networks during secondary metabolite formation in cocultures have been reported. One consists of specific interaction between *Aspergillus nidulans* and *Streptomyces rapamycinicus*. Physical contact of *A. nidulans* (producer) with *S. rapamycinicus* (inducer) led to the selective activation of silent PKS and NRPS gene clusters in the fungus. Activation of a silent fungal gene cluster and production of novel compounds was mediated by bacterial manipulation of chromatin-based regulation (i.e., acetylation of chromatin) in the fungus. Another example is the production of red pigment by *S. lividans*, which is induced by coculturing *S. lividans*

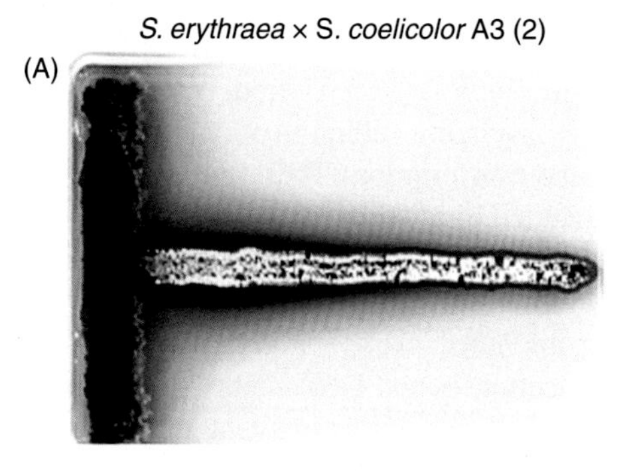

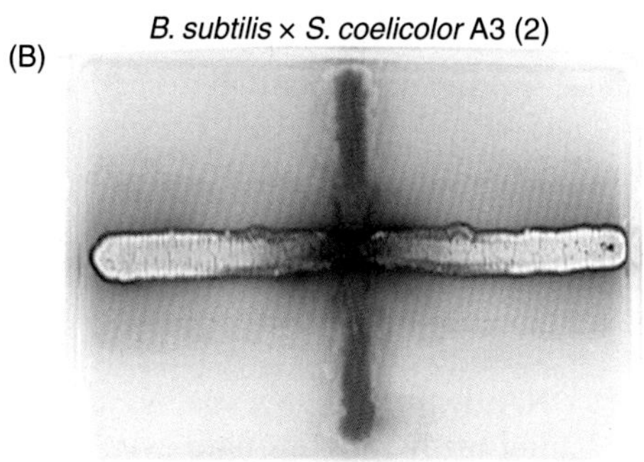

FIGURE 9.5 Cross-culture of *S. coelicolor* A3(2) with *S. erythraea* or *B. subtilis*. *S. coelicolor* A3(2) 1147 (*horizontal line*) was cross-cultured with *S. erythraea* NRRL 2338 (A) or *B. subtilis* 168 (B) (*vertical line*) on MR5 agar at 30°C for 8 days. *Blue color* represents the antibiotic actinorhodin.

(producer) with *Tsukamurella pulmonis* (inducer), a mycolic acid-containing bacterium. Importantly, coculture of *T. pulmonis* with *Streptomyces* strains isolated from soil altered the synthesis of natural products in 88% of these bacterial strains, with new secondary metabolites produced in 37% and metabolite production increased in 55%. Thus, mycolic acid-containing bacteria can influence, at very high frequencies, the biosynthesis of cryptic natural products in *Streptomyces* spp. Coculture of *T. pulmonis* with *Streptomyces endus* led to the identification of a novel antibiotic, alchivemycin A. The addition of mycolic acid to the medium of pure *S. lividans* cultures, however, had no effect on antibiotic production, suggesting that direct cell-cell contact between *S. lividans* and *T. pulmonis* was necessary. The cross talk between inducers and producers during cocultivation of microorganisms appears to be a highly dynamic process. Although the functional mechanisms underlying this method remain unclear, this approach is easily scalable for the production of cryptic antibiotics, as it only involves the addition of a mycolic acid-containing bacterium to a pure culture of an actinomycete (Abdelmohsen et al., 2015; Bertrand et al., 2014; Marmann et al., 2014; Netzker et al., 2015; Onaka et al., 2011; Shank and Kolter, 2009).

Epigenetics

Histone modifications, such as acetylation, methylation, phosphorylation, ADP-ribosylation, glycosylation, ubiquitination, and sumoylation, have been shown to regulate gene transcription. These modifications alter histone structure, thus affecting histone binding to DNA and altering its transcription. Several chromatin modifiers, which regulate secondary metabolite biosynthesis in fungi, have been identified to date. Histones or their associated DNA may be modified by treating microorganisms with epigenetic modifiers, including inhibitors of histone deacetylases and DNA methyl-transferases. These modifications may also initiate the transcription of silent genes. Epigenetic modification to access cryptic natural product gene clusters has been evaluated primarily in fungi. Inhibitors of histone deacetylation in fungi can induce changes in chromatin structure, increasing promoter accessibility, and inducing the production of secondary metabolites. Analogously, elicitors, such as histone deacetylation inhibitors were found to induce the expression of otherwise silent gene clusters in *Streptomyces coelicolor*. Treatment of *Aspergillus niger* with the epigenetic modifier suberohydroxamic acid upregulated the expression of 48 of 55 gene clusters.

The best understood method used by bacteria to alter their epigenome is the methylation of DNA. Different DNA methytransferases target unique DNA sequences, attaching methyl groups onto cytosine bases. Methylation close to gene promoter regions can alter gene expression. Treatment with drugs, such as 5-azacytidine and procaine, which inhibit DNA methyltransferases, may therefore access silent natural product gene clusters in bacteria. Epigenetics may also play a role in triggering the activation of silent fungal genes in response to cocultivation with bacteria (Marmann et al., 2014; Netzker et al., 2015).

HETEROLOGOUS EXPRESSION OF SILENT GENES

Apart from homologous expression of cryptic genes, an additional promising alternative for genome mining is heterologous expression of complex biosynthetic pathways in prioritized host organisms that allow fast growth and good production. The development of cloning methods for large DNA fragments has made possible the cloning of entire biosynthetic clusters, some of which are very large. By modifying these clones to include site-specific integration sites, it may be possible to transfer them into strains for heterologous expression. Use of such approaches with gene clusters in other organisms requires the development of a good molecular genetic system for that organism or secondary metabolism genes or the introduction of gene clusters from that organism into appropriate heterologous host organisms. *Streptomyces lividans* and *Streptomyces albus* J1074 were originally used in this process because of their low secondary metabolite output and limited restriction barriers. More recently, *S.*

coelicolor and *S. avermitilis* have been developed for the expression of heterologous metabolites with very exciting results. For example, *S. coelicolor* M1154 was constructed by deleting four gene clusters (actinorhodin, prodiginine, calcium-dependent antibiotic, and cryptic polyketide), followed by the introduction into the *rpoB* and *rpsL* genes of point mutations that enhance specialized metabolite production (see ribosome engineering). Expression of the gene clusters for chloramphenicol and congocidine in this strain led to 40- and 30-fold increases in production, respectively, when compared with *S. coelicolor* M145.

Another surrogate host was developed from the industrial strain *S. avermitilis*, a producer of avermectin. Genome-minimized strains were used to express heterologous secondary metabolite pathways. For example, wild-type *S. avermitilis* produced ~30 mg/L streptomycin, whereas the SUKA5 recombinant produced ~180 mg/L. These host strains may offer a general solution to the production of high levels of compounds of interest. In fungi, heterologous expression of genes from the proposed aspyridone biosynthetic cluster of *Aspergillus nidulans* in the host *Aspergillus oryzae* led to production of eight different compounds in addition to aspyridone A 1, one of the previously observed products. More recently, a new high-performance heterologous fungal expression system was described in *Aspergillus terreus*. This system was found to be particularly suitable for high level expression of polyketides in heterologous hosts. However, despite some successes, challenges remain and molecular-based methods utilizing heterologous expression systems continue to be limited by the location and cloning of genes, by difficulties with gene transformation and inactivation, and by host incompatibilities. The gene clusters encoding putative bioactive compounds are typically large (10–100 kb) and often contain many genes. Therefore, the cloning and expression of these clusters is significantly more challenging than cloning of a single gene and expressing its product. Also, these clusters/genes may not be expressed in heterologous hosts due to the lack of signals naturally present in the native producer.

Synthetic biology is an emerging strategy to awaken cryptic metabolic gene clusters in actinomycetes. In this strategy, the identified silent biosynthetic gene clusters are subjected to PCR-amplification or chemical synthesis, followed by gene cluster reconstruction using suitable heterologous promoters. Once reconstructed, these gene clusters are assembled in yeast. The assembled constructs are subsequently isolated from yeast, used to retransform *E. coli.* Following verification of the identity of the assembled constructs, they are transferred to a heterologous host. Use of synthetic biology led to the activation of the cryptic macrolactam biosynthetic gene cluster in *S. griseus*, as well as the awakening of the silent spectinabilin pathway in *Streptomyces orinoci* (Baltz, 2010; Breitling and Takano, 2015; Gomez-Escribano and Bibb, 2014; Kim et al., 2015; Komatsu et al., 2010; Rebets et al., 2014; Yaegashi et al., 2014).

METABOLISM REMODELING AND METABOLIC ENGINEERING

Although activation of biosynthetic genes at the transcriptional level is of prime importance in exploiting useful metabolites, perturbation of biosynthesis by modulating, for example, the supply of precursors ("metabolism remodeling" as named by Justin Nodwell) may also be a promising approach to enhancing a cell's ability to produce secondary metabolites (Fig. 9.1). The yields of secondary metabolites may be enhanced by small molecules that modulate fatty acid biosynthesis. For example, screening of a large number (>30,000) of small molecules, seeking candidates that may alter the yield of actinorhodin in *S. coelicolor*, identified 19 compounds that induce or elevate actinorhodin production. As fatty acid and polyketide synthesis pathways share the precursors acetyl-CoA and malonyl-CoA, partial inhibition of fatty acid synthesis may recruit these acyl-CoAs preferentially to polyketide biosynthesis. Theoretically, this approach may be enhanced by combination with transcription-associated approaches (Fig. 9.1). These studies suggest that optimally allocating metabolic resources between primary and secondary metabolic pathways should engineer primary metabolism such that the pool of building blocks is enhanced.

Metabolic engineering involves the genetic modification of producer organisms to enhance the available levels of certain metabolic precursors, such as acetyl-CoA, molonyl-CoA, and methylmalonyl-CoA. This precursor pool can be improved by manipulating biochemical pathways involved in fatty acid biosynthesis and degradation, branched chain amino-acid degradation, and glucose metabolism (Craneym et al., 2012; Olano et al., 2008; Weber et al., 2015).

CONCLUDING REMARKS

Apart from the technology, it is important to determine why cryptic genes are not expressed under laboratory fermentation conditions. In nature, secondary metabolites were thought to function in the defense of producer habitats by inhibiting the growth of competing organisms. Secondary metabolites may have evolved because microorganisms used them as chemical signals for communication between cells of the same species or different species. It is unclear whether silent genes are expressed under special, as yet-unknown environmental conditions, or if the cryptic secondary metabolites play an intrinsic biological role(s) in producing organisms under these special environmental conditions. Determination of any special environmental conditions, as well as understanding the mechanism(s) underlying the silencing of cryptic genes, would help to fully utilize microbial gene clusters for secondary metabolism. Studies showing the effects of REEs on cryptic gene activation may be a clue to address these questions, as REEs are distributed ubiquitously throughout the world, albeit at very low concentrations. Microorganisms may have acquired the ability to respond to low levels of these elements as an abiotic stress, possibly to adapt to prevailing conditions.

Another clue may be provided by cryptic gene activation through cell-to-cell interactions. The presence of "natural" mutant type RNAPs, either as polymorphisms or nonpolymorphisms, may also be used to resolve questions about cryptic genes. Although only <0.1% of actinomycetes in the soil have a natural mutant-type RNAP gene, their cryptic genes are apparently expressed during late growth phase. Therefore, it is likely that, in the absence of the natural mutant-type RNAP gene, any cryptic genes in actinomycetes would be expressed under as yet unknown, specific environmental conditions and that activation of these secondary metabolite-biosynthetic pathways may enhance the survival advantages of these organisms.

The availability of genome sequence information of various microorganisms and newly developed methods of activating silent and poorly expressed genes suggest that natural product research has entered a promising new era. Particular attention should be paid to combined approaches, which include the transcription-activation of key genes and metabolism-remodeling. Reinforcement of biosynthetic processes by efficient substrate supply may be synergistic with transcription-activation, eventually leading to the efficient discovery of new secondary metabolites. It is also of interest to identify the target(s) of REEs responsible for the activation of cryptic genes. As the RNAP and ribosome are both involved in cryptic gene activation, REEs may bind to RNAP and/or the ribosome, eventually leading to functional alteration of these macromolecules. Rare earth microbiology (Ochi et al., 2014) may thus offer new insights into entirely unknown regulatory events that occur in all organisms.

REFERENCES

Abdelmohsen, U.R., Grkovic, T., Balasubramanian, S., Kamel, M.S., Quinn, R.J., Hentschel, U., 2015. Elicitation of secondary metabolism in actinomycetes. Biotechnol. Adv. 33, 798–811.

Alifano, P., Palumbo, C., Pasanisi, D., Tala, A., 2015. Rifampicin-resistance, *rpoB* polymorphism and polymerase genetic engineering. J. Biotechnol. 202, 60–77.

Baltz, R.H., 2010. *Streptomyces* and *Saccharopolyspora* hosts for heterologous expression of secondary metabolite gene clusters. J. Ind. Microbiol. Biotechnol. 37, 759–772.

Baltz, R.H., 2011. Strain improvement in actinomycetes in the postgenomic era. J. Ind. Microbiol. Biotechnol. 38, 657–666.

Bentley, S.D., Chater, K.F., Cerdeno-Tarraga, A.M., Challis, G.L., Thomson, N.R., James, K.D., Harris, D.E., Quail, M.A., Kieser, H., Harper, D., Bateman, A., Brown, S., Chandra, G., Chen, C.W., Collins, M., Cronin, A., Fraser, A., Goble, A., Hidalgo, J., Hornsby, T., Howarth, S., Huang, C.H., Kieser, T., Larke, L., Murphy, L., Oliver, K., O'Neil, S., Rabbinowitsch, E., Rajandream, M.A., Rutherford, K., Rutter, S., Seeger, K., Saunders, D., Sharp, S., Squares, R., Squares, S., Taylor, K., Warren, T., Wietzorrek, A., Woodward, J., Barrell, B.G., Parkhill, J., Hopwood, D.A., 2002. Complete genome sequence of the model actinomycete *Streptomyces coelicolor* A3(2). Nature 417, 141–147.

Bertrand, S., Bohni, N., Schnee, S., Schumpp, O., Gindro, K., Wolfender, J.L., 2014. Metabolite induction via microorganism co-culture: a potential way to enhance chemical diversity for drug discovery. Biotechnol. Adv. 32, 1180–1204.

Breitling, R., Takano, E., 2015. Synthetic biology advances for pharmaceutical production. Curr. Opin. Microbiol. 35, 46–51.

Challis, G.L., 2014. Exploitation of the *Streptomyces coelicolor* A3(2) genome sequence for discovery of new natural products and biosynthetic pathways. J. Ind. Microbiol. Biotechnol. 41, 219–232.

Chiang, Y.M., Chang, S.L., Oakley, B.R., Wang, C.C., 2011. Recent advances in awakening silent biosynthetic gene clusters and linking orphan clusters to natural products in microorganisms. Curr. Opin. Chem. Biol. 15, 137–143.

Choi, S.S., Kim, H.J., Lee, H.S., Kim, P., Kim, E.S., 2015. Genome mining of rare actinomycetes and cryptic pathway awakening. Proc. Biochem. 50, 1184–1193.

Craneym, A., Ozimok, C., Pimentel-Elardo, S.M., Capretta, A., Nodwell, J.R., 2012. Chemical perturbation of secondary metabolism demonstrates important links to primary metabolism. Chem. Biol. 19, 1020–1027.

Fedorenko, V., Genilloud, O., Horbal, L., Marcone, G.L., Marinelli, F., Paitan, Y., Ron, E.Z., 2015. Antibacterial discovery and development: from gene to product and back. BioMed. Res. Int. 2015, 16. Available from: http://dx.doi.org/10.1155/2015/591349

Gomez-Escribano, J.P., Bibb, M.J., 2014. Heterologous expression of natural product biosynthetic gene clusters in *Streptomyces coelicolor*: from genome mining to manipulation of biosynthetic pathways. J. Ind. Microbiol. Biotechnol. 41, 425–431.

Hosaka, T., Ohnishi-Kameyama, M., Muramatsu, H., Murakami, K., Tsurumi, Y., Kodani, S., Yoshida, M., Fujie, A., Ochi, K., 2009. Antibacterial discovery in actinomycetes strains with mutations in RNA polymerase or ribosomal protein S12. Nat. Biotechnol. 27, 462–464.

Ikeda, H., Ishikawa, J., Hanamoto, A., Shinose, M., Kikuchi, H., Shiba, T., Sakaki, Y., Hattori, M., Omura, S., 2003. Complete genome sequence and comparative analysis of the industrial microorganism *Streptomyces avermitilis*. Nat. Biotechnol. 21, 526–531.

Imai, Y., Sato, S., Tanaka, Y., Ochi, K., Hosaka, T., 2015. Lincomycin at subinhibitory concentrations potentiates secondary metabolite production by *Streptomyces* spp. Appl. Environ. Microbiol. 81, 3869–3879.

Kim, E., Moore, B.S., Yoon, Y.J., 2015. Reinvigorating natural product combinatorial biosynthesis with synthetic biology. Nat. Chem. Biol. 11, 649–659.

Komatsu, M., Uchiyama, T., Omura, S., Cane, D.E., Ikeda, H., 2010. Genome-minimized *Streptomyces* host for the heterologous expression of secondary metabolism. Proc. Natl. Acad. Sci. USA 107, 2646–2651.

Laureti, L., Song, L., Huang, S., Corre, C., Leblond, P., Challis, G.L., Aigle, B., 2011. Identification of a bioactive 51-membered macrolide complex by activation of a silent polyketide synthase in *Streptomyces ambofaciens*. Proc. Natl. Acad. Sci. USA 108, 6258–6263.

Marmann, A., Aly, A.H., Lin, W., Wang, B., Proksch, P., 2014. Co-cultivation–a powerful emerging tool for enhancing the chemical diversity of microorganisms. Marine Drugs 12, 1043–1065.

Müller, R., Wink, J., 2014. Future potential for anti-infectives from bacteria- how to exploit biodiversity and genomic potential. Int. J. Med. Microbiol. 304, 3–13.

Netzker, T., Fischer, J., Weber, J., Mattern, D.J., Konig, C.C., Valiante, V., Schroeckh, V., Brakhage, A.A., 2015. Microbial communication leading to the activation of silent fungal secondary metabolite gene clusters. Front. Microbiol. 6, 299.

Ochi, K., Hosaka, T., 2013. New strategies for drug discovery: activation of silent or weakly expressed microbial gene clusters. Appl. Microbiol. Biotechnol. 97, 87–98.

Ochi, K., Tanaka, Y., Tojo, S., 2014. Activating the expression of bacterial cryptic genes by *rpoB* mutations in RNA polymerase or by rare earth elements. J. Ind. Microbiol. Biotechnol. 41, 403–414.

Ohnishi, Y., Ishikawa, J., Hara, H., Suzuki, H., Ikenoya, M., Ikeda, H., Yamashita, A., Hattori, M., Horinouchi, S., 2008. Genome sequence of the streptomycin-producing microorganism *Streptomyces griseus* IFO 13350. J. Bacteriol. 190, 4050–4060.

Olano, C., Lombo, F., Mendez, C., Salas, J.A., 2008. Improving production of bioactive secondary metabolites in actinomycetes by metabolic engineering. Metab. Eng. 10, 281–292.

Oliynyk, M., Samborskyy, M., Lester, J.B., Mironenko, T., Scott, N., Dickens, S., Haydock, S.F., Leadlay, P.F., 2007. Complete genome sequence of the erythromycin-producing bacterium *Saccharopolyspora erythraea* NRRL23338. Nat. Biotechnol. 25, 447–453.

Onaka, H., Mori, Y., Igarashi, Y., Furumai, T., 2011. Mycolic acid-containing bacteria induce natural-product biosynthesis in *Streptomyces* species. Appl. Environ. Microbiol. 77, 400–406.

Rebets, Y., Brotz, E., Tokovenko, B., Luzhetskyy, A., 2014. Actinomycetes biosynthetic potential: how to bridge in silico and in vivo? J. Ind. Microbiol. Biotechnol. 41, 387–402.

Reen, F.J., Romano, S., Dobson, A.D.W., O'Gara, F., 2015. The sound of silence: activating silent biosynthetic gene clusters in marine microorganisms. Marine Drugs 13, 4754–4783.

Seyedsayamdost, M.R., 2014. High-throughput platform for the discovery of elicitors of silent bacterial gene clusters. Proc. Natl. Acad. Sci. USA 111, 7266–7271.

Shank, E.A., Kolter, R., 2009. New developments in microbial interspecies signaling. Curr. Opin. Microbiol. 12, 205–214.

Tala, A., Wang, G., Zemanova, M., Okamoto, S., Ochi, K., Alifano, P., 2009. Activation of dormant bacterial genes by *Nonomuraea* sp. strain ATCC 39727 mutant-type RNA polymerase. J. Bacteriol. 191, 805–814.

Tanaka, Y., Kasahara, K., Hirose, Y., Murakami, K., Kugimiya, R., Ochi, K., 2013. Activation and products of the cryptic secondary metabolite biosynthetic gene clusters by rifampin resistance (*rpoB*) mutations in actinomycetes. J. Bacteriol. 195, 2959–2970.

Weber, T., Charusanti, P., Musiol-Kroll, E.M., Jiang, X., Tong, Y., Kim, H.U., Lee, S.Y., 2015. Metabolic engineering of antibiotic factories: new tools for antibiotic production in actinomycetes. Trends Biotechnol. 33, 15–26.

Yaegashi, J., Oakley, B.R., Wang, C.C.C., 2014. Recent advances in genome mining of secondary metabolite biosynthetic gene clusters and the development of heterologous expression systems in *Aspergillus nidulans*. J. Ind. Microbiol. Biotechnol. 41, 433–442.

Zhu, H., Sandiford, S.K., van Wezel, G.P., 2014. Triggers and cues that activate antibiotic production by actinomycetes. J. Ind. Microbiol. Biotechnol. 41, 371–386.

Chapter 10

The Nagoya Protocol Applied to Microbial Genetic Resources

Philippe Desmeth
Belgian Science Policy Office, Brussels, Belgium

THE CONVENTION ON BIOLOGICAL DIVERSITY

The Convention on Biological Diversity entered into force on 29 December 1993, with three objectives: (1) the conservation of biological diversity, (2) the sustainable use of the components of biological diversity, and (3) the fair and equitable sharing of the benefits arising out of the utilization of genetic resources. The implementation of the third objective is ruled by the Nagoya Protocol (NP) on Access and Benefit Sharing (ABS) http://www.cbd.int/abs/ which entered into force on 12 October 2014. This chapter highlights the need for sector specific regulations adapted to the technical specificities of microorganisms and to the legal, ethical, and social impacts of their uses. It shows the role of CCs in the effective implementation of the "ABS" *quid pro quo*: benefit sharing with providers in return for facilitated access for users.

Fifty per cent of the living biomass on the planet is said to be microbial nature and the microorganisms have the potential to become detrimental or beneficial such as providing solutions in agriculture, industry, plant, animal and human health and several other biotechnological applications. The empirical use of microorganisms existed for millennia, among others for brewing and preservation of food and is the basis of traditional knowledge source of vivid cultural traditions and fundamentals for craft industries. But the systematic study of the microbial realm suspected to exist in the 17th century when Antoni van Leeuwenhoek opened a window on this unseen world, has really started in the late 19th century with Pasteur, Cohn, Hansen and their contemporaries. It was rapidly understood that sustainable use and control of beneficial or detrimental microorganisms are possible if secure access to microbial raw material is effective. The cultivation of microorganisms in the laboratories became a routine and nowadays it is a necessity to sustain biotech developments. The conceiving of specialized infrastructures dedicated to the long-term preservation of microbial

Microbial Resources. http://dx.doi.org/10.1016/B978-0-12-804765-1.00010-2

raw material evolved in a similar way, from *ad hoc* structure to high-tech dedicated specialized infrastructure.

CULTURE COLLECTIONS: PROFESSIONALS UNDERPINNING MICROBIAL REALM EXPLOITATION

Comprehensive exploration and structured study of the microbial diversity implies access to huge numbers of specimens. "Type strains" are the archetypes of a species and constitute the primary elements of taxonomy. "Reference strains" are representative of a lineage with well identified properties; they are essential parts of coherent R&D processes (Federhen et al., 2016). These assets of fundamental scientific importance must be conserved and provided with the highest level of reliability to ensure cumulative research, to build microbiology on solid ground. This requires that scientists routinely deposit significant strains in CCs on the one hand, and on the other a worldwide coordinated effort of the CCs to optimize the long-term preservation of their holdings (Stackebrandt, 2010).

CCs have decades of expertise in long-term ex situ conservation of living microbial material and related electronic data processing. These specialized infrastructures provide for facilitated access to technically and legally fit for use microbiological resources of consistent quality. CCs have evolved from mere centers of conservation and distribution of microbiological material to "Biological Resources Centres (BRC) (http://www.oecd.org/dataoecd/26/19/31685725.pdf, http://www.oecd.org/dataoecd/6/27/38778261.pdf, http://www.wfcc.info/guidelines/)," conceived as basic infrastructures, sources of all essentials for Research and Innovation in "Knowledge Base Bio-Economy". KBBE can be defined as "transforming life sciences knowledge into new, sustainable, ecoefficient, and competitive products." "Knowledge based" refers to the increasing amount of data on biological material produced as research outputs, and processed by analytical tools. The term "bio-economy" includes all industries and economic sectors that produce, manage, and exploit biological resources (agriculture, food, pharmaceutical, cosmetic, and other bio-based industries). Advanced biotechnology is breaking new ground in understanding microbial diversity and bioprocesses that could lead to valuable bioproducts and biomaterials. See *New Perspectives on the Knowledge-Based Bio-Economy*, Conference Report, European Commission, Brussels, 2005 (https://ec.europa.eu/research/conferences/2005/kbb/pdf/kbbe_conferencereport.pdf).

The concept of BRC was thought up as early as 1946, at UNESCO, on the set up of the MIRCEN—*Microbial Resources Centres Network*—program. In 1999, the *Organization for Economic Co-operation and Development* (OECD) expert Task Force on BRC initiated the development of the concept into the 21st century, pointing out the crucial roles of BRCs for human life and the biosphere, underlining the necessity to provide the adequate support to enable the BRCs to meet the increasing challenges of biodiversity and genomics. While the emphasis was previously put on the biological resources conserved in specialized facilities, at present a BRC is conceived as a functional unit having all the necessary components to study, preserve, and use biological diversity. It integrates appropriate infrastructure, human, financial and technical resources, skills related to information production, processing and diffusion as well as legal, administrative, management, and quality control systems. Since 2013, the Technical Committee 276 of the International Organization for Standardization (ISO) calls CCs "Microbial Biobanks," by which it recognizes the increasing role of collections as key socioeconomic actors in biotechnology.

This evolution of the socioeconomic role of collections goes with technical and scientific change. In the 19th century and until the middle of the 20th century, the exploration of the microbial realm was mainly carried out in wet laboratory, based on growing pure cultures of microorganisms in laboratory. The development of bioinformatics and the inclusion of a more complex reality, apprehending the microbiomes, have revolutionized microbiology where digital tools play a central role. The microbial research infrastructures, including the microbial Biobanks, have to integrate this new research paradigm and growing institutional and legal constraints into their business plan to meet the societal demands.

SPECIFICITIES OF MICROORGANISMS, OWNERSHIP OF MICROBIOLOGICAL MATERIAL

On the entry into force of the CBD, when providers of biological resources started to question the apportionment of profit generated by exploitation of biological resources, the issue of ownership of biological diversity was brought in the foreground, with divergent opinion about who are the holders of biological resources (Dedeurwaerdere et al., 2007). This legitimate question was unfortunately raised in a confused manner, mixing tangible and intangible ownership. Such misty atmosphere creates an unfavorable context for biodiversity exploitation but has the merit to urge for further clarity, if necessary through innovative approach. In a context where ownership is not squarely defined by law, providers of biological material are reluctant to distribute biological items, to loose the tie with the risk of losing unforeseen profit. Users of biological material are hesitant to invest considerable sums in bioprospecting, not knowing if their gathering will be usable legally, the results of lengthy and expensive R&D jeopardized by late claiming. Intermediaries and facilitators like CCs are caught between these two ends, and look for answers that facilitate access to and sustainable use of biological resources.

First, one must kill a myth: the recognition in the CBD of "the sovereign rights of states over their natural resources" (CBD Art 15.1) does not mean per se that a State owns genetic resources. It merely says that each State governs its biological resources according to national laws. The State decides on the regime of tangible ownership applicable to its natural resources. If the State is one of the 163 members (as of July 2016) of the World Trade Organization, it will enforce the intangible rights according to the TRIPS agreement. That is for the general legal framework.

Then, one may examine the ownership issue from the legal or natural person perspective, with innovative thinking. Property is not a monolithic right, it has various components—tangible and intangible that can be owned by only one exclusive, or shared nonexclusive. In short, there is a "bundle of rights" attached to biological resources which must be regulated by laws and managed through agreement and contracts. The present concept of ownership seems ill adapted to the requirements of the knowledge economy. The question: "Who owns what?" must be modified in: "Who has what rights?" on microbiological resources.

And yes, CCs have rights on the material they manage and preserve. Having these rights, they also can specify the conditions under which they distribute their material. So ownership, appropriately understood, allows the CCs community to define when biological material should be shared on a nonexclusive basis or, conversely, when a restrictive licensing policy is justified. In that way, the "microbial commons" on the one hand (shared use) and the "Intellectual Property Rights" on the other (exclusive use) are not to be opposed, but are two complementary tools at the disposition of the CCs community.

Well-defined property rights play a key role in enhancing the economic innovation and the provision of services of general interest. Their most important contribution is their stimulation of long-term investment in the continuous renewal and adaptation of the institutional rules, adapting them to new technologies, external checks, and internal dynamics in order to sustain indefinite growth.

However, well-defined property rights do not necessarily imply full ownership or a fortiori private ownership. For example, well-defined rights to a good, such as natural resource can include exclusion and management rights being attributed to a private organization while the resource itself remains in state ownership. In a similar way, data sharing through a data portal can imply the exercise of management and exclusion rights by an organization without full ownership of the original database necessarily being transferred to this entity.

More precisely ownership can be analyzed as a "bundle" of use and decision rights that are attributed to a number of economic agents. This model of "bundle" can be applied to the networks of BRCs for the distribution of data and resources based on common management and exclusion rights. This concept of "bundle of rights" is a dynamic model of ownership management moving away from the static concept of ownership toward a flexible allotment of rights.

Ownership constitutes a "bundle" of use and decision rights that are attributed to a number of stakeholders/economic agents. It is a set of operational and collective choice rights defining, respectively who decides upon the use that one can make of a resource, and who decides upon the future exercise of the rights on the resource. Such scheme allows multiownership of a gradual level of use and decision rights. These rights can begin with basic access rights, encompassing research delivering outputs to the public domain, distribution on to third parties, exploitation rights to develop intellectual property, and its ownership which may include reach through rights.

Full ownership implies the full bundle of rights. It thus includes exclusive property (exclusive use and decision rights) and the right to sell or lease this exclusive property and all the other rights to another party. This concept of full ownership has important but limited application to microbiological resources. Indeed, it only applies poorly at the level of the biological resources itself—for instance, the same material may be owned by different CCs—and in specific case only can be applied at the level of the associated knowledge, should there be a nonobvious innovation with direct use value that satisfies the conditions for patentability. Moreover, innovation in the life sciences is characterized by a diffuse process of identification, purification, and transformation of the microbiological resources and by a distributed and accumulative process of generation of information. Forms of nonexclusive property, such as the sharing of resources among public and/or private research institutions or collaborative databases are thus common in the intermediary stages of the innovation process.

The application of the "bundle of rights" makes possible the enforcement of the "sovereign rights of States over their natural resources" without prejudice to private rights. Unambiguous allotment of rights in advance will facilitate rightful benefit sharing "at the end of the pipe (Dedeurwaerdere, 2005).

CONTINUUM IN LIFE SCIENCES R&D—EXTENDED ROLE OF CULTURE COLLECTIONS

In microbiology like in other disciplines, the pattern of complementary distinctive basic and applied sciences is replaced by a continuum of upstream and downstream researches producing flows of data and information handled via bioinformatics (Stokes, 1997). In "Knowledge Based Bio-Economy," the various players in this R&D flow are closely tangled in an interdisciplinary fabric where CCs are providers of raw material as well as infrastructures underpinning scientific activities.

Optimal collaboration between the providers and the users of raw material is necessary to produce valuable—exploitable—research outcomes. It requires also appropriate policies regarding Intellectual Property Rights management as required by the TRIPS Agreement (Agreement on Trade-Related Aspects of Intellectual Property Rights. The TRIPS Agreement is Annex 1C of the Marrakesh Agreement Establishing the World Trade Organization, signed in Marrakesh, Morocco on 15 April 1994. See World Trade Organisation website http://www.wto.org) and material distribution policies in accordance with the provisions of the Nagoya Protocol under the Convention on Biological Diversity. That requires mastering the public and proprietary laws interface, as well as expertise in managing tangible and intangible rights related to microorganisms and related data. The role of CCs is not anymore limited to providing for microbial material and related data. Now CCs are expected to provide also for legal and administrative services, even advising lawmakers and decision makers on research policy. But while the roles and duties of CCs extend the funding does not.

BUILDING A COMMUNITY OF CONNECTED CENTERS OF EXPERTISE

CCs must set a strategy to fulfil their role under the evolving biosciences paradigm, in an era of fewer funds and more duties. One way to meet the challenges is to combine their strength into structured and more or less integrated networks, at national, regional, or international level. Such functional networking requires adapted Information and Communication Technology capacities to optimize collaboration, minimum quality standards, and technical level to facilitate scientific cooperation.

CCs are established all around the world, many are members of the World Federation for Culture Collections (WFCC) http://www.wfcc.info. Created in 1970, WFCC has developed a pioneering databases system managed by the World Data Centre for Microorganisms (WDCM) http://www.wdcm.org. The WFCC registers its members through a unique acronym and numerical identifier. There are currently more than 700 CCs in more than 70 countries registered in the official WFCC directory of collections CCINFO http://www.wfcc.info/ccinfo/home/

The CCINFO database records data on the organization, management, services, and fields of expertise of the collections. It is one of four major complementary, interconnected tools forming the information system of WDCM.

LIFE SCIENCES PRACTITIONERS IN THE ARENA OF POLITICAL NEGOTIATIONS

It took 16 years from the start of the negotiations initiated at COP 4 in 1998 to the entry into force of the Nagoya Protocol in October 2014. The Protocol on Access and Benefit Sharing was adopted in October 2010 in Nagoya when negotiators confronted with the risk of political silting reached compromise on a text finalized in urgency.

The slow pace of the process and its abrupt outcome can be explained by the subjective negotiating climate and the lack of preexisting legislation and expertise in the field of ABS. On the one hand, some problems have been exaggerated and oversimplified. Although real, facts were so distorted that they betrayed the field reality. For instance, the word "biopiracy" (refers to commercial exploitation or monopolization of biological or genetic material, without compensating the indigenous peoples or countries from which the material or relevant knowledge is obtained) was created to depict the stereotype of the exploitation of poor providers of biological material in southern nonindustrial countries by greedy users working for wealthy companies in opulent countries. Confirmed or alleged cases of biopiracy were reported by media and spread by lobby groups to raise the issue of equitable sharing of the fruits of economic growth. Examples of leonine contracts about the exploitation of biological resources by wealthy industries, and cases of stolen knowledge and raw material used by greedy companies exist but are over mediatized. These exceptions are presented as common practices. Such Manichean view was not conducive to smooth and fruitful negotiations anchored in facts.

On the other hand, issues of primary importance were neglected in the Nagoya Protocol and must be dealt with afterward, via pragmatic, applicable implementing regulation at regional and national levels.

First, the complexity of the R&D processes leading from microbiodiversity to socioeconomic outcomes is underestimated. This oversimplified technical apprehension of the scientific approach produced legal solutions inappropriate to field complexity. Then, the cost-benefit and return on investment ratio of an ABS system was overlooked, where the measures may cost more than the benefits it intends to redistribute. Cost-benefit analysis in life sciences is difficult for it has to take all parameters, the multiple actors, multiple factors aspects as well as the duration of research programmes into account. Also, key concepts, such as commercialization, placing on the market, R&D are not defined non-equivocally from the upset, that doesn't help reduce legal uncertainty.

LACK OF AWARENESS, IGNORANCE, MISTRUST, DISDAIN

Notwithstanding extended studies in legal and social sciences and the big media fuzz about ABS, this concept is still largely ignored by microbiologists because of a lack of information and a lack of awareness.

From the life sciences practitioner's perspective, several aspects of the debate concerning the fair and equitable sharing of the benefits seem not particularly relevant. Some case studies are not representative of the reality in the field. Certain problems are exacerbated while others are neglected.

The misperception of the situation by the lawmakers and inappropriate expertise input during and around the negotiations led to inappropriate solutions that are seen as not relevant and thus rejected because not credible by actors in the field.

The actors in life sciences and the experts of human sciences are equally at fault. The first for having ignored the ABS concept and the Nagoya Protocol, because it seemed not of their concern; the latter for having oversimplified the complexity of apprehending the biological diversity, in particular, the complexity of research processes and the concept of ownership impractical to microbial diversity. Ignorance, misperception of the reality and more importantly lack of communication between disciplines tainted the climate of negotiation with mutual mistrust at least, disdain at worst. Yet, even imperfect, the Nagoya Protocol is a valuable legal framework in a new field of international law. As for the CBD, the NP set principles that have to be converted into feasible regulation at national and/or regional levels.

Common sense and the need for workable solutions to keep the cost/benefit ratio beneficial and the administrative workload manageable recognize the usefulness of best practices and guidelines or standards. These are complementary ways to set the standards, set the regulations and finally enable *bona fide* stakeholders to abide by the law at affordable costs. The European Regulation 511/2014 (Regulation (EU) No 511/2014 of the European Parliament and of the Council of 16 April 2014 on compliance measures for users from the Nagoya Protocol on Access to Genetic Resources and the Fair and Equitable Sharing of Benefits Arising from their Utilization in the Union. And its implementing act the Commission Implementing Regulation (EU) 2015/1866 of 13 October 2015 laying down detailed rules for the implementation of Regulation (EU) No 511/2014 of the European Parliament and of the Council as regards the register of collections, monitoring user compliance and best practices), is a good example of legal framework matching a complex and multifaceted reality with feasible solutions. It creates possible solutions in an imperfect system. The regulation sets up the concept of registered collections of biological resources where genetic resources can be accessed with an increased level of legal certainty. It also organizes an ABS Consultation Forum where pragmatism prevails and stakeholders are brought to compromise.

THE INITIATIVES OF THE CULTURE COLLECTIONS RELATED TO ABS AND THE NAGOYA PROTOCOL

On the contrary of general microbiologists, CCs were aware of the matters at stake from the onset. At this point, it is important to highlight paragraph 2 of CBD article 15 as the key stake in the Access and Benefit Sharing process: facilitated access to genetic resources as the prerequisite for any advancement in life sciences.

The proactive contribution of CCs to the Convention on Biological diversity and the Nagoya Protocol dated back from September 1997 when the Belgian Coordinated Collections of Microorganisms (BCCM) launched the concerted action MOSAICC with the support of the Directorate General for Science, Research, and Development of the European Commission.

MOSAICC (Concerted action # BIO4-CT97-2206 (DGXII - SSMI). http://bccm.belspo.be/projects/mosaicc) stands for «Micro-Organisms Sustainable use and Access regulation International Code of Conduct ». It is a voluntary Code of Conduct focused on the implementation in microbiology of the Convention on Biological Diversity (CBD) and other international applicable rules. Among others the Budapest Treaty on the International Recognition of the Deposit of Microorganisms for the Purposes of Patent Procedure (28 April, 1977, amended on 26 September, 1980 and Regulations) and the Agreement on Trade-Related aspects of Intellectual Property Rights (TRIPS Agreement, Marrakech, 15 April, 1994). See also CBD article 22 ruling the relationship with other international conventions. MOSAICC was first issued in spring ’99, two years before the Bonn Guidelines (see Bonn Guidelines on Access to Genetic Resources and Fair and Equitable Sharing of the Benefits Arising out of their Utilization (see Convention on Biological Diversity – Conference of Parties 6 Decision VI/24. http://www.cbd.int/decision/cop/?id=7198).

MOSAICS http://bccm.belspo.be/projects/mosaics followed and focused on three needs previously identified by MOSAICC: an estimate of the economic value of microbiological resources, models of documents to be used in validated procedures and a system to manage and convey the deposit in CCs and the subsequent distribution by CCs of microbiological resources. MOSAICS stands for Micro-Organisms Sustainable use and Access management Integrated Conveyance System.

During these projects, the "bundle of rights" concept was adapted to the biological diversity and the design of "microbial commons" (this development is complementary to national regulations on ABS and to existing IPR laws, as it constitutes a demarcated space where material and information are relatively freely accessible provided that the outputs are injected back into this open space, to be shared again. Inside this space access and benefit-sharing are "commonly shared." Outside this demarcated space, access and benefit-sharing will be ruled through ordinary national and international laws, including IPR and specific CBD inspired regulations. A practical development similar to this model is the NIEMA system. See http://www.thecommonsjournal.org/index.php/ijc/article/view/215/144) for the exchange of (micro) biological material provided for basic common use principles for access to both material and information. These outcomes developed already before the Nagoya Protocol have been reviewed and updated to match the latest legal developments in Access and Benefit Sharing.

MOSAICS and MOSAICC are the forerunners of TRUST, the latest initiative of WFCC collections, a global system to implement ABS in microbiology, in an efficient way and at affordable cost.

WDCM, FROM PIONEERING TO BREAKTHROUGH, FROM CCINFO TO GCM

Created on the onset of WFCC, the World Data Centre for Microorganisms (WDCM) fulfils several roles. The WDCM website is a communication tool and the data hub of the CCs community. Its CCINFO directory provides information about the collections. The collections' catalogs display their holdings. WDCM becomes a "one-stop-shop" portal where scientists can find information about microbial resources. In addition, WDCM proposes training and software to help WFCC members to catalog their microbiological resources.

To create a catalog of microorganisms, taxonomists define the Minimum Data Set (MDS) necessary for accurate identification of microbiological material while bioinformatics professionals structure the database of microbiological material. The database is published as online or printed catalog.

Combining the CC's acronym recorded in CCINFO with the numbering of every strain in the database's catalog creates a code called strain number. Combining this WDCM code system with electronic markers called "Globally Unique Identifiers" (GUIDs) http://bccm.belspo.be/documents/files/projects/mosaics/ics_report.pdf links every strains to potentially all kinds of data stored in various databases in different institutions: scientific, technical, administrative, legal, etc., for any kind of use: research, conveyance, resources conservation, resources exploitation, etc. The catalogs entries of CCs are thus the interfaces between wet lab products and the dry lab data. In effect, once a microorganism is deposited in a WFCC member collection and is assigned a number in the online catalog it can be traced in all online scientific and technical literature, including patent files.

WDCM manages the ground breaking Global Catalog of Microorganisms (GCM) Global catalog of microorganisms (GCM): a comprehensive database and information retrieval, analysis, and visualization system for microbial resources; Wu et al. (2013) which combines as many online catalogs as possible, creating a robust system connecting the strain catalogs information with all kinds of data, such as nucleic acids sequences, protein, reference, citations, etc. and making the information accessible through one portal. It acts as information broker between the online catalog entries of the CCs and the multiple users.

GCM can play a central role in the implementation of ABS in microbiology as it also may convey the transfers of microbiological items, helping track the flow of resources, for the purpose of the access and benefit sharing deal.

The system can potentially retrieve all kinds of information about microbiological resources, including information related to the possession, location, transfers, and use of microbial strains, including country of origin, existence of official documents such as PIC permit and contractual terms MAT, the creation of derived patents and all associated scientific publications. The flow of ABS related information generated by GCM will be connected to the ABS Clearing House Mechanism via machine-processed link.

Beginning 2017 GCM already involved 110collections from more than 40 countries and information on more than 385,000 strains from nearly 50,000 species. That makes GCM a transparent, safe, and sustainable data management system of ex situ microbial diversity worldwide. With GCM, WDCM strengthens the bridging role of WFCC between providers and users of microbiological resources. GCM is a powerful scientific tool as well as a way to build safe, ethical and socioeconomically balanced ABS processes at global level. Systems like GCM are automated and thus more cost effective; yet nonnegligible amount of investments are and will be necessary to manage the flow of data generated by ABS requirements.

GAINING TRUST, BUILDING TRUST

Because trust is a prerequisite for lasting cooperation in science, prominent CCs have decided to build TRUST, literally and practically. TRUST stands for TRansparent User-friendly System of Transfer, for Science & Technology.

TRUST is an example of sector specific ABS best practices. Yet it is more than a set of best practices since it provides also for data management of ABS related information.

It is a cost-efficient, simple, and fast system that can be used by multiple users for various tasks of R&D management as well as for administrative purposes. It combines the Code of Conduct MOSAICC and the MOSAICS Integrated Conveyance System for the best practices part with the Global Catalog of Microorganisms for the technical part. This construct constitutes the general management system facilitating access to microbial genetic resources stored in CCs.

It is an initiative launched on 6 December, 2012, at the occasion of the 10th Anniversary of NITE BRC Symposium focusing on "Addressing Public function of BRC and the Nagoya Protocol on ABS." Following the suggestion of the WFCC President, representatives of major collections and some associations (include BCCM, Belgian Coordinated Collections of Microorganisms; BIOTEC, Thailand National Center for Genetic Engineering and Biotechnology; CBS, Netherlands Centraalbureau Schimmelcultures; JBA, Japan Bio-industry Association; KCTC, Korean Collection for Type Cultures; NBRC, Japan NITE Biological Resource Center; UNU, United Nations University; WDCM, World Data Centre for Microorganisms, NHML, Natural History Museum London; WFCC, World Federation for Culture Collections) decided to revisit MOSAICC to answer efficiently the technical and legal challenges posed by the Nagoya Protocol. It aims at mitigating the impact of the CBD and the Nagoya Protocol by limiting the additional work on the scientific, technical, and administrative activities of CCs, incorporating at less cost the legal obligations into the daily life of microbiologists.

The ABS *quid pro quo* applied to microbiology may be expressed as such: on the one hand access to microbial genetic resources is a prerequisite for the advancement of microbiology; on the other hand, monitoring the transfer of MGRs is necessary to identify the individuals or groups that are entitled to be scientifically or financially rewarded for their contribution to the conservation and sustainable use of the microbial genetic resources.

The purpose of TRUST is to facilitate accesses and transfers as well as to monitor the scientific outcomes and their authors, contributing to the appropriate sharing of benefits as covered by the Nagoya Protocol.

As cited in regulation EU 511/2014 on ABS, the general principle that prevails is the obligation for all users to "exercise due diligence to ascertain whether genetic resources [...] have been accessed in accordance with applicable legal [...] and to ensure that, [...], benefits are fairly and equitably shared." To that end, all users will seek, keep, and transmit information relevant for access and benefit-sharing. TRUST proposes to manage the due diligence data, basically information about *what*-what is accessed, *when*-when is it accessed, *where*-where it is accessed, and *by whom*. In addition, transfers are monitored, to answer the question *to whom* are the resources transferred. A system based on trustworthy accurate information about these five questions increases transparency.

Along the timeline of an ABS process there are three critical events:

1. At the start, the in situ origin of a sample is identified via a procedure of notification and/or authorization for sampling called Prior Informed Consent-PIC in CBD's jargon. The in situ origin of the microorganism is recorded then.

   In TRUST the information collected in the GCM will be connected to the ABS Clearing House (ABSCH) via a cost-effective machine-processed link. The ABSCH is the official international information hub of the secretariat of the Convention on Biological Diversity for ABS issues.
2. When the microorganism is deposited in a CC, it receives a catalog entry number, a Global Unique Identifier (GUID). This code is kept throughout transfers and connected to all relevant scientific or administrative data.
   The Deposit of a microorganism into a CC is covered by a Material Accession Agreement (MAA) which on the one hand records basic data such as place and date of sampling, etc., and on the other hand specifies the rights and duties of depositor and collection. Data compiled in catalogs are usually publicly accessible, at least the Minimum Data Set.
3. The transfers of a microorganism are recorded in CCs' confidential customer databases and occur under Material Transfer Agreement (MTA) which terms are mutually agreed by recipient and provider. MTA is a generic term that covers very short shipment document, simple standard delivery notice, standard invoice containing minimal standard requirements, or more detailed specific contract including tailor-made mutually agreed terms. According to the use and intended distribution of the MGRs, mutually agreed terms ("mutually agreed terms" is considered here as a general expression defining terms of a contract, for instance terms like these made under private laws. It is not limited to the definition of articles 6 and 7 of the Nagoya Protocol in the context of PIC deliverance) can be short or very detailed.

In summary, origin of samples are recorded in both the ABSCH, the official database of the CBD, and in catalogs of CCs accepting the microorganisms isolated from samples. The catalogs' data are compiled into GCM. The GCM data are connected to all scientific and administrative data. These data show the transfers, studies, and uses made for every microorganism.

The system underpins cumulative research by linking scientific publication, microbial material, and data via referencing of microbial strains. When looking to apportion the benefit arising out of the use of microbiological genetic resources, one will go up the electronic data chain, from the end of the pipe result up to the sample(s) of origin. Every piece of added value and socioeconomic profit out of microbiological material can be traced back to its original provider.

CC may provide this service to microbiologists, to country officials, and other providers and users. Yet this cannot be achieved without cooperation with all stakeholders, particularly the publishers. These are part of the equation of referencing. Not only referencing articles to authors, but also linking the biological material cited in publication indirectly to the legal authorization by requesting from the authors the CC's reference of the material described in the publication. Doing so will close the circle, as the best way to achieve ABS with effective socioeconomic benefits is to build on existing procedures, to make the appropriate linkages between the various actors. Common sense tells us that providing for appropriate ICT tools to make life easier for all stakeholders is more effective than setting up coercive measures or penalties to impose additional regulation. Also scientific excellence should be properly rewarded in terms of financial sustainability of research groups. Excessive regulation may lead to situation where the most advanced scientific team is not equally financially successful, when the strongest legal team prevails. To be competitive, public and private research infrastructures need to combine efforts in three directions: a strong scientific strategy supported by progressive legal policy and sustained investment.

REFERENCES

Dedeurwaerdere, T., 2005. Understanding ownership in the knowledge economy: the concept of the bundle of rights. BCCM News Edition 18—Autumn 2005.

Dedeurwaerdere, T., Desmeth, P., Green, P., 2007. Microbiological resources, who owns what? Building the microbial commons. Communication ICCC11, Goslar, Germany.

Federhen, S., Rossello-Mora, R., Klenk, H.P., Tindall, B.J., Konstantinidis, K.T., Whitman, W.B., Brown, D., Labeda, D., Ussery, D., Garrity, G.M., Colwell, R.R., 2016. Meeting report: GenBank microbial genomic taxonomy workshop (12-13 May, 2015). Stand. Genomic Sci. 11 (1), 1, http://standardsingenomics.biomedcentral.com/articles/10.1186/s40793-016-0134-1.

Stackebrandt, E., 2010. Diversification and focusing: strategies of microbial culture collections. Trends Microbiol. 18, 283–287.

Stokes, D.E., 1997. Pasteur's Quadrant: Basic Science and Technological Innovation. The Brookings Institution, Washington, USA.

Wu et al., 2013. Global catalogue of microorganisms (GCM): a comprehensive database and information retrieval, analysis, and visualization system for microbial resources. BMC Genomics 14, 933. Available from: http://www.biomedcentral.com/1471-2164/14/933 and http://gcm.wfcc.info/

Chapter 11

Fungal Genetic Resources for Biotechnology

Kevin McCluskey
Kansas State University, Manhattan, KS, United States

INTRODUCTION

Living microbial collections of many types support the development of therapeutics for human, animal, and plant health management, the progress of food and fiber production, and a myriad of industrial and chemical processes. Among these some collections emphasize the diversity of microorganisms in the environment, or in a specific geographic or politically defined region. Other collections emphasize the diversity of genetics of a few well studied organisms, such as the Fungal Genetics Stock Center, or the *E. coli* Stock Center. Each has much to offer and the complementarity of these collections is essential to the orderly progress of the scientific endeavor. Fungal collections and collections holding fungi are well represented among the World Federation for Culture Collections membership (http://www.wdcm.org/). Collections holding many mutants of few species are fewer in number, even when one includes genetic collections of bacteria and eukaryotic microorganisms.

Genetic Collection Holdings

Living microbe collections typically emphasize the biological diversity of one ecological region, sovereign country, or taxonomic category. With the exception of taxonomy, genetic collections are not typically circumscribed by the same boundaries. Indeed, genetic collections are typically tied to the community of researchers who utilize the specific research organism in the collection. Key among these repositories holding characterized genetic mutants and related strains are the collections historically supported by the US National Science Foundation Living Stock Collection program (Table 11.1).

Genetic resource collections are similar to type culture collections by virtue of their shared mission of preserving and distributing microbial resources for

Microbial Resources. http://dx.doi.org/10.1016/B978-0-12-804765-1.00011-4

TABLE 11.1 Fungi with Genetic Strains in the FGSC Collection

Fungi	Number of loci/markers[a]	Status of FGSC mutant collection	References
Aspergillus	1,769	Active	Clutterbuck (1997)
Neurospora	2,000	Active	Perkins et al. (2001)
Schizophyllum	<100	Active	Frankel and Ellingboe (1977)
Fusarium	<100	Active	Jurgenson et al. (2002); Xu and Leslie (1996)
Magnaporthe	<100	Active/Archival	Skinner et al. (1993)
Ustilago	<100	Archival	Perkins (1949)
Allomyces	<100	Archival	Ingraham and Emerson (1954)

[a]*The number of loci is from classical genetic analyses and is 15%–20% of all loci predicted in genome sequences.*

research and commercial development. Unlike type culture collections, however, genetic resource collections, also called stock centers, typically house hundreds to thousands of strains of one species, or several closely related species. The Fungal Genetics Stock Center (FGSC, http://www.fgsc.net/) is the largest public repository of genetically characterized fungi and yeast and holds diverse and authoritative resources for genetics, cell biology, medicine, and agriculture (McCluskey, 2011).

With expanded holdings of mutants from the classical genetics era to gene deletion mutants from the postgenomics era, as well as reference strains from whole genome sequencing programs, diverse tools for working with fungi are provided (Fig. 11.1) and the FGSC holdings are utilized by researchers around the world. Many international repositories hold fungal resources, but these are utilized in different research areas of mycology compared to fungal genetics. For example, the NITE BRC reports that only two fungal strain are among the most widely requested to be used as reference strains for antifungal testing. Similarly, fungal resources at the Centraalbureau voor Schimmelcultures (CBS) collection emphasize taxonomic diversity, as do the holdings of many fungal collections. The NRRL collection maintained by the US Department of Agriculture holds a variety of biological materials including fungi from industrial practices, agriculture, and patent strains. Research resource repositories, tied to their historical research communities or host institutions, often grow from the project of one researcher or a community of researchers. Some of these become organized as distributing collections, usually in response to the pressure on the research program by the demands to distribute large numbers of strains or other materials. For example, the hybridoma collection at the University of

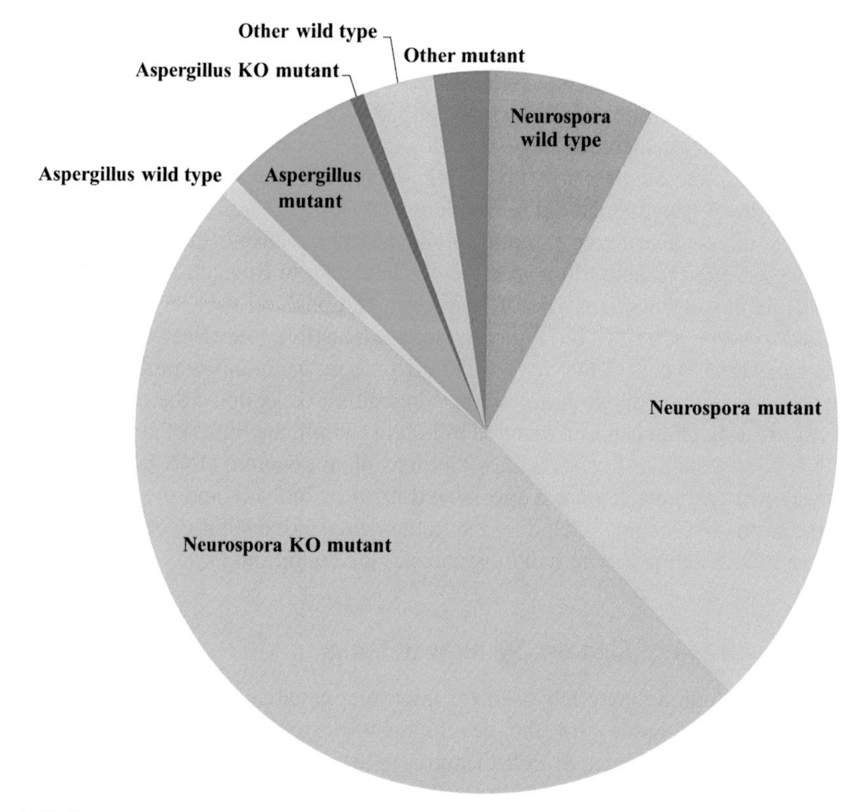

FIGURE 11.1 **Holdings of different types of strains in the FGSC collection.**

Iowa, (USA) which began as a small collection targeting developmental biology resources, is now a self-supporting resource with diverse reagents utilized in cell and developmental biology research (http://dshb.biology.uiowa.edu/). Similarly, the Center for Fungal Genetics Resources in Korea grew out of research programs in plant microbe interactions (Dean et al., 2005) and has benefitted greatly by the significant commitment to living microbe collections by the Korean National Research Resource Center.

Also, the Yeast Resource Center at the University of Washington (USA) provides limited resources but also gives access to expertise in six core technology areas including mass spectrometry, phenotyping, fluorescence microscopy, and computational biology. The collection holds plasmids and datasets for work with *Saccharomyces*.

Most strains in the FGSC collection are *Neurospora* strains with a majority being gene deletion and classic mutants. Smaller number of *Aspergillus* mutants and of *Neurospora* and *Aspergillus* wild type strains fill out the collection. Among the other wild type strains, reference strains from whole genome sequencing programs predominate.

Related Collections

Historical collections include herbaria and other natural history museums. The US National Fungus Collection (http://www.ars.usda.gov/Services/docs.htm?docid=9397) includes large mycological herbaria (Verkley et al., 2015). These specimens are primarily taxonomic in emphasis and attempts to use them as a source of genetic material for modern analyses has had limited impact both because of nondestructive sampling policies and because of harsh treatments for preservation of herbarium specimens (Gaudeul and Rouhan, 2013).

Biobanks are another type of related collections and they support a variety of research activities but typically hold exhaustible resources (*Cooperative Human Tissue Network* http://www.chtn.nci.nih.gov/). In this regard, microbial research resource collections are more like culture collections: the materials in a culture collection can be expanded infinitely (within the limits of practicality). DNA banks, such as the RBG Kew Gardens plant genomic DNA bank (http://apps.kew.org/dnabank/) are a specialized form of biobank and they share the same limits of biobanks including the fundamental exhaustibility of the materials and the vulnerability to handling and storage conditions.

Establishment of Genetic Systems in Fungi

Among the unique materials used for microbial genetics, mutant strains were the defining resources for the development of modern molecular genetic practices. From the first *E. coli* (Tatum, 1945) and *Neurospora* (Beadle and Tatum, 1941) mutants of the 1940s to the systematic gene deletion mutant programs of the postgenomics era (Baba et al., 2006; Colot et al., 2006) mutants in diverse traits have been useful in propelling genetic understanding from simple haploid microorganisms to complex diploid and polyploid eukaryotes. The first genetic traits that were evaluated was mating type for fungi and phage resistance for bacteria. This was rapidly followed by the production of biochemical, morphological, or sexual variants which were used in the construction of genetic maps for diverse organisms. Genetic mapping strains include strains with known characteristics in various combinations which allow rapid association of a newly identified phenotypic trait to a particular chromosome. Mapping strains also allow evaluation of allelism for mutations that may be in the same gene. For example, in *Neurospora* a variety of phenotypes including arginine auxotrophy and temperature sensitive lethality are found in different mutants at the *spe-1* locus (Pitkin and Davis, 1990; Pomraning et al., 2011). Interestingly, the most densely populated genetic maps based purely on phenotypic characteristics have allowed identification of only 15%–20% of genes (open reading frames) from genome sequence analysis. This is recapitulated by the observation that gene deletion projects typically only produce visible or biochemical phenotypes in the same small fraction of potential mutants (Park et al., 2011; Winzeler et al., 1999).

Another category of genetically characterized strains is those used in the characterization of unknown strains. These include reference strains with defined mating types, such as the bipolar mating system in eukaryotic model organisms *Saccharomyces cerevisiae* (Lindegren and Lindegren, 1943) and *Neurospora crassa* (Dodge, 1932). The use of defined mating type testers was applied very early to diverse systems and soon revealed that in some fungi mating type serves as a form of vegetative incompatibility (Papazian, 1950). This references the useful vegetative incompatibility reference strains held in genetic collections, including *Fusarium* (Puhalla, 1985), and other important plant and human pathogenic fungi. These and related studies led to the description of a tetrapolar mating system in *Ustilago maydis* (Holliday, 1961) as well as the multiallelic tetrapolar system for *Schizophyllum commune* (Raper and Miles, 1958). Reference strains for the distinct mating types are present, for example, in the collection of the Fungal Genetics Stock Center and these include reference mating type strains for a diverse group of fungi.

During the early classical genetics era, special strains were generated, typically by sexual crossing, to allow for increased success with mutagenesis. Among these, strains auxotrophic for production of inositol were especially useful (Lester and Gross, 1959). Additional strains with spore color, spore surface protein, and ultimately strains for the evaluation of mutagenicity were also generated (De Serres, 1968) and many found impactful use in genetics, cell biology, and as models for more complicated systems. Since the advent of DNA mediated transformation for fungi, special strains for transformation have been generated in a number of organisms. These include strains with defined auxotrophies, which allow selection by complementation of the deficiency, for transformation of *Neurospora* (Case et al., 1979; Legerton and Yanofsky, 1985; Mishra and Tatum, 1973), *Aspergillus* (Turner and Ballance, 1985; Yelton et al., 1984), and numerous fungi (Bennett and Lasure, 1991). While the development of the hygromycin selectable marker for fungi made the use of auxotrophic strains less important (Punt et al., 1987), targeted transformation remained a challenge with most fungi having only 1%–5% homologous integration.

Genetic Maps

Several species of filamentous fungi were important as research systems during the classical genetics era and each was utilized for its specific advantages in one area or another. Genetic map construction was an important goal during this era as it allowed advances in understanding of the nature of inheritable characters, the physical nature of the material of inheritance, and the mutability, stability, and divisibility of genetic characters. After the demonstration of the heritable nature of biochemical characters using mutants of *Neurospora* (Beadle and Tatum, 1941), genetic maps were constructed in a number of different research systems including *E. coli* (Lederberg, 1947), *A. nidulans* (Pontecorvo et al., 1954), Drosophila (Sturtevant, 1951), and Maize (Anderson and Randolph, 1945). Researchers working in each of these systems generated and shared strains for mapping and several were formally established in living collections. The FGSC was first established in 1960 with holdings including *Neurospora* and *Aspergillus* mutants (Barratt et al., 1965; Murray, 1965). Incidentally, the decision in 1960 to assign parallel numbers to strains from different species continues to challenge investigators and collection staff, and only the prefix "A" differentiates *Aspergillus* strains from strains from over 120 other species in the FGSC collection (McCluskey, 2000). The genetic maps made for *Neurospora* (Barratt et al., 1954) and *Aspergillus* (Clutterbuck, 1994) were regularly publicized in the Fungal Genetic Newsletter which was published by the FGSC and insured the greatest access and impact during an era when rapid publication was not the norm. The *Neurospora* genetic map included over 1,200 phenotypic markers and additional genetic or cytological markers such as the centromeres, telomeres, and intra and interchromosomal rearrangements (Perkins et al., 1982, 2000). Similarly, the *Aspergillus* genetic map, originally constructed using para-sexual genetics (Pontecorvo et al., 1954), eventually included biochemical, morphological, and physiological traits (Clutterbuck, 1997). Other research systems were being exploited from the classical genetics era into the molecular genetic era and genetic maps were generated for *Allomyces* (Ingraham and Emerson, 1954), *Schizophyllum* (Raper and Miles, 1958), *Ustilago* (Perkins, 1949), *Fusarium* (Snyder et al., 1975), and other plant pathogenic fungi (Hulbert et al., 1988; Michelmore and Hulbert, 1987; Valent, 1990). Because genetic mapping was time consuming and depended on stabile genetic systems, various schemes were organized for simplifying genetic mapping. In *Neurospora*, these included the development of a variety strains specifically for genetic mapping including strains with classical markers on many or all of the seven chromosomes to allow rapid assignment of a novel marker to a linkage group by genetic crossing, and strains with cytoplasmic (mitochondrial) markers. Additional strains for genetic analysis include diverse species (mating) testers, strains carrying reciprocal translocations allowing rapid identification of linkage (Perkins et al., 1969), multiply inbred strains for stability and fecundity, as well as strains with unique characteristics that make them suitable for utilization in student laboratories such as conidial separation mutant strains (Selitrennikoff, 1974) or strains with easily wettable spores (Lauter et al., 1992). While classical genetic mapping has been very robust, it is limited to the identification of genes for which mutation has a visible, biochemical, or physiological phenotype. The ability to generate molecular genetic maps where the markers are phenotypically neutral has been very significant. Molecular mapping tools have been generated for a number of organisms and they range from the RFLP probes for *Neurospora* (Vollmer and Yanofsky, 1986), *Magnaporthe* (Dioh et al., 2000), and *Fusarium* (Xu and Leslie, 1996), to a set of primers for amplification of fragments associated with distributed Single Nucleotide Polymorphisms in *Neurospora* (Lambreghts et al., 2009), although now whole genome sequence analysis makes this all somewhat quaint (McCluskey et al., 2011; Nowrousian et al., 2010; Pomraning et al., 2011).

Molecular Tools

Among the resources held in genetic strains collections, perhaps the plasmid resources are the most widely utilized. This utilization spans the historical delimitation by research system. Plasmids may be used in the intended host, in related host strains, or even across diverse taxonomic lineages. Indeed, the molecular genetic resources of the FGSC collection have been used by a diverse group of researchers well beyond the narrow community originally served by the collection. This diversification has mirrored the concomitant diversification of the meetings hosted by the community. The largest fungal genetics meeting, originally the *Neurospora Information Conference*, has become the most diverse and well attended fungal genetics conference worldwide (Heitman and Howlett, 2008). The diversity of the meeting is mirrored by the diversity of resources in the FGSC collection, most acutely among the molecular plasmids (Fig. 11.2). The number of plasmids for protein tagging for microscopy and purification, and the diverse organisms in which these materials are utilized (McCluskey, 2015) is a proxy for the success of the model organisms in which the tools were developed. The first plasmid deposited into the FGSC collection carried the transposable element, Tad from *Neurospora* (Cambareri et al., 1994) and this plasmid collection has grown to include diverse tools of limited impact,

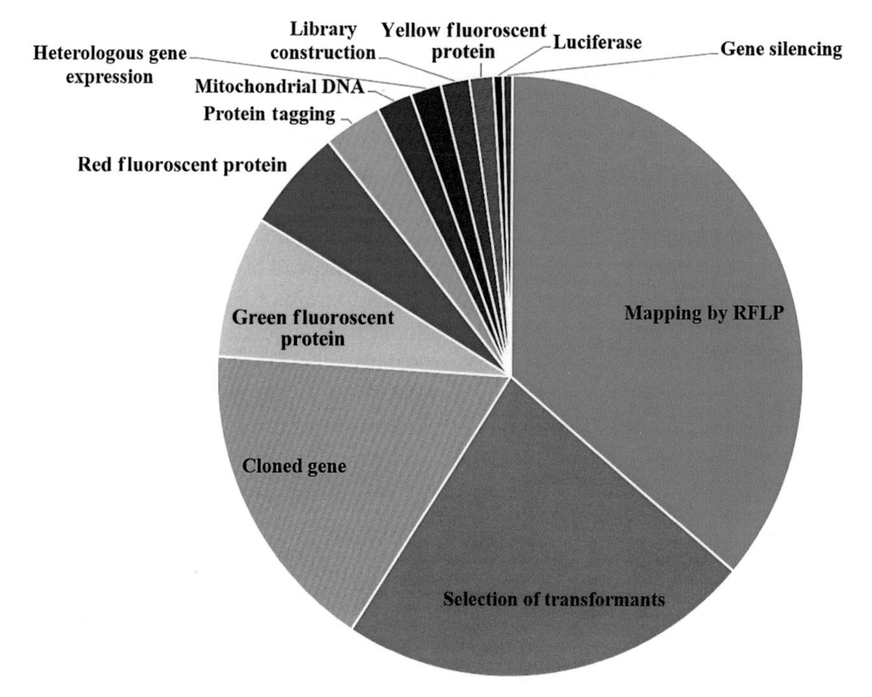

FIGURE 11.2 **Molecular reagents in the FGSC collection.**

such as RFLP probes, as well as tools of enduring impact, such as selectable markers for transformation and protein tags for visualization or purification.

As a genetic resource center, the FGSC has held a variety of tools that were built by the community and shared to promote the rapid advance of the field. Beginning with ordered gene libraries (Vollmer and Yanofsky, 1986) and advancing to chromosome specific (Brody et al., 1991) and ultimately genome linked cosmid, fosmid, and BAC libraries for *Neurospora* (Orbach, 1994), *Aspergillus nidulans* (Galagan et al., 2005), and *Fusarium graminearum* (Cuomo et al., 2007) as well as phage pools containing genomic or cDNA libraries for diverse organisms. These resources were distributed hundreds of times in the pregenome sequence era (McCluskey, 2013) and were utilized in the assembly of numerous fungal genomes (Cuomo et al., 2007; Dean et al., 2005; Galagan et al., 2003, 2005). The public availability of the location on the genome of cosmids used in the assembly was exploited by a number of researchers to simplify cosmid walking approaches to gene identification (Dieterle et al., 2010; Kazama et al., 2008). Similarly, and as a representation of the collegiality of the *Neurospora* and *Aspergillus* research communities, the identity of cosmids carrying specific genes was published in the FGSC catalog for many years, (McCluskey and Plamann, 2006) simplifying gene identification and other molecular characterizations.

Among diverse tools for molecular manipulation of yeast and filamentous fungi, selectable markers for fungal transformation and fluorescent tags for protein localization are the most diverse and widely utilized.

As genomics impacted the filamentous fungal genetics community (Galagan et al., 2003, 2005) tools took on a new character. Among these, genome linked cosmid (Brody et al., 1991; Kelkar et al., 2001) and fosmid (Clutterbuck and Farman, 2007) libraries, and cDNA libraries (Nelson et al., 1997) contributed to rapid dissemination of techniques and tools from model systems like *Neurospora* and *Aspergillus* to nonmodel research systems.

The demonstration that DNA based transformation of strains deficient for nonhomologous end joining of DNA sequences allowed highly efficient targeting led to a renaissance of utilization of mutant strains for characterization of gene function (Ninomiya et al., 2004), although gene deletion mutants are not the same as single nucleotide variants, or strains carrying small (1–15) base insertions or deletions (McCluskey et al., 2011). The ability to target transformation for gene deletion or replacement has allowed the use of filamentous fungi to investigate questions not accessible with simpler systems (such as yeast) which have greatly reduced genome content (Goffeau et al., 1996) and minimal morphology, or more complex, but economically important systems which struggle with instable genomes (Dean et al., 2005), or which are diploid (Jones et al., 2004) or aneuploid (McCluskey et al., 1994; VanEtten et al., 1994), complicating interpretation and assembly.

These limits notwithstanding, the use of systematic gene deletion resources has had a great impact on fungal genetics and gene deletion mutant sets are available commercially for some organisms (such as *Saccharomyces*) (Tong et al., 2001) or through noncommercial laboratories, such as the FGSC which holds complete or partial gene deletion sets for *Neurospora* (Colot et al., 2006), *Candida* (Homann et al., 2009; Nobile and Mitchell, 2009; Noble et al., 2010), and *Cryptococcus* (Baker et al., 2007; Liu et al., 2008). Additional gene deletion resources were made available from the FGSC including over 10,000 gene deletion cassettes (as naked DNA) and a small number of gene deletion mutants for *A. nidulans*. Rounding out the suite of strains and tools are diverse strains carrying the mutation necessary to enable targeted transformation, although regulatory requirements mean that plant pathogen strains which are genetically modified through recombinant DNA technology have limited distribution. The unfortunate end-result of this regulation is that people are making gene deletions in their favorite strain thereby limiting comparisons of results from individuals working in different laboratories.

Genomics as a Proxy for Genetics

While traditional genetic stock centers were built upon the foundation of classical genetic mutants, modern resource collections do not depend on the observation of unique genetically segregating phenotypes to define useful materials. Indeed, modern gene deletion projects are not dependent on observation of visible or biochemical phenotypes and thus allow access to the majority of genes for which no phenotype is observed (Dunlap et al., 2007; Liu et al., 2008; Nobile and Mitchell, 2009). The definition of genetics has had to be reevaluated in this context and even "wild-type" strains are genetic when they are accompanied by genome sequence. In this context, genome sequence associated with living microbial resources define a paradigm shift where we no longer depend on phenotypes to identify and follow genes in crosses or populations but rather use genomics to identify quantitative traits (Reeves et al., 2007). Moreover, the ability to use genomics to compare isolates that are otherwise phenotypically identical will have profound impact both on the utilization of fungal genetic resources, but also on the ability to identify traits of economic importance and associate them with their geographic or temporal origins which may be highly significant in the context of Access and Benefit Sharing under the Nagoya Protocol (Dedeurwaerdere et al., 2016) (Chapter 10).

Because genome biology has changed the nature of fungal genetics, it is important to consider the relationship between genetic repositories and genome databases. While the Broad Institute invested in fungal genome resources, beginning with their elucidation of the genome of the *Neurospora crassa* strain FGSC 2489 (Galagan et al., 2003), their support for maintaining these resources was temporary and in 2015, they stopped investing in their fungal genomics resources. Some fungal genomes are hosted at the *US National Center for Biotechnology Information* and others are hosted at the *US Department of Energy* (DOE) Mycocosm server. Many of these strains are also available from the

FGSC or from the *Centraalbureau voor Schimmelcultures* (CBS) collection in the Netherlands. Recognizing the value of genome sequence for fungi, the US DOE launched the 1000 fungal genomes program with the intent of providing context for metagenomics studies (Spatafora et al., 2013).

Trajectory of a Genetic Repository

The growth and persistence of culture collections, and therefore of genetic resource collections, follows a typical trajectory. Many collections grow organically in the laboratory of a researcher with a specific interest. A research community may advocate for the formation of a shared repository, and once this happens a formal collection grows in two different modes. The first is incremental growth and the second is by accretion (Fig. 11.3). Incremental growth occurs by individual deposit, either solicited by the curator or offered by the depositor, either out of generosity, or to benefit from the established increase in citation associated with shared materials (Furman and Stern, 2011). Organized growth would occur typically upon the retirement of researchers who had developed useful or valuable strain collections, or more recently, as the result of systematic gene deletion or manipulation projects (Colot et al., 2006). This recognizes that genetic collections can be ephemeral as the research questions they were developed to address are elucidated. They are typically based upon model organisms

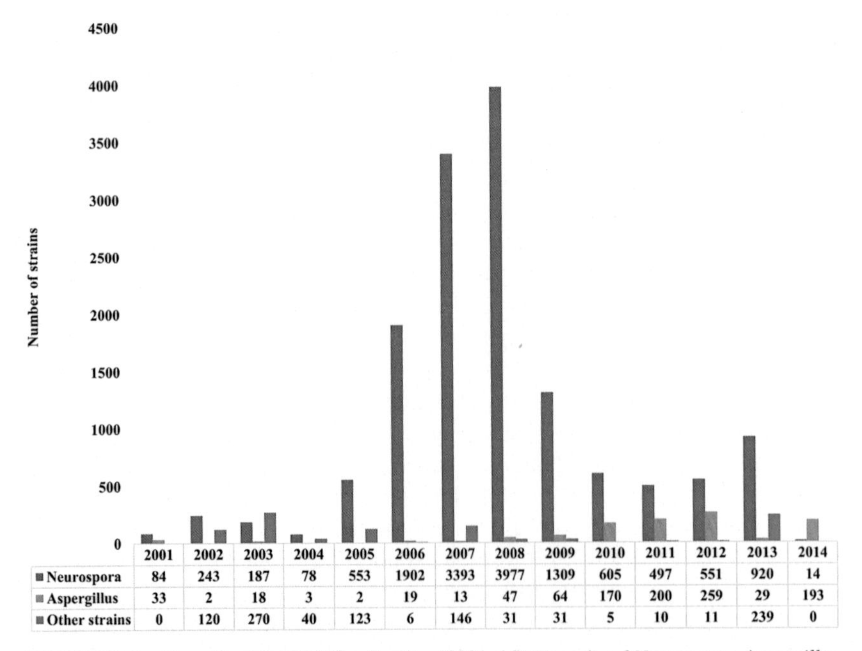

	2001	2002	2003	2004	2005	2006	2007	2008	2009	2010	2011	2012	2013	2014
■ Neurospora	84	243	187	78	553	1902	3393	3977	1309	605	497	551	920	14
■ Aspergillus	33	2	18	3	2	19	13	47	64	170	200	259	29	193
■ Other strains	0	120	270	40	123	6	146	31	31	5	10	11	239	0

FIGURE 11.3 **Growth of the FGSC collection (2001–14).** Deposits of *Neurospora, Aspergillus,* and other fungi into the FGSC collection show the impact of large scale deposits.

which have no intrinsic value. While genome biology offers unique insight, it is not clear that it will replace historical genetics. This emphasizes the point made elsewhere that classic mutants and wild types may have sequence polymorphisms not available in deletion mutants. Indeed, as modern genome biology is demonstrating that nature conducts "mutation screens," synthetic biology may represent the future of genetic resources. This of course relies on the ability to manipulate and characterize strains and points out that wild type collections may be inherently more durable.

CONCLUSIONS

Genetic resource collections are complimentary to type collections and biodiversity collections. Within their limited taxonomic mandate, genetic collections often have diverse holdings, not limited to strains, but rather including diverse materials serving a specific research community. Examples include the Fungal Genetics Stock Center as well as collections of *E. coli*, *Chlamydomonas*, *Bacillus*, as well as higher eukaryotic organisms. Medically relevant collections have access to funding not available to environmental, industrial, taxonomic, or genetic collections. Some genetic collections have tremendous numbers of isolates, sometimes including hundreds of mutants at the same genetic locus and these may outlive their utility as the questions they were intended to answer are better understood. Similarly, the ability to generate targeted and genome-wide molecular genetic mutant sets has led to the rapid growth in living collection holdings, unfortunately at the same time that financial support of these collections is being shifted to the end users. Because support for basic research using microorganisms has dwindled, the shift of the expense to the end user creates a downward pressure on both the research community (which may prefer peer-to-peer exchanges because of the lower perceived expense) and on the collection, which has higher expenses by virtue of the necessity to follow established regulations regarding shipping, intellectual property rights, and even international treaty obligations.

Microbial genetic resource collections are culture collections, but by virtue of their distinct holdings, define a different category of culture collection. Along with type collections, biodiversity collections, and industrial collections, genetic resource collections share the fate of all culture collections. While medically relevant collections have unique funding available, all microbial genetic resource collections face similar challenges and share characteristics with other culture collections. They differ in that the resources held in microbial genetic resource collections may become obsolete as technology changes and the questions they were developed to answer are resolved. They are similar in having large numbers of quality assured living microbial strains utilized in academic and commercial research, in distributing living microorganisms around the world, and in implementing best practices for quality control, for regulatory compliance, and for client service. As support for living biodiversity and type collections waxes and wanes, so does support for genetic microbial resource collections. With the growing awareness of reproducibility in science (Nosek et al., 2015), the need for robust and well supported living microorganism collections becomes more obvious. Combined with the demonstrated impact of living microorganism collections, support for these valuable infrastructure resources must be prioritized and acknowledged for the benefit they provide to research, industry, and to society.

ACKNOWLEDGMENT

This article is publication number 16-252-B from the Kansas Agricultural Experiment Station.

REFERENCES

Anderson, E., Randolph, L., 1945. Location of the centromeres on the linkage maps of maize. Genetics 30 (6), 518.

Baba, T., Ara, T., Hasegawa, M., Takai, Y., Okumura, Y., Baba, M., Datsenko, K.A., Tomita, M., Wanner, Barry L., Mori, H., 2006. Construction of *Escherichia coli* K-12 in-frame, single-gene knockout mutants: the Keio collection. Mol. Syst. Biol. 2 (1), 2006.

Baker, L.G., Specht, C.A., Donlin, M.J., Lodge, J.K., 2007. Chitosan, the deacetylated form of chitin, is necessary for cell wall integrity in *Cryptococcus neoformans*. Eukaryot. Cell 6 (5), 855–867.

Barratt, R.W., Newmeyer, D., Perkins, D.D., Garnjobst, L., 1954. Map construction in *Neurospora crassa*. Adv. Genet. 6, 1–93.

Barratt, R., Johnson, G., Ogata, W., 1965. Wild-type and mutant stocks of *Aspergillus nidulans*. Genetics 52 (1), 233–246.

Beadle, G.W., Tatum, E.L., 1941. Genetic control of biochemical reactions in *Neurospora*. Proc. Natl. Acad. Sci. USA 27 (11), 499–506.

Bennett, J.W., Lasure, L.L., 1991. More Gene Manipulations in Fungi. Academic Press, San Diego, 558 pp.

Brody, H., Griffith, J., Cuticchia, A.J., Arnold, J., Timberlake, W.E., 1991. Chromosome-specific recombinant DNA libraries from the fungus *Aspergillus nidulans*. Nucleic Acids Res. 19 (11), 3105–3109.

Cambareri, E.B., Helber, J., Kinsey, J.A., 1994. Tad1-1, an active LINE-like element of *Neurospora crassa*. Mol. Gen. Genet. 242 (6), 658–665.

Case, M.E., Schweizer, M., Kushner, S.R., Giles, N.H., 1979. Efficient transformation of *Neurospora crassa* by utilizing hybrid plasmid DNA. Proc. Natl. Acad. Sci. 76 (10), 5259–5263.

Clutterbuck, A.J., 1994. Linkage map and locus list. Prog. Ind. Microbiol. 29, 791–824.

Clutterbuck, A., 1997. The validity of the *Aspergillus nidulans* linkage map. Fungal Genet. Biol. 21 (3), 267–277.

Clutterbuck, A.J., Farman, M., 2007. *Aspergillus nidulans* linkage map and genome sequence: closing gaps and adding telomeres. CRC Press, Boca Raton, FL, pp. 57–74.

Colot, H.V., et al., 2006. A high-throughput gene knockout procedure for *Neurospora* reveals functions for multiple transcription factors. Proc. Natl. Acad. Sci. USA 103 (27), 10352–10357.

Cuomo, C.A., Guldener, U., Xu, J.R., Trail, F., Turgeon, B.G., Di Pietro, A., et al., 2007. The *Fusarium graminearum* genome reveals a link between localized polymorphism and pathogen specialization. Science 317 (5843), 1400–1402.

De Serres, F., 1968. Genetic analysis of the extent and type of functional inactivation in irreparable recessive lethal mutations in the ad-3 region of *Neurospora crassa*. Genetics 58 (1), 69.

Dean, R.A., Talbot, N.J., Ebbole, D.J., Farman, M.L., Mitchell, T.K., Orbach, M.J., Thon, M., Kulkarni, R., Xu, J.-R., Pan, H., 2005. The genome sequence of the rice blast fungus *Magnaporthe grisea*. Nature 434 (7036), 980–986.

Dedeurwaerdere, T., Melindi-Ghidi, P., Broggiato, A., 2016. Global scientific research commons under the Nagoya Protocol: towards a collaborative economy model for the sharing of basic research assets. Environ. Sci. Policy 55, 1–10.

Dieterle, M.G., Wiest, A.E., Plamann, M., McCluskey, K., 2010. Characterization of the temperature-sensitive mutations un-7 and png-1 in *Neurospora crassa*. PLoS One 5 (5), e10703.

Dioh, W., Tharreau, D., Notteghem, J.L., Orbach, M., Lebrun, M.H., 2000. Mapping of avirulence genes in the rice blast fungus, *Magnaporthe grisea*, with RFLP and RAPD markers. Mol. Plant Microbe Interact. 13 (2), 217–227.

Dodge, B.O., 1932. The non-sexual and the sexual functions of microconidia of *Neurospora*. Torrey Botanical Society, KS, United States, pp. 347–360.

Dunlap, J.C., Borkovich, K.A., Henn, M.R., Turner, G.E., Sachs, M.S., Glass, N.L., McCluskey, K., Plamann, M., Galagan, J.E., Birren, B.W., Weiss, R.L., Townsend, J.P., Loros, J.J., Nelson, M.A., Lambreghts, R., Colot, H.V., Park, G., Collopy, P., Ringelberg, C., Crew, C., Litvinkova, L., DeCaprio, D., Hood, H.M., Curilla, S., Shi, M., Crawford, M., Koerhsen, M., Montgomery, P., Larson, L., Pearson, M., Kasuga, T., Tian, C., Basturkmen, M., Altamirano, L., Xu, J., 2007. Enabling a community to dissect an organism: overview of the *Neurospora* functional genomics project. Adv. Genet. 57, 49–96.

Frankel, C., Ellingboe, A.H., 1977. New mutations and a 7-chromosome linkage map of *Schizophyllum commune*. Genetics 85 (3), 417–425.

Furman, J.L., Stern, S., 2011. Climbing atop the shoulders of giants: the impact of institutions on cumulative research. Am. Econ. Rev. 101 (5), 1933–1963.

Galagan, J.E., Calvo, S.E., Borkovich, K.A., Selker, E.U., Read, N.D., Jaffe, D., FitzHugh, W., Ma, L.J., Smirnov, S., Purcell, S., Rehman, B., Elkins, T., Engels, R., Wang, S., Nielsen, C.B., Butler, J., Endrizzi, M., Qui, D., Ianakiev, P., Bell-Pedersen, D., Nelson, M.A., Werner-Washburne, M., Selitrennikoff, C.P., Kinsey, J.A., Braun, E.L., Zelter, A., Schulte, U., Kothe, G.O., Jedd, G., Mewes, W., Staben, C., Marcotte, E., Greenberg, D., Roy, A., Foley, K., Naylor, J., Stange-Thomann, N., Barrett, R., Gnerre, S., Kamal, M., Kamvysselis, M., Mauceli, E., Bielke, C., Rudd, S., Frishman, D., Krystofova, S., Rasmussen, C., Metzenberg, R.L., Perkins, D.D., Kroken, S., Cogoni, C., Macino, G., Catcheside, D., Li, W., Pratt, R.J., Osmani, S.A., DeSouza, C.P., Glass, L., Orbach, M.J., Berglund, J.A., Voelker, R., Yarden, O., Plamann, M., Seiler, S., Dunlap, J., Radford, A., Aramayo, R., Natvig, D.O., Alex, L.A., Mannhaupt, G., Ebbole, D.J., Freitag, M., Paulsen, I., Sachs, M.S., Lander, E.S., Nusbaum, C., Birren, B., 2003. The genome sequence of the filamentous fungus *Neurospora crassa*. Nature 422 (6934), 859–868.

Galagan, J.E., Calvo, S.E., Cuomo, C., Ma, L.J., Wortman, J.R., Batzoglou, S., Lee, S.I., Basturkmen, M., Spevak, C.C., Clutterbuck, J., Kapitonov, V., Jurka, J., Scazzocchio, C., Farman, M., Butler, J., Purcell, S., Harris, S., Braus, G.H., Draht, O., Busch, S., D'Enfert, C., Bouchier, C., Goldman, G.H., Bell-Pedersen, D., Griffiths-Jones, S., Doonan, J.H., Yu, J., Vienken, K., Pain, A., Freitag, M., Selker, E.U., Archer, D.B., Penalva, M.A., Oakley, B.R., Momany, M., Tanaka, T., Kumagai, T., Asai, K., Machida, M., Nierman, W.C., Denning, D.W., Caddick, M., Hynes, M., Paoletti, M., Fischer, R., Miller, B., Dyer, P., Sachs, M.S., Osmani, S.A., Birren, B.W., 2005. Sequencing of *Aspergillus nidulans* and comparative analysis with *A. fumigatus* and *A. oryzae*. Nature 438 (7071), 1105–1115.

Gaudeul, M., Rouhan, G., 2013. A plea for modern botanical collections to include DNA-friendly material. Trends Plant Sci. 18 (4), 184–185.

Goffeau, A., Barrell, B.G., Bussey, H., Davis, R.W., Dujon, B., Feldmann, H., Galibert, F., Hoheisel, J.D., Jacq, C., Johnston, M., Louis, E.J., Mewes, H.W., Murakami, Y., Philippsen, P., Tettelin, H., Oliver, S.G., 1996. Life with 6000 Genes. Science 274 (5287), 546–567.

Heitman, J., Howlett, B., 2008. Fungal horizons: The Asilomar Fungal Genetics Conference 2007. Fungal Genet. Biol. 45 (2), 77–83.

Holliday, R., 1961. The genetics of *Ustilago maydis*. Genet. Res. 2 (02), 204–230.

Homann, O.R., Dea, J., Noble, S.M., Johnson, A.D., 2009. A phenotypic profile of the *Candida albicans* regulatory network. PLoS Genet. 5 (12), e1000783.

Hulbert, S., Ilott, T., Legg, E., Lincoln, S., Lander, E., Michelmore, R., 1988. Genetic analysis of the fungus, *Bremia lactucae*, using restriction fragment length polymorphisms. Genetics 120 (4), 947–958.

Ingraham, J.L., Emerson, R., 1954. Studies of the nutrition and metabolism of the aquatic phycomycete, *Allomyces*. Am. J. Bot. 41, 146–152.

Jones, T., Federspiel, N.A., Chibana, H., Dungan, J., Kalman, S., Magee, B.B., Newport, G., Thorstenson, Y.R., Agabian, N., Magee, P.T., 2004. The diploid genome sequence of *Candida albicans*. Proc. Natl. Acad. Sci. USA 101 (19), 7329–7334.

Jurgenson, J.E., Bowden, R.L., Zeller, K.A., Leslie, J.F., Alexander, N.J., Plattner, R.D., 2002. A genetic map of *Gibberella zeae* (*Fusarium graminearum*). Genetics 160 (4), 1451–1460.

Kazama, Y., Ishii, C., Schroeder, A.L., Shimada, H., Wakabayashi, M., Inoue, H., 2008. The *Neurospora crassa* UVS-3 epistasis group encodes homologues of the ATR/ATRIP checkpoint control system. DNA Repair 7 (2), 213–229.

Kelkar, H.S., Griffith, J., Case, M.E., Covert, S.F., Hall, R.D., Keith, C.H., Oliver, J.S., Orbach, M.J., Sachs, M.S., Wagner, J.R., 2001. The *Neurospora crassa* genome: cosmid libraries sorted by chromosome. Genetics 157 (3), 979–990.

Lambreghts, R., Shi, M., Belden, W.J., Decaprio, D., Park, D., Henn, M.R., Galagan, J.E., Basturkmen, M., Birren, B.W., Sachs, M.S., Dunlap, J.C., Loros, J.J., 2009. A high-density single nucleotide polymorphism map for *Neurospora crassa*. Genetics 181 (2), 767–781.

Lauter, F.-R., Russo, V., Yanofsky, C., 1992. Developmental and light regulation of eas, the structural gene for the rodlet protein of *Neurospora*. Genes Dev. 6 (12a), 2373–2381.

Lederberg, J., 1947. Gene recombination and linked segregations in *Escherichia coli*. Genetics 32 (5), 505.

Legerton, T.L., Yanofsky, C., 1985. Cloning and characterization of the multifunctional his-3 gene of *Neurospora crassa*. Gene 39 (2), 129–140.

Lester, H.E., Gross, S., 1959. Efficient method for selection of auxotrophic mutants of *Neurospora*. Science 129 (3348), 572.

Lindegren, C.C., Lindegren, G., 1943. Selecting, inbreeding, recombining, and hybridizing commercial yeasts. J. Bacteriol. 46 (5), 405–419.

Liu, O.W., Chun, C.D., Chow, E.D., Chen, C., Madhani, H.D., Noble, S.M., 2008. Systematic genetic analysis of virulence in the human fungal pathogen *Cryptococcus neoformans*. Cell 135 (1), 174–188.

McCluskey, K., 2000. A relational database for the FGSC. Fungal Genet. Newslett. 47, 74–78.

McCluskey, K., 2011. From genetics to genomics: fungal collections at the Fungal Genetics Stock Center. Mycology 2 (3), 161–168.

McCluskey, K., 2013. Biological resource centers provide data and characterized living material for industrial biotechnology. Ind. Biotechnol. 9 (3), 117–122.

McCluskey, K., 2015. Boosting research and industry by providing extensive resources for fungal research. In: Schmoll, M., Dattenböck, C. (Eds.), Gene Expression Systems in Fungi: Advancements and Applications. Springer, Berlin.

McCluskey, K., Plamann, M., 2006. Fungal Genetics Stock Center Catalog of Strains, eleventh ed. Fungal Genetics Newsletter, 53S.

McCluskey, K., Agnan, J., Mills, D., 1994. Characterization of genome plasticity in *Ustilago hordei*. Curr. Genet. 26 (5–6), 486–493.

McCluskey, K., Wiest, A., Grigoriev, I.V., Lipzen, A., Martin, J., Schackwitz, W., Baker, S.E., 2011. Rediscovery by whole genome sequencing: classical mutations and genome polymorphisms in *Neurospora crassa*. Genes Genomes Genet. 1 (4), 303–316.

Michelmore, R., Hulbert, S., 1987. Molecular markers for genetic analysis of phytopathogenic fungi. Annu. Rev. Phytopathol. 25 (1), 383–404.

Mishra, N., Tatum, E., 1973. Non-Mendelian inheritance of DNA-induced inositol independence in *Neurospora*. Proc. Natl. Acad. Sci. 70 (12), 3875–3879.

Murray, N.E., 1965. Cysteine mutant strains of *Neurospora*. Genetics 52 (4), 801.

Nelson, M.A., Kang, S., Braun, E.L., Crawford, M.E., Dolan, P.L., Leonard, P.M., Mitchell, J., Armijo, A.M., Bean, L., Blueyes, E., 1997. Expressed Sequences from conidial, mycelial, and sexual stages of *Neurospora crassa*. Fungal Genet. Biol. 21 (3), 348–363.

Ninomiya, Y., Suzuki, K., Ishii, C., Inoue, H., 2004. Highly efficient gene replacements in Neurospora strains deficient for nonhomologous end-joining. Proc. Natl. Acad. Sci. USA 101 (33), 12248–12253.

Nobile, C.J., Mitchell, A.P., 2009. Large-scale gene disruption using the UAU1 cassette. Methods Mol. Biol. 499, 175–194, Available from: http://doi.org/10.1007/978-1-60327-151-6_17.

Noble, S.M., French, S., Kohn, L.A., Chen, V., Johnson, A.D., 2010. Systematic screens of a *Candida albicans* homozygous deletion library decouple morphogenetic switching and pathogenicity. Nat. Genet. 42 (7), 590–598.

Nosek, B.A., Alter, G., Banks, G.C., Borsboom, D., Bowman, S.D., Breckler, S.J., Buck, S., Chambers, C.D., Chin, G., Christensen, G., Contestabile, M., 2015. Promoting an open research culture: Author guidelines for journals could help to promote transparency, openness, and reproducibility. Science 348 (6242), 1422.

Nowrousian, M., Stajich, J.E., Chu, M., Engh, I., Espagne, E., Halliday, K., Kamerewerd, J., Kempken, F., Knab, B., Kuo, H.C., Osiewacz, H.D., Poggeler, S., Read, N.D., Seiler, S., Smith, K.M., Zickler, D., Kuck, U., Freitag, M., 2010. De novo assembly of a 40 Mb eukaryotic genome from short sequence reads: *Sordaria macrospora*, a model organism for fungal morphogenesis. PLoS Genet. 6 (4), e1000891.

Orbach, M.J., 1994. A cosmid with a HyR marker for fungal library construction and screening. Gene 150 (1), 159–162.

Papazian, H., 1950. Physiology of the incompatibility factors in *Schizophyllum commune*. Botanical Gazette 112, 143–163.

Park, G., Colot, H.V., Collopy, P.D., Krystofova, S., Crew, C., Ringelberg, C., Litvinkova, L., Altamirano, L., Li, L., Curilla, S., Wang, W., Gorrochotegui-Escalante, N., Dunlap, J.C., Borkovich, K.A., 2011. High-throughput production of gene replacement mutants in *Neurospora crassa*. Methods Mol. Biol. 722, 179–189.

Perkins, D.D., 1949. Biochemical mutants in the smut fungus *Ustilago maydis*. Genetics 34 (5), 607–626.

Perkins, D.D., Newmeyer, D., Taylor, C.W., Bennett, D.C., 1969. New markers and map sequences in *Neurospora crassa*, with a description of mapping by duplication coverage, and of multiple translocation stocks for testing linkage. Genetica 40 (1), 247–278.

Perkins, D.D., Radford, A., Newmeyer, D., Bjorkman, M., 1982. Chromosomal loci of *Neurospora crassa*. Microbiol. Rev. 46 (4), 426–570.

Perkins, D.D., Radford, A., Sachs, M.S., 2000. The *Neurospora compendium*: Chromosomal Loci. Academic Press, San Diego, CA.

Perkins, D.D., Radford, A., Sachs, M.S., 2001. The *Neurospora compendium*: Chromosomal Loci. Academic press, San Diego, CA.

Pitkin, J., Davis, R.H., 1990. The genetics of polyamine synthesis in *Neurospora crassa*. Arch. Biochem. Biophys. 278 (2), 386–391.

Pomraning, K.R., Smith, K.M., Freitag, M., 2011. Bulk segregant analysis followed by high-throughput sequencing reveals the *Neurospora* cell cycle gene, ndc-1, to be allelic with the gene for ornithine decarboxylase, spe-1. Eukaryot. Cell 10 (6), 724–733.

Pontecorvo, G., Gloor, E.T., Forbes, E., 1954. Analysis of mitotic recombination in *Aspergillus nidulans*. J. Genet. 52 (1), 226–237.

Puhalla, J.E., 1985. Classification of strains of *Fusarium oxysporum* on the basis of vegetative compatibility. Can. J. Bot. 63 (2), 179–183.

Punt, P.J., Oliver, R.P., Dingemanse, M.A., Pouwels, P.H., van den Hondel, C.A., 1987. Transformation of *Aspergillus* based on the hygromycin B resistance marker from *Escherichia coli*. Gene 56 (1), 117–124.

Raper, J.R., Miles, P.G., 1958. The genetics of *Schizophyllum commune*. Genetics 43 (3), 530.

Reeves, P.A., He, Y., Schmitz, R.J., Amasino, R.M., Panella, L.W., Richards, C.M., 2007. Evolutionary conservation of the FLOWERING LOCUS C-mediated vernalization response: evidence from the sugar beet (*Beta vulgaris*). Genetics 176 (1), 295–307.

Selitrennikoff, C., 1974. Use of conidial separation-defective strains. Neurospora. Newslett. 21, 22.

Skinner, D., Budde, A., Farman, M., Smith, J., Leung, H., Leong, S., 1993. Genome organization of Magnaporthe grisea: genetic map, electrophoretic karyotype, and occurrence of repeated DNAs. Theor. Appl. Genet. 87 (5), 545–557.

Snyder, W.C., Georgopoulos, S., Webster, R.K., Smith, S.N., 1975. Sexuality and Genetic Behavior in the Fungus Hypomyces *(Fusarium) solani* f. sp. *cucurbitae*. Hilgardia 43, 161–185.

Spatafora, J., Stajich, J., Grigoriev, I., 2013. 1000 fungal genomes project. Phytopathology 103 (6), 137.

Sturtevant, A., 1951. A map of the fourth chromosome of *Drosophila melanogaster*, based on crossing over in triploid females. Proc. Natl. Acad. Sci. USA 37 (7), 405–407.

Tatum, E., 1945. X-ray induced mutant strains of *Escherichia coli*. Proc. Natl. Acad. Sci. USA 31 (8), 215.

Tong, A.H.Y., Evangelista, M., Parsons, A.B., Xu, H., Bader, G.D., Pagé, N., Robinson, M., Raghibizadeh, S., Hogue, C.W.V., Bussey, H., 2001. Systematic genetic analysis with ordered arrays of yeast deletion mutants. Science 294 (5550), 2364–2368.

Turner, G., Ballance, D., 1985. Cloning and transformation in *Aspergillus*. In: Bennett, J.W. (Ed.), Gene Manipulations in Fungi. Academic Press, London, pp. 259–278.

Valent, B., 1990. Rice blast as a model system for plant pathology. Phytopathology 80 (1), 33–36.

VanEtten, H., Funnell-Baerg, D., Wasmann, C., McCluskey, K., 1994. Location of pathogenicity genes on dispensable chromosomes in *Nectria haematococca* MPVI. Antonie Van Leeuwenhoek 65 (3), 263–267.

Verkley, G.J., Rossman, A., Crouch, J.A., 2015. The role of herbaria and culture collections. In: McLaughlin, Spatafora (Eds.), Systematics and Evolution,. second ed. Springer-Verlag, Berlin, pp. 205–225.

Vollmer, S.J., Yanofsky, C., 1986. Efficient cloning of genes of *Neurospora crassa*. Proc. Natl. Acad. Sci. USA 83 (13), 4869–4873.

Winzeler, E.A., Shoemaker, D.D., Astromoff, A., Liang, H., Anderson, K., Andre, B., Bangham, R., Benito, R., Boeke, J.D., Bussey, H., 1999. Functional characterization of the *S. cerevisiae* genome by gene deletion and parallel analysis. Science 285 (5429), 901–906.

Xu, J., Leslie, J.F., 1996. A genetic map of *Gibberella fujikuroi* mating population A (*Fusarium moniliforme*). Genetics 143 (1), 175–189.

Yelton, M.M., Hamer, J.E., Timberlake, W.E., 1984. Transformation of *Aspergillus nidulans* by using a trpC plasmid. Proc. Natl. Acad. Sci. 81 (5), 1470–1474.

Chapter 12

Industrial Culture Collections: Gateways from Microbial Diversity to Applications

Olga Genilloud
Fundación MEDINA, Health Sciences Technology Park, Granada, Spain

INTRODUCTION

Microbial resources have been intensively exploited by the pharmaceutical and biotechnology industry and are at the origin of a broad diversity of commercial products derived from microbial natural products. Whereas only a small fraction of existing microbial diversity is hold in more than 704 culture collections registered worldwide in the *World Directory of Culture Collections* (WDCM, 2014), the 2,500,000 strains of bacteria and fungi contained in these collections still represent immense reservoirs of well characterized microorganisms that are maintained and distributed to support research in the academic and private sectors. Public collections today play a key role in the conservation and sustainable use of microbial resources, preserving and describing new strains, and providing pure cultures required by academic institutions and the industry. Many of them have been evolving from their original role of public repositories of well characterized new isolated strains to a more specialized model based on the expertize of their curators, providing microbiological services a more important focus (Stackebrandt et al., 2015).

In contrast, most industrial microbial collections were built in the second half of the 20th century as part of the big pharma high throughput screening programs to ensure the access to large reservoirs of talented strains in the production of bioactive compounds or by industrial processes to produce food, additives, aminoacids, or enzymes. These strains were used as sources of novel drugs and other biotechnology products, and the production of a vast diversity of commercially important molecules obtained from microbial strains (bioactive

Microbial Resources. http://dx.doi.org/10.1016/B978-0-12-804765-1.00012-6

small molecules, enzymes, primary metabolism compounds, colorants, flavors, etc.) that ensured the development of industrial processes based on the biosynthesis of microbial metabolites that continue in these days. Strains in industrial collections were selected to capture the microbial metabolic diversity from the environment and to respond to the needs of intensive screening programs that required to be continuously fed with new isolates belonging to highly productive or specialized microbial groups.

The chapter will focus on the differential key aspects and criteria that were followed and guided the construction of industrial culture collections to ensure the discovery of potential new drugs and high value biotechnology products derived from microbial sources, and the current added value of these collections in the new public-private research paradigm.

ACCESS TO MICROBIAL DIVERSITY: THE CONVENTION OF BIOLOGICAL DIVERSITY AND THE NAGOYA PROTOCOL

From an industrial perspective, the access to the microbial diversity in the environment was traditionally focused on intensive collecting activity from a wide diversity of geographical locations and habitats. *The Convention on Biological Diversity* (Convention, 2016) established a turning point in 1993 in the way biological resources were obtained and exploited by industry. Few years before Merck and Co. had been pioneering in this field and had already shown the path that, although initially criticized, would be followed by many others to address the access to biological resources and the lack of existing regulations to obtain materials in most of the countries of origin. The agreement between Merck and Co and the INBIO institute in Costa Rica in 1991 was a landmark and one of the leading and most cited best examples cases (Blum, 1993). For more than 7 years, this agreement regulated the access and provided upfront resources ensuring research training, technology transfer, and the conservation of Costa Rican biodiversity, and established a framework for the fair future return in the case a commercial product was developed from these sources. The new legal framework established by the CBD to regulate the access and benefit sharing associated with the use of genetic resources permitted the establishment of a voluntary code of conduct and the ground for more formal collaboration with research institutes and universities in third countries. Agreements were negotiated with local institutes and government agencies, and permitted working closely with organizations in the source countries, and with local scientists that supported the identification of local species. All these efforts were also confronted to the lack of local regulations and designed competent authorities, limiting the possibilities to obtain sampling permits and to establish legal sourcing contracts. The lack of common regulations that would satisfy all parties required a legal framework to secure the access and future development of any biological resource. It is only in October 2014, after more than 20 years, that the

Nagoya Protocol on Access to Genetic Resources and the Fair and Equitable Sharing of Benefits Arising from their Utilization (Convention, 2016), entered into force providing increased legal certainty and transparency on the procedures for providers and users of genetic resources, but at the same time enforcing the traceability of the biological resources in all the value chain, including future commercial exploitations.

Paralleling this evolving legal landscape, millions of strains have been isolated and screened over the decades in industrial laboratories. During the golden era of antibiotics, environmental samples were processed by thousands following general isolation methods, aiming at isolating new strains most of them included among predominant species of bacteria and fungi, without putting much emphasis in the intensive exploitation of the existing microbial diversity (Walsh and Wencewicz, 2014). In most cases this approach led to the recurrent isolation of these predominant species easily isolated from many habitats despite of the biogeography of the samples used. Many of these strains were producing well-known and structurally related molecules that were identified and discarded as early as possible in drug discovery programs.

The need to identify new microbial strains producing novel compounds and to improve the possibility of success of finding new chemical entities as leads for the development of drug candidates triggered the development of new approaches to exploit the environment microbial diversity. To expand the required microbial diversity in industrial screening programs, the exploration of the environment was extended to a broader variety of sources that were shown later to contain untapped communities with the potential to deliver novel and unique microbial diversity. Today it is well accepted that the distribution of some microbial species presents biogeographic patterns mostly determined by microenvironmental conditions and that these differential behaviors could be translated into novel products (Fiedler et al., 2005; Martiny et al., 2006; Thornburg et al., 2010).

New species are being systematically described that might be the source of novel chemistry with potential new activities. There is extended evidence suggesting that minor species or genetically distinct strains of many bacterial taxa that have not yet been cultured under laboratory conditions occur in most environments, reinforcing the interest in a continued effort to apply new technologies to uncover novel bioactive compounds (Davies, 2011; Stackebrandt et al., 2015). Whereas many novel species have been deposited in public culture collections worldwide, there is still a broad diversity of microbial strains preserved in industrial private collections in Europe, the United States, and Japan, without any means to assess the diversity level in these collections. Some of these assets have found their way to public or academic institutions, and in some cases some compendia have been built aiming at collecting this knowhow (Wink, 2016).

BUILDING AN INDUSTRIAL CULTURE COLLECTION: EXPANDING THE DIVERSITY OF CULTURED MICROBIAL SOURCES

The discovery of novel bioactive natural products in industrial screenings is directly dependent on the chemical diversity of the extract libraries derived from selected cultured microbial strains (Donadio et al., 2010; Genilloud et al., 2011). To respond to the required microbial diversity in culture collections, exploring a broader variety of sources from terrestrial and marine origin, and extending the scope to unique assemblages, plant host associations and marine environments, have also shown that untapped communities in the microbial source have the potential to deliver novel and unique diversity (Aly et al., 2010; Bredholt et al., 2008; Bugni and Ireland, 2004; González et al., 2005; Janso and Carter, 2010; Montalvo et al., 2005; Pathom-aree et al., 2006).

The extraordinary metabolic richness of bacterial and fungal strains triggered the interest in the development of specific isolation approaches targeted at enriching microbial collections in novel strains from these species, with special emphasis on actinomycetes and endophytic fungi from a large diversity of specific and unique microbial communities (Bills et al., 2009; Genilloud et al., 2011). Other bacterial groups, such as myxobacteria, pseudomonads or cyanobacteria, were also actively explored by different research groups as potential sources of new bioactive compounds (Dittmann et al., 2015; Raaijmakers and Mazzola, 2012; Wenzel and Müller, 2009).

The access to still untapped ecosystems has required to introduce the use of novel isolation methods targeting the cultivation of potentially minor species frequently underrepresented or previously not cultivated under standard laboratory conditions (Collado et al., 2007; Davis et al., 2005; Hamaki et al., 2005; Janssen et al., 2002; Suzuki, 2001; Taechowisan et al., 2003; Tamaki et al., 2005). This ensured the isolation of novel or less frequently occurring species that are becoming common members of industrial screening collections (Janssen et al., 2002; Suzuki, 2001).

In the case of bacteria, and especially actinomycetes, a large diversity of approaches were developed for the selective isolation of members of these different taxa combining the use of poor nutritional media devoid of carbon sources with extended incubation periods, and in some cases, subinhibitory concentrations of antibiotics enabling the development of slow-growing members of these microbial communities after weeks of incubation. Many specific enrichments in certain taxa have been described in the last 15 years (Hayakawa et al., 2000; Kizuka et al., 1997), but the most productive approaches resulted from systematic isolation of colonies from nutrient-poor and slow-growing isolation plates (Janssen et al., 2002) and more recently on the domestication of previously uncultured species using specially designed diffusion chambers (Lewis et al., 2010; Pascual et al., 2016). In the case of fungi, the approaches have included methods involving high throughput culturing of new strains using

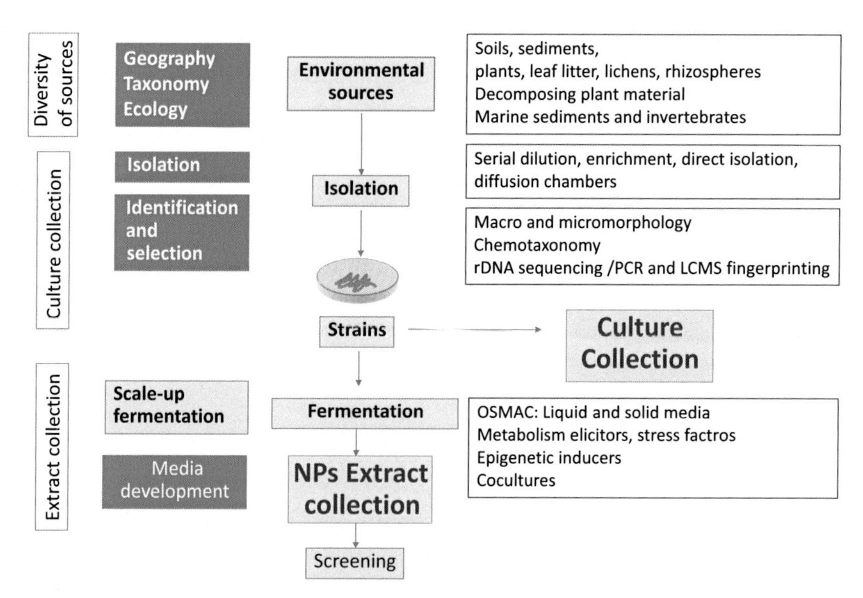

FIGURE 12.1 **Key stages in the construction of industrial culture collections.**

serial dilutions and extinction methods, methods that were easily applied to washed particles obtained from plant materials and other types of substrates, as well as direct isolation of individual endophytes from plant substrates (Collado et al., 2007; Bills et al., 2009).

The continued study of the wide diversity of terrestrial and marine habitats including endophytic plant species and symbiotic assemblages of bacteria and fungi, has shown that the capacity to produce novel compounds by secondary metabolism is broadly distributed across many bacterial and fungal taxa, emphasizing the need to apply key selection criteria to ensure that the strains of major interest and biosynthetic potential are efficiently selected (Bills et al., 2009, 2012; Crawford and Clardy, 2011; Jensen et al., 2007; Singh et al., 2010; Zotchev, 2012) (Fig. 12.1).

Despite these efforts to diversify the microbial sources, only a small fraction of microbial species has been isolated and tested for bioactivity (Bérdy, 2012). Traditional cultivation-based methods often fail to isolate the dominant members of microbial communities and the lack of cultivability of the majority of microbial species in laboratory conditions is therefore limiting the characterization of a large diversity of microbial sources remaining largely underexplored (Rappe and Giovannoni, 2003). New methods have been developed for the isolation and domestication of new strains as part of complex microbial communities that cannot be cultured in standard laboratory conditions or require the interaction with other species for survival or the production of secondary metabolites (Lewis, 2013; Pascual et al., 2016). In some cases, isolation efforts that were targeted at unusual microbial niches have relied on the understanding of the natural interactions between species that can lead to culturability, and they succeeded in the isolation of novel compounds from the entomopathogenic bacteria *Photorhabdus luminescens* and *Xenorhabdus nematophila,* or strict anaerobe *Clostridium* species (Fuchs et al., 2011; Lincke et al., 2010). The use of miniaturized incubation chambers exposed to environment signals, and the presence of other community members required for growth in artificial media have shown new ways to cultivate previously uncultured groups (Kaeberlein et al., 2002; Lewis, 2013; Lewis et al., 2010; Pascual et al., 2016; Stewart, 2012). These approaches still represent a serious challenge and from a cultured-based approach they are being overlooked using adapted liquid production media and long incubation periods, the potential of these new groups to produce novel compounds.

STRAIN SELECTION AND DEREPLICATION TOOLS

Building any screening microbial collection is also limited by the number of strains that can be selected and maintained, and requires a rigorous selection process to ensure that the broadest diversity of strains is included in the collection. In addition, it is well known that despite the intensive exploitation of a substrate to identify and isolate novel lineages from a taxonomic perspective, the efforts invested in observing this diversity do not always guarantee the production of novel compounds. Horizontal transfer events may be at the origin of microbial evolution, and convergence of biosynthetic pathways within a microbial community can determine the dispersion of biosynthetic genes among different members of the microbial community (Egan et al., 2001; Morningstar et al., 2006).

Different methods have been developed to be used as selection criteria including from the most immediate macroscopic observation of the colony growth and the micromorphological description providing specific characters of the new strains, the comparison to other isolates from the same substrate, to more effective early selection tools based on chemotaxonomic, physiologic, and molecular analyses. All these tools have been applied to ensure the highest microbial diversity in the collections (Table 12.1). To be efficiently applied to large numbers of strains, these methods required to be rapid and reproducible to permit the comparison with existing species data bases in the collections. Whereas morphological characterization of fungi has been critical for the identification of the isolates, the lack of specific characters in most of the bacterial groups required to implement alternative high throughput techniques to ensure the distinction of morphologically similar isolates. One of the first methods most intensively used for the dereplication of bacteria was the determination of their fatty acid methyl esters profiles. These phenotypic biomarkers permitted the comparison and clustering of new isolates with strains already in the collection (Bugni and Ireland, 2004).

Molecular tools, such as rDNA and ITS sequencing has been systematically applied as selection criteria of strains in the collection. They permitted to

TABLE 12.1 Bacterial Dereplication Tools

Character	Technology	Coverage	Output
Morphology	Macro and micromorphology	Phenotype	Differentiation structures
Chemotaxonomy	Chemical fingerprinting	Intraspecific	GC chemo markers: fatty acids
	Metabolic fingerprinting	Whole metabolome intraspecific	LC-MS metabolite profiling: small molecules
	Protein fingerprinting		MALDI-TOF MS (BIOTYPER)
Molecular tools	DNA fingerprinting	Intraspecific, whole genome	Ribosomal primers: polymorphism ribosomal spacers,
		Intraspecific, whole genome	tRNA PCR REP primers, (repetitive sequences): BOX, REP, ERIC, GTG5
	Metabolic fingerprinting	Metabolic pathways	PCR profiling
	DNA sequencing	Phylogenetic studies	Individual genes: rDNA sequencing and house-keeping genes
		Full annotation	Whole genome

confirm not only the taxonomic affiliation of the new strains but as well to infer individual phylogenetic relationships among strains, and reveal their taxonomic position in a given niche. Despite of the high value of these sequences, this information cannot be used upfront as selection criterion to infer the diversity of the strains within a given clade. The intraspecies heterogeneity in microbial taxa cannot be resolved in inner branches, requiring alternative dereplication tools to assess the uniqueness of potential new strains. The intraspecific variability between strains of the same species isolated from the same environment is frequent and requires additional tools for the dereplication of the strains, such as PCR fingerprinting methods. Many different molecular fingerprinting approaches have been described and used according to their level of resolution: amplified fragment length polymorphism (AFLPs), random amplified polymorphic DNA (RAPDS), repetitive sequence-based PCR (REP-PCR) fingerprinting, random BOX-PCR amplification (PCR based on primers targeting the highly conserved repetitive DNA sequences of the BOX element of *Streptococcus pneumoniae*). BOX-PCR amplification was shown to be one of the most useful tools to detect variability within actinomycetes species and to establish the relatedness of bacterial isolates (Ayuso et al., 2005; Lanoot et al., 2005; Versalovic et al., 1994; Vos et al., 1995; Welsh and McClelland, 1990).

The advent of matrix-assisted laser desorption/ionization (MALDI) time-of-flight mass spectrometry (TOF MS) as a tool for microbial characterization has permitted to implement a new profiling method of bacterial proteins allowing strain resolution level that could be directly applied on isolates without precultivation (Sandrin et al., 2012).

Alternative molecular fingerprints were also developed on the basis of the restriction pattern of polyketide synthase (PKS) and nonribosomal peptide synthetase (NRPS) biosynthetic genes that might provide additional information on their potential metabolic diversity and permitted to generate strain-specific profiles after enzymatic restriction of these PCR products (Ayuso-Sacido and Genilloud, 2005; Yarbrough et al., 1993).

CULTURE-BASED APPROACHES AND GENERATION OF LIBRARIES OF MICROBIAL EXTRACTS

The exploitation of the extraordinary richness in biosynthetic genes of different microbial groups requires the cultivation and expression of secondary metabolites. The generation of library of extracts is always limited in numbers and is determined by the combination of the number of strains and the production conditions that can be selected, making always critical the selection criteria applied. Traditionally, production of most known compounds has been triggered just by empirical changes in cultivation parameters and a limited number of stress conditions generally accepted as sufficient to exploit the production of new secondary metabolites (Hantke, 2001).

One of the first challenges when dealing with wild-type strains with little or no knowledge of their physiological characteristics has been to address how to manipulate the cultivation conditions by modulating culture growth parameters and introducing stress factors that might influence drastically their secondary metabolism. Microbial metabolism is greatly influenced by many factors related to media composition, oxygen limitation, temperature, pH, osmolarity, redox stress, and CO_2 concentration are well known factors affecting NPs production. The effect of many of these factors has been well documented in previous works involving modifications of these factors (Hesketh et al., 2009; Martin, 2004; Rigali et al., 2008; Ruiz et al., 2010). The effect on NPs production on the availability of carbon sources, phosphate depletion, disruption of zinc and iron homeostasis, or the presence of polysaccharide intermediate, such as chitin derived GlcNAc have been well documented and are routinely included in media formulations (Bentley et al., 2002; Hesketh et al., 2009; Martin, 2004; Ruiz et al., 2010).

Genome sequence analyses of most important secondary metabolite producers have shown that a large fraction of the microbial genomes still remains unexploited, and that switching on silent pathways might reveal the production of novel molecules (Donadio et al., 2007; Ikeda et al., 2003; Oliynyk et al., 2007; Rutledge and GL Challis, 2015; Udwary et al., 2007). Today, despite the numerous whole-genome studies confirming the presence of this huge biosynthetic potential, in most of the cases only revealed as cryptic pathways, we still cannot predict the nutritional requirements and physiology of most of the groups of strains screened and the signals involved in the regulation of their secondary metabolite production. The introduction of multiple nutritional conditions with miniaturized parallel fermentation devices has been intensively exploited in industrial laboratories to expand the possibilities of triggering the production of new secondary metabolites from large numbers of strains of actinomycetes and fungi (Duetz and Witholt, 2001; Duetz, 2007). All major taxonomic groups of actinomycetes have been cultivated in a large variety of complex and synthetic liquid media, even with long incubations, with a parallel testing of a large number of nutritional conditions opening the opportunity to explore minor groups of isolates and understand their requirements. The fermentation of a large collection of new isolates using a multiple production media in microwell plate format following an OSMAC approach has shown the increase in the number of antibiotic activities identified against a broad panel of human bacterial and fungal pathogens (Bills et al., 2008; Genilloud et al., 2011). These results provided new evidence supporting that proper manipulation of media conditions can further promote the biosynthetic potential of microorganisms and increase the chances of detecting novel molecules from active extracts selected to be pursued in chemical isolation projects (Bode et al., 2002; Genilloud et al., 2011).

Different classes of chemical inducers, such as metabolism intermediates, siderophores, and rare earths have been used to stimulate the production of novel bacterial secondary metabolites (Kawai et al., 2007; van Wezel and McDowall, 2011; Yamanaka et al., 2005). Small diffusible bacterial hormone-like molecules, such as the γ-butyrolactones, and other widely distributed butenolides have been used to enhance antibiotic production in many species of the genus *Streptomyces* (Corre et al., 2008; Nishida et al., 2007; Tezuka and Ohnishi, 2014). Avenolide, a butenolide-type autoregulator from *Streptomyces avermitilis*, is one of the most recent examples of this class of signaling molecules and it has been shown to be critical to elicit avermectin production (Kitani et al., 2011). Other elicitors of secondary metabolism, such as N-acetylglucosamine or ribosomal mutations resulting in altered ppGpp biosynthesis and catabolite repression, have been used to activate cryptic pathways and the modulation of precursors has been shown as another efficient way to elicit the synthesis of previously undetected compounds (Craney et al., 2012; Hosaka et al., 2009; Ruiz et al., 2010).

Epigenetic modulation by histone acetylation and methylation is another gene expression regulatory level that has shown to determine dramatic changes on the production of bioactive compounds in fungi (Brakhage and Schroeckh, 2011). Small-molecule epigenetic modifiers used to inhibit histone deacetylase (HDAC) or DNA methyltransferase (DMAT) can be used to activate unpredicted silent natural product pathways in different fungal species (De La Cruz et al., 2012; González-Menéndez et al., 2016; Henrikson et al., 2009; Williams et al., 2008. Similarly to fungi in their control of biosynthetic pathways, HDAC orthologues have been identified to be broadly distributed not only among *Streptomyces* but in members of *Pseudonocardia, Saccharopolyspora*, and *Amycolatopsis*, prolific taxa for production of secondary metabolites (Moore et al., 2012).

LC-MS PROFILING AND METABOLITE ANNOTATION

The application of liquid chromatography-mass spectrometry (LC-MS) to study the production of metabolites has been extremely useful to monitor the performance of strains in many different conditions. In addition, the use of early dereplication methods to identify known bioactive compounds have been essential tools to select the most diverse strains from culture collections (Genilloud et al., 2011; Moore et al., 2012). Frequently produced NPs are synthetized by a broad diversity of extensively screened taxonomic groups of actinomycetes and fungi, whereas many novel compounds have been shown to be species-specific (Genilloud et al., 2011; Ziemert et al., 2004). Recent diversity studies aimed at understanding the distribution of biosynthetic gene clusters have shown that the production of many compounds is concentrated in certain species (Genilloud et al., 2011). Many NPs have been identified to be produced only once by unique strains isolated from very specific habitats. The limited characterization of the microbial metabolic diversity that remains to be exploited and the richness in microbial biosynthetic pathways observed from genetic studies with many of the newly explored taxa provide a measure of the need to expand the search for novel compounds among poorly explored taxa and microbial assemblages not-yet cultivated involving new bacterial and fungal species. Early dereplication of known molecules has been ensured through systematic analysis of combined LC-MS and Ultra Violet (UV) profiles of each of the active extracts, and screening of these data against a proprietary LC-MS library of known antibiotic compounds. In spite of the wide occurrence of known molecules, and the challenge of identifying these compounds early in the process, microbial strains may produce interesting unknown compounds continuously masked by potent or highly produced bioactive molecules. New approaches to identify novel compounds from industrial microbial collections take advantage of the continued expanding knowledge related with the genome and metabolome of microbial strains (Alam et al., 2010; Challis, 2008; Wagenaar, 2008), semiautomated generation of enriched semifractionated samples in screening, and high-resolution analytical LC-MS/NMR techniques to identify the presence of scaffolds in partially purified samples (Koehn, 2008; Wagenaar, 2008), and generation of three-dimensional (3D) databases that will foster future virtual mining of natural products (Lin et al., 2008; Rollinger, 2009). Frequently unrelated species can produce the same compound, supporting the wide distribution of many of these biosynthetic clusters in the microbial community.

To complement the LC-MS annotation of the active extracts, the strains taxonomic affiliation is determined by ribosomal DNA sequences analysis for all the active strains and mapping the known metabolites produced in all these taxa, trying to correlate the phylogenetic data with the compounds produced. Besides the partial information reported in public culture collections about the secondary production of the strains, few studies have been performed describing the different bacterial and fungal species known to produce bioactive molecules. Contrary to what has been observed in filamentous fungi (Bills et al., 2008), with small molecules associated to some extent with specific families of fungi, the production of frequently occurring compounds studied in several actinomycetes families is widespread across the different lineages and cannot be related to any specific taxon, probably reflecting horizontal gene transfer events that have been reported as frequently occurring among actinomycetes (Egan et al., 2001; Thomas and Nielsen, 2005) in the environment.

IMPACT OF GENOMICS AND GENOME MINING TO EXPLOIT MICROBIAL CULTURE COLLECTIONS

The increase in available genome sequences from public databases has confirmed the large number and diversity of biosynthetic gene clusters involving NPs. Unnumbered cases of silent biosynthetic genes pathways derived from genetic analysis of these strains have unveiled the potential production of previously unknown NPs, and have changed the perception of the strains, especially fungi and actinomycetes, as efficient secondary metabolite producers and untapped reservoirs of NPs (Lin et al., 2008; Moore et al., 2012; Rollinger, 2009). As many as 5,031 full bacterial genome sequences and 671 fungal genomes are available in public databases in 2016 (http://www.ncbi.nlm.nih.gov/genome/browse/), and comparative genome analysis has revealed that the genes involved in the production and regulation of secondary metabolites biosynthesis are much higher than those anticipated although most of them remain silent in the absence of induction signals. This metabolic potential is under represented in microbial natural products libraries and the need to exploit these resources has set up the basis for a new NP discovery paradigm based on genome mining (Challis, 2008; Nett et al., 2009; Zerikly and Challis, 2009). Industrial microbial collections are still underexploited to discover novel metabolic products. Current efforts are aimed at revealing new biosynthetic pathways and increase our understanding of the mechanisms of NPs synthesis, including new NRPS and PKS, the diversity of starter and extender units, terpenes synthases, or other assembly lines, using biosynthetic pathways engineering tools focused to increase chemical diversity. Most NP industrial microbial collections derived from large pharma NPs discovery programs or any other biotechnology application have shown to be extremely prolific in delivering novel compounds and represent an untapped source of metabolic diversity from which novel genes and novel pathways that have been overlooked by traditional empiric screens can be identified by genome mining. This new paradigm that involves the exploitation of these unique resources to express new pathways in laboratory conditions is backed up by the development of new synthetic biology tools, included specialized engineered chassis to overcome limitations due to low expression rates, inappropriate culture conditions, and lack of inducing signals.

APPLICATIONS IN BIOTECHNOLOGY AND ROLE IN FUTURE NPS DISCOVERY

The most intensive research in microbial natural products has been traditionally focused on the discovery of new drug leads for many different therapeutic areas, and natural products discovery is mostly associated to this field. Nevertheless, bacteria and fungi are one of the richest and most productive factories optimized by nature to produce a broad diversity of molecules and enzymes with application in all sectors of biotechnology (Fig. 12.2).

Microbial products are seeing an emerging interest in agriculture where new applications are being developed as part of the interest in developing more sustainable crop protection products. Compounds produced by bacterial and fungal strains have been developed as nature friendly biopesticides and agents stimulating plant growth and control regulators. Many of these bioactive compounds have been found to be produced by a broad variety of proteobacteria, bacilli and actinomycetes species isolated from rhizospheres, and endophytic fungal species. The agro industry is very much interested in exploring these strains as products given their potential to produce compounds that will help crops to resist the high salinity conditions and temperature stresses of many countries of the world. Another key role of some groups of microbial strains in this bioeconomy is their capacity to synthetize primary metabolites that are produced in high scale to respond to the market needs. These involve the industry scale production of aminoacids, sugars, vitamins, as well as colorants, flavors, food additives that are directly commercialized and consumed or used as starting materials for industrial processes. The cosmetic sector is also looking for attractive strains from exotic niches to be used in the production of flavors and volatile compounds, but there is a high interest in bioactive compounds that could be developed as antioxidants and skin care products. Similarly, actinomycetes and fungi, as saprophytic organisms, are able to grow on almost any substrate and produce hydrolytic enzymes degrading recalcitrant substrates derived from wood and wastes. These microorganisms are key elements in many of the new bioprocesses that are being developed to generate new products from waste recycling as well as other enzymes of high interest to be added to detergents and cleaning products. Microalgae-based

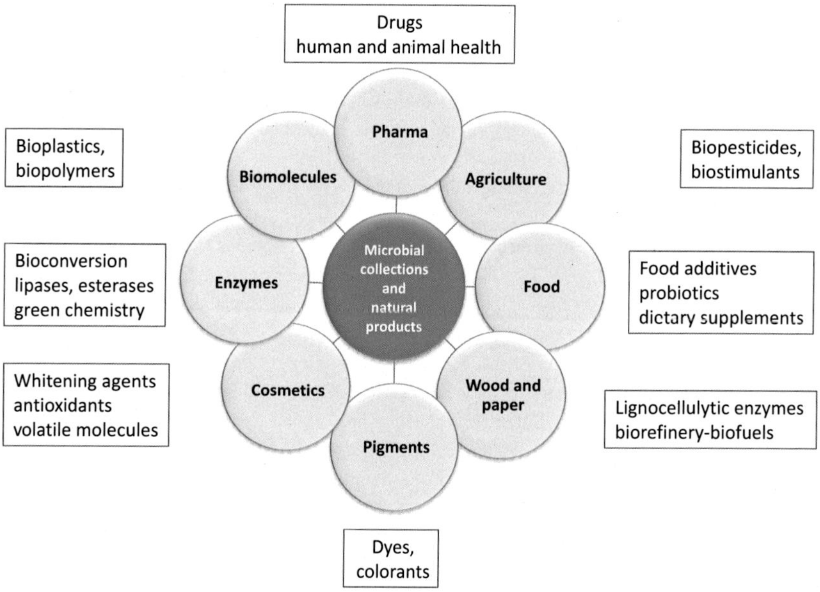

Drugs
human and animal health
Pharma
Biomolecules
Agriculture
Bioplastics,
biopolymers
Biopesticides,
biostimulants
Bioconversion
lipases, esterases
green chemistry
Enzymes
Microbial collections and natural products
Food
Food additives
probiotics
dietary supplements
Whitening agents
antioxidants
volatile molecules
Cosmetics
Wood and paper
Lignocellulytic enzymes
biorefinery-biofuels
Pigments
Dyes,
colorants

FIGURE 12.2 **Biotechnology sectors with industrial applications of microbial derived products.**

bioenergy plants are exploiting the capacity of these bacteria to grow in tanks and capture solar energy to be transformed in new bioproducts. In summary, the examples referred earlier support the role of microbial strains molecules in a wide diversity of industrial applications and in almost any field of biotechnology. All these sectors are sustained by a research community focused at continuing to explore new microbial diversity and identifying novel sources microbial natural products.

The renovated interest in natural product research as sources for new drug discovery will impact as well in the development of culture collections. In this sense, as more biotechnology products are developed relying on a microorganism-based process, microbial collections will represent unique resources and will continue to play a key role in the emerging new field of bioeconomy and the human health sector.

REFERENCES

Alam, M.T., Merlo, M.E., Takano, E., Breitling, R., 2010. Genomebased phylogenetic analysis of Streptomyces and its relatives. Mol. Phylogenet. Evol. 54, 763–772.

Aly, A.H., Debbab, A., Kjer, J., Proksch, P., 2010. Fungal endophytes from higher plants: a prolific source of phytochemicals and other bioactive natural products. Fungal Divers. 41, 1–16.

Ayuso, A., Clark, D., Gonzalez, I., Salazar, O., Anderson, A., Genilloud, O., 2005. A novel actinomycete strain de-replication approach based on the diversity of polyketide synthase and nonribosomal peptide synthetase biosynthetic pathways. Appl. Microbiol. Biotechnol. 67, 795–806.

Ayuso-Sacido, A., Genilloud, O., 2005. New PCR primers for the screening of NRPS and PKS-I systems in actinomycetes: detection and distribution of these biosynthetic gene sequences in major taxonomic groups. Microbiol. Ecol. 49, 10–24.

Bentley, S.D., Chater, K.F., Cerdeno-Tarraga, A.M., Challis, G.L., Thomson, N.R., James, K.D., Harris, D.E., et al., 2002. Complete genome sequence of the model actinomycete Streptomyces coelicolor A3(2). Nature 417, 141–147.

Bérdy, J., 2012. Thoughts and facts about antibiotics: where we are now and where we are heading. J. Antibiot. (Tokyo) 65, 385–395.

Bills, G.F., Platas, G., Fillola, A., Jimenez, M.R., Collado, J., Vicente, F., Martin, J., Gonzalez, A., Bur-Zimmermann, J., Tormo, J.R., Pelaéz, F., 2008. Enhancement of antibiotic and secondary metabolite detection from filamentous fungi by growth on nutritional arrays. J. Appl. Microbiol. 104, 1644–1658.

Bills, G., Overy, D., Genilloud, O., Pelaez, F., 2009. Contributions of pharmaceutical antibiotic and secondary metabolite discovery to the understanding of microbial defense and antagonism. In: White, J., Torres, M.S. (Eds.), Defensive Mutualism in Microbial Symbiosis. Dekker, New York.

Bills, G.F., González-meéndez, V., Platas, G., 2012. *Kabatiella bupleuri* sp. nov.(Dothideales), a pleomorphic epiphyte and endophyte of the Mediterranean plant *Bupleurum gibraltarium* (*Apiaceae*). Mycologia 104, 962–973.

Blum, E., 1993. Making biodiversity profitable: a case study of the Merck/INBio agreement. Environment 35, 38–45.

Bode, H.B., Bethe, B., Hofs, R., Zeeck, A., 2002. Big effects from small changes: possible ways to explore nature's chemical diversity. Chembiochem 3, 619–627.

Brakhage, A.A., Schroeckh, V., 2011. Fungal secondary metabolites. Strategies to activate silent gene clusters. Fungal Genet. Biol. 48, 15–22, 2011.

Bredholt, H., Fjærvik, E., Johnsen, G., Zotchev, S.B., 2008. Actinomycetes from sediments in the Trondheim Fjord, Norway: diversity and biological activity. Mar. Drugs 6, 12–24.

Bugni, T.S., Ireland, C.M., 2004. Marine-derived fungi: a chemically and biologically diverse group of microorganisms. Nat. Prod. Rep. 21, 143–163.

Challis, G.L., 2008. Mining microbial genomes for new natural products and biosynthetic pathways. Microbiology 154, 1555–1569.

Collado, J., Platas, G., Paulus, B., Bills, G.F., 2007. High-throughput culturing of fungi from plant litter by a dilution-to-extinction technique. FEMS Microbiol. Ecol. 60, 521–533.

Convention on Biological Diversity, 2016. Available from: https://www.cbd.int/abs/ 2016.

Corre, C., Song, L., O'Rourke, S., Chater, K.F., Challis, G.L., 2008. 2-Alkyl-4-hydroxymethylfuran-3-carboxylic acids, antibiotic production inducers discovered by Streptomyces coelicolor genome mining. Proc. Natl. Acad. Sci. USA 105, 17510–17515.

Craney, A., Ozimok, C., Pimentel-Elardo, S.M., Capretta, A., Nodwell, J.R., 2012. Chemical perturbation of secondary metabolism demonstrates important links to primary metabolism. Chem. Biol. 19, 1020–1027.

Crawford, J.M., Clardy, J., 2011. Bacterial symbionts and natural products. Chem. Commun. 47, 7559–7566.

Davies, J., 2011. How to discover new antibiotics: harvesting the parvome: havesting the parvome. Curr. Opin. Chem. Biol. 15, 5–10.

Davis, K.E.R., Joseph, S.J., Janssen, P.H., 2005. Effects of growth medium, inoculum size, and incubation time on culturability and isolation of soil bacteria. Appl. Environ. Microbiol. 71, 826–834.

De La Cruz, M., Martín, J., González-Menéndez, V., Pérez-Victoria, I., Moreno, C., Tormo, J.R., El Aouad, N., Guarro, J., Vicente, F., Reyes, F., et al., 2012. Chemical and physical modulation of antibiotic activity in Emericella species. Chem. Biodivers. 9, 1095–1113.

Dittmann, E., Gugger, M., Sivonen, K., Fewer, D.P., 2015. Natural product biosynthetic diversity and comparative genomics of the cyanobacteria. Trends Microbiol. 23, 642–652.

Donadio, S., Monciardini, P., Sosio, M., 2007. Polyketide synthases and nonribosomal peptide synthetases: the emerging view from bacterial genomics. Nat. Prod. Rep. 24, 1073–1109.

Donadio, S., Maffioli, S., Monciardini, P., Sosio, M., Jabes, D., 2010. Antibiotic discovery in the twenty-first century: current trends and future perspectives. J. Antibiot. (Tokyo) 63, 423–430.

Duetz, W.A., 2007. Microtiter plates as mini-bioreactors: miniaturization of fermentation methods. Trends Microbiol. 15, 469–475.

Duetz, W.A., Witholt, B., 2001. Effectiveness of orbital shaking for the aeration of suspended bacterial cultures insquare-deepwell microtiter plates. Biochem. Eng. J. 7, 113–115.

Egan, S., Wiener, P., Kallifidas, D., Wellington, E.M.H., 2001. Phylogeny of Streptomyces species and evidence for horizontal transfer of entire and partial antibiotic gene clusters. Antonie van Leeuwenhoek 79, 127–133.

Fiedler, H.-P., Bruntner, C., Bull, A.T., Ward, A.C., Goodfellow, M., Potterat, O., Puder, C., Mihm, G., 2005. Marine actinomycetes as a source of novel secondary metabolites. Antonie van Leeuwenhoek 87, 37–42.

Fuchs, S.W., Proschak, A., Jaskolla, T.W., Karas, M., Bode, H.B., 2011. Structure elucidation and biosynthesis of lysine-rich cyclic peptides in Xenorhabdus nematophila. Org. Biomol. Chem. 9, 3130–3132.

Genilloud, O., Gonzalez, I., Salazar, O., Martin, J., Tormo, J.R., Vicente, F., 2011. Current approaches to exploit actinomycetes as a source of novel natural products. J. Ind. Microbiol. Biotechnol. 38, 375–389.

González, I., Ayuso, A., Anderson, A., Genilloud, O., 2005. Actinomycetes isolated from lichens: evaluation of their diversity and detection of biosynthetic gene sequences. FEMS Ecol. Microbiol. 54, 401–415.

González-Menéndez, V., Pérez-Bonilla, M., Pérez-Victoria, I., Martín, J., Muñoz, F., Reyes, F., Tormo, J.R., Genilloud, O., 2016. Multicomponent Analysis of the Differential Induction of Secondary Metabolite Profiles in Fungal Endophytes. Molecules 21, 234.

Hamaki, T., Suzuki, M., Fudou, R., Jojima, Y., Kajiura, T., Tabuchi, A., Sen, K., Shibai, H., 2005. Isolation of novel bacteria and actinomycetes using soil-extract agar medium. J. Biosci. Bioeng. 99, 485–492.

Hantke, K., 2001. Iron and metal regulation in bacteria. Curr. Opin. Microbiol. 4, 172–177.

Hayakawa, M., Otoguro, M., Takeuchi, T., Yamazaki, T., Iimura, Y., 2000. Application of a method incorporating differential centrifugation for selective isolation of motile actinomycetes in soil and plant litter. Antonie Van Leeuwenhoek 83, 107–116.

Henrikson, J.C., Hoover, A.R., Joyner, P.M., Cichewicz, R.H., 2009. A chemical epigenetics approach for engineering the in situ biosynthesis of a cryptic natural product from Aspergillus niger. Org. Biomol. Chem. 7, 435–438.

Hesketh, A., Kock, H., Mootien, S., Bibb, M., 2009,. The role of absC, a novel regulatory gene for secondary metabolism, in zinc-dependent antibiotic production in *Streptomyces coelicolor* A3(2). Mol. Microbiol. 74, 1427–1444.

Hosaka, T., Ohnishi-Kameyama, M., Muramatsu, H., Murakami, K., Tsurumi, Y., Kodani, S., Yoshida, M., Fujie, A., Ochi, K., 2009. New strategies for drug discovery: activation of silent or weakly expressed microbial gene clusters. Nat. Biotechnol. 27, 462–464.

Ikeda, H., Ishikawa, J., Hanamoto, A., Shinose, M., Kikuchi, H., Shiba, T., Sakaki, Y., Hattori, M., Omura, S., 2003. Complete genome sequence and comparative analysis of the industrial microorganism Streptomyces avermitilis. Nat. Biotechnol. 21, 526–531.

Janso, J.E., Carter, G.T., 2010. Phylogenetically unique andophytic Actinomycetes from tropical plants possess great biosynthetic potential. Appl. Environ. Microbiol. 76, 4377–4386.

Janssen, P.H., Yates, P.S., Grinton, B.E., Taylor, P.M., Sait, M., 2002. Improved culturability of soil bacteria and isolation in pure culture of novel members of the divisions *Acidobacteria, Actinobacteria, Proteobacteria,* and *Verrucomicrobia*. Appl. Environ. Microbiol. 68, 2391–2396.

Jensen, P.R., Williams, P.G., Oh, D.-C., Zeigler, L., Fenical, W., 2007. Species specific secondary metabolite production in marine actinomycetes of the genus *Salinispora*. Appl. Environ. Microbiol. 73, 1146–1152.

Kaeberlein, T., Lewis, K., Epstein, S.S., 2002. Isolating uncultivable microorganisms in pure culture in a simulated natural environment. Science 296, 1127–1129.

Kawai, K., Wang, G., Okamoto, S., Ochi, K., 2007. The rare earth, scandium, causes antibiotic overproduction in *Streptomyces spp.* FEMS Microbiol. Lett. 274, 311–315.

Kitani, S., Miyamoto, K.T., Takamatsu, S., Herawati, E., Iguchi, H., Nishitomi, K., Uchida, M., Nagamitsu, T., Omura, S., Ikeda, H., Nihira, T., 2011. Avenolide, a Streptomyces hormone controlling antibiotic production in *Streptomyces avermitilis*. Proc. Natl. Acad. Sci. USA 108, 16410–16415.

Kizuka, M., Enokita, R., Takahashi, K., Okazaki, T., 1997. Distribution of the actinomycetes in the Republic of South Africa investigated using a newly developed isolation method. Actinomycetologica 11, 54–58.

Koehn, F.E., 2008. High impact technologies for natural products screening. In: Petersen, F., Amstutz, R. (Eds.), Progress in Drug Research 65, Natural Products as Drugs I. Basel, Birkhauser, pp. 177–210.

Lanoot, B., Vancanneyt, M., Van Schoor, A., Liu, Z., Swings, J., 2005. Reclassification of *Streptomyces nigrifaciens* as a later synonym of *Streptomyces flavovirens*; *Streptomyces citreofluorescens, Streptomyces chrysomallus* subsp. chrysomallus and *Streptomyces fluorescens* as later synonyms of *Streptomyces anulatus; Streptomyces chibaensis* as a later synonym of *Streptomyces corchorusii; Streptomyces flaviscleroticus* as a later synonym of *Streptomyces minutiscleroticus*; and *Streptomyces lipmanii, Streptomyces griseus* subsp. alpha, *Streptomyces griseus subsp. cretosus and Streptomyces willmorei as later synonyms of Streptomyces microflavus.* Int. J. Syst. Evol. Microbiol. 55, 729–731.

Lewis, K., 2013. Platforms for antibiotic discovery. Nat. Rev. Drug Discov. 12, 371–387.

Lewis, K., Epstein, S., D'Onofrio, A., Ling, L.L., 2010. Uncultured microorganisms as a source of secondary metabolites. J. Antibiot. 63, 468–476.

Lin, Y., Schiavo, S., Orjala, J., Vouros, P., Kautz, R., 2008. Microscale LC-MS-NMR platform applied to the identification of active cyanobacterial metabolites. Anal. Chem. 80, 8045–8054.

Lincke, T., Behnken, S., Ishida, K., Roth, M., Hertweck, C., 2010. Closthioamide: an unprecedented polythioamide antibiotic from the strictly anaerobic bacterium *Clostridium cellulolyticum*. Angew. Chem. Int. Ed. Engl. 49, 2011–2013.

Martin, J.F., 2004. Phosphate control of the biosynthesis of antibiotics and other secondary metabolites is mediated the PhoR-PhoP system: an unfinished story. J. Bacteriol. 186, 5197–5201.

Martiny, J.B.H., Bohannan, B.J.M., Brown, J.H., Colwell, R.K., Fuhrman, J.A., Green, J.L., Horner-Devine, M.C., Kane, M., Krumins, J.A., Kuske, C.R., Morin, P.J., Naeem, S., Øvrea°s, L., Reysenbach, A.-L., Smith, V.H., Staley, J.T., 2006. Microbial biogeography: putting microorganisms on the map. Nat. Rev. Microbiol. 4, 102–112.

Montalvo, N.F., Mohamed, N.M., Enticknap, J.J., Hill, R.T., 2005. Novel actinobacteria from marine sponges. Antonie van Leeuwenhoek 87, 29–36.

Moore, J.M., Bradshaw, E., Seipke, R.F., Hutchings, M.I., McArthur, M., 2012. Methods in enzymology, natural product biosynthesis by microorganisms and plants, Part C. Hopwood, D.

(Ed.), Use and Discovery of Chemical Elicitors that Stimulate Biosynthetic Gene Clusters in Streptomyces Bacteria, vol. 517, Academic Press, Oxford, pp. 368–385.

Morningstar, A., Gaze, W.H., Tolba, S., Wellington, E.M., 2006. Evolving gene clusters in bacteria. In: Logan, A., Lappin-Scott, H.M., Oyston, P.C. (Eds.), Prokaryotic Diversity: Mechanisms and Significance. Cambridge University Press, Cambridge, pp. 201–222.

Nett, M., Ikeda, H., Moore, B.S., 2009. Genomic basis for natural product biosynthetic diversity in the actinomycetes. Nat. Prod. Rep. 26, 1362–1384.

Nishida, H., Ohnishi, Y., Beppu, T., Horinouchi, D., 2007. Evolution of c-butyrolactone synthases and receptors in Streptomyces. Environ. Microbiol. 9, 1986–1994.

Oliynyk, M., Samborskyy, M., Lester, J.B., Mironenko, T., Scott, N., Dickens, S., Haydock, S.F., Leadlay, P.F., 2007. Complete genome sequence of the erythromycin-producing bacterium *Saccharopolyspora erythraea* NRRL23338. Nat. Biotechnol. 25, 447–453.

Pascual, J., García-López, M., Bills, G.F., Genilloud, O., 2016. *Longimicrobium terrae* gen. nov., sp. nov., a novel oligotrophic bacterium of the underrepresented phylum *Gemmatimonadetes* isolated through a system of miniaturized diffusion chambers. Int. J. Syst. Evol. Microbiol. 66, 1976–1985.

Pathom-aree, W., Stach, J.E.M., Ward, A.C., Horikoshi, K., Bull, A.T., Goodfellow, M., 2006. Diversity of actinomycetes isolated from challenger deep sediment (10,898 m) from the Mariana Trench. Extremophiles 10, 181–189.

Raaijmakers, J.M., Mazzola, M., 2012. Diversity and natural functions of antibiotics produced by beneficial and plant pathogenic bacteria. Ann. Rev. Phytopathol. 50, 403–424.

Rappe, M.S., Giovannoni, S.J., 2003. The uncultured microbial majority. Ann. Rev. Microbiol. 57, 369–394.

Rigali, S., Titgemeyer, F., Barends, S., Mulder, S., Thomae, A.W., Hopwood, D.A., van Wezel, G.P., 2008. Feast or famine: the global regulator DasR links nutrient stress to antibiotic production by *Streptomyces*. EMBO Rep. 9, 670–675.

Rollinger, J.M., 2009. Accessing target information by virtual parallel screening—the impact on natural product research. Phytochem. Lett. 2, 53–58.

Ruiz, B., Chavez, A., Forero, A., Garcia-Huante, Y., Romero, A., Sanchez, M., Rocha, D., Sanchez, B., Rodriguez-Sanoja, R., Sanchez, S., Langley, E., 2010. Production of microbial secondary metabolites: regulation by the carbon source. Crit. Rev. Microbiol. 36, 146–167.

Rutledge, P.J., GL Challis, G.L., 2015. Discovery of microbial natural products by activation of silent biosynthetic gene clusters. Nat. Rev. Microbiol. 13, 509–523.

Sandrin, T.R., Goldstein, J.E., Schumaker, S., 2012. MALDI TOF MS profiling of bacteria at the strain level: a review. Mass Spectrom. Rev. 32, 188–217.

Singh, S.B., Genilloud, O., Pelaez, F., 2010. NP structural diversity II—secondary metabolite sources, evolution and selected molecular structures: terrestrial micro-organisms—bacteria. In: Moore, B., Crews, P. (Eds.), Comprehensive Natural Products Chemistry II. Elsevier, Oxford.

Stackebrandt, E., Schüngel, M., Dunja Martin, D., Smith, D., 2015. The microbial resource research infrastructure MIRRI: strength through coordination. Microorganisms 3, 890–902.

Stewart, E.J., 2012. Growing unculturable bacteria. J. Bacteriol. 194, 4151–4160.

Suzuki, S., 2001. Establishment and use of gellan gum media for selective isolation and distribution survey of specific rare actinomycetes. Actinomycetologica 15, 55–60.

Taechowisan, T., Peberdy, J.F., Lumyong, S., 2003. Isolation of endophytic actinomycetes from selected plants and their antifungal activity. World J. Microbiol. Biotechnol. 19, 381–385.

Tamaki, H., Sekiguchi, Y., Hanada, S., Nakamura, K., Nomura, N., Matsumura, M., Kamagata, Y., 2005. Comparative analysis of bacterial diversity in freshwater sediment of a shallow eutrophic lake by molecular and improved cultivation-based techniques. Appl. Environ. Microbiol. 71, 2162–2169.

Tezuka, T., Ohnishi, Y., 2014. Microbial Hormones as a master switch for secondary metabolism in Streptomyces. In: Anazawa, H., Shimizu, S. (Eds.), Microbial Production. From Genome Design to Cell Engineering. Springer, Japan, pp. 179–190.

Thomas, C.M., Nielsen, K.M., 2005. Mechanisms of, an barriers to horizontal gene transfer between bacteria. Nat. Rev. Microbiol. 3, 711–721.

Thornburg, C.C., Zabriskie, T.M., McPhail, K.L., 2010. Deep-sea hydrothermal vents: potential hot spots for natural products discovery? J. Nat. Prod. 73, 489–499, 2010.

Udwary, D.W., Zeigler, L., Asolkar, R.N., Singan, V., Lapidus, A., Fenical, W., Jensen, P.R., Moore, B.S., 2007. Genome sequencing reveals complex secondary metabolome in the marine actinomycete Salinispora tropica. Proc. Natl. Acad. Sci. USA 104, 10376–10381.

van Wezel, G.P., McDowall, K.J., 2011. The regulation of the secondary metabolism of Streptomyces: new links and experimental advances. Nat. Prod. Rep. 28, 1311–1333.

Versalovic, J., Schneider, M., De Bruijn, F.J., Lupski, J.R., 1994. Genomic fingerprinting of bacteria using repetitive sequence based polymerase chain reaction. Methods Mol. Cell Biol. 5, 25–40.

Vos, P., Hogers, R., Bleecker, M., Reijens, M., Lee, J., Hornes, M., Friders, A., Pot, J., Palemann, J., Kaiper, M., Zabean, M., 1995. AFLP: a new technique for DNA fingerprinting. Nucl. Acids Res. 23, 4407–4414.

Wagenaar, M.M., 2008. Pre-fractionated microbial samples—the second generation natural products library at Wyeth. Molecules 13, 1406–1426.

Walsh, C.T., Wencewicz, T.A., 2014. Prospects for new antibiotics: a molecule-centered perspective. J. Antibiot. 67, 7–22.

Welsh, J., McClelland, M., 1990. Finger printing of genomes using PCR with arbitrary primers. Nucl. Acids Res. 18, 7213–7218.

Wenzel, S.C., Müller, R., 2009. Molecular BioSystems, Myxobacteria—'microbial factories' for the production of bioactive secondary metabolites. Mol. BioSyst. 5, 567–574.

Williams, R.B., Henrikson, J.C., Hoover, A.R., Lee, A.E., Cichewicz, R.H., 2008. Epigenetic remodeling of the fungal secondary metabolome. Org. Biomol. Chem. 6, 1895–1897.

Wink, J., 2016. Atlas of Actinomycetes, DSMZ. Available from: https://www.dsmz.de/bacterial-diversity/compendium-of-actinobacteria.html

World Directory of Culture Collections (WDCM), Sixth version, 2014. Available from: http://www.wfcc.info/ccinfo/index.php/home/statistics/home/

Yamanaka, K., Oikawa, H., Ogawa, H.O., Hosono, K., Shinmachi, F., Takano, H., Sakuda, S., Beppu, T., Ueda, K., 2005. Desferrioxamine E produced by *Streptomyces griseus* stimulates growth and development of *Streptomyces tanashiensis*. Microbiology 151, 2899–2905.

Yarbrough, G.G., Taylor, D.P., Rowlands, R.T., Crawford, M.S., Lasure, L.L., 1993. Screening microbial metabolites for new drugs. Theoretical and practical issues. J. Antibiot. 46, 535–544.

Zerikly, M., Challis, G.L., 2009. Strategies for the discovery of new natural products by genome mining. Chembiochem. 10, 625–663.

Ziemert, N., Lechner, A., Wietz, M., Millán-Aguiñaga, N., Chavarria, K.L., Jensen, P.R., 2004. Diversity and evolution of secondary metabolism in the marine actinomycete genus *Salinispora*. Proc. Natl. Acad. Sci. USA 111, E1130–E1139.

Zotchev, S., 2012. Marine actinomycetes as an emerging resource for the drug development. J. Biotechnol. 159, 168–175.

FURTHER READING

Bode, H.B., 2009. Entomopathogenic bacteria as a source of secondary metabolites. Curr. Opin. Chem. Biol. 13, 224–230.

Butler, M.S., 2004. The role of natural product chemistry in drug discovery. J. Nat. Prod. 67, 2141–2153.

Piel, J., 2004. Metabolites from symbiotic bacteria. Nat. Prod. Rep. 21, 519–538.

Chapter 13

An Overview of Biological Resource Center-Maintenance of Microbial Resources and Their Management

Ken-ichiro Suzuki
Tokyo University of Agriculture, Setagaya-ku, Tokyo, Japan

INTRODUCTION

Different from higher organisms, such as the animals and plants, microorganisms need culture collections for ex situ conservation and to be shared among the scientific communities. A culture collection is an infrastructure collecting, preserving, and providing microbial strains. A microbial "strain" is an operating unit of research similar to the "individual" of higher organisms. "Strain" is defined as a mass of cells originated from single cell isolated from nature and designated by a certain number or symbol, such as a-1, AU 101, etc. by researchers or organizations. When public culture collections accept the deposition of strains from scientists, they give accession numbers to the strains, preserve the strain with the allocated numbers, and then inform the depositor of the accession number by issuing a certificate. The type strains used for taxonomical level characterization of bacteria must be assigned with the allocated strain accession number and used to confirm the identity of a new closely related species are proposed. In the *International Code of the Nomenclature of Prokaryotes*, the type strains for species of prokaryotes must be living cultures available for common use from these public culture collections. Whereas yeasts and filamentous fungi which are governed by the *International Code for Botany* and can assign the taxonomic type by only examining the dried and dead specimen. However, needs of their active cultures are increasing following the identification studies for industrial reasons.

In a published paper, information on the strain designations and their accession numbers issued by the public culture collections are currently indicating the availability of materials for the replication of the study. In additions, current

Microbial Resources. http://dx.doi.org/10.1016/B978-0-12-804765-1.00013-8

culture collections are providing cell components, such as the whole or a part of the chromosomal DNA as well as other related genetic materials. This activity is supporting molecular biologists who want to use the DNA of hard-to-cultivate microorganisms, such as the extremophiles but who may not have the in vitro capabilities to cultivate such microorganisms.

The role of culture collections is not only for scientific research but they can also provide resources for quality control purposes, such as the reference strains used in standardized tests in various fields like food safety, performance of antibiotics, etc. For such tests, the strains to be used are mostly specified at the strain level and the quality requirement is specified for each test.

Microorganisms are controlled for their access and use by various laws and regulations, relevant to the protection from infectious diseases, plant quarantine, etc. Recently, Convention on Biological Diversity have made it difficult to share biological materials for international collaboration. Correct identification of microorganisms with taxonomic background is an essential factor for the compliance with these laws and regulations.

Biological resource center is an advanced level institution of a culture collection, which manages biological materials with high quality and well documented to support the research activities of scientists who use microorganisms for their research.

Culture Collection

Around 1890, a culture collection, which collects, preserves, and distributes microorganism, was firstly established in Prague, Czech Republic by Dr. Frantisek Kral (1846–1911). This collection was aimed for education in clinical bacteriology. Microorganisms, especially bacteria cannot be easily recognized due to their smaller morphology by naked eyes and even by a microscope. First, he prepared a display of cultures as a museum. Then he recognized the importance of supply of living cultures for scientists. Since then, culture collections have been established in the world, such as *Centraalbureau voor Schimmelcultures* (CBS) in the Netherlands in 1904, *American Type Culture Collection* (ATCC) in United States of America in 1925, and *Institute for Fermentation, Osaka* (IFO, now NBRC) in Japan in 1944. Originally the function of these culture collections was to collect microbial strains from the depositors and deposit them with the accompanying information attached by the depositors. This means that the depositors were responsible for the microorganisms in the collections. However, in the current practice the collections are responsible for the quality of the deposited strains, which includes the identity and the related information on the strain once it is provided to public. Culture collections are increasingly becoming the important infrastructure for microbiological research, biotechnology and industrial applications. In addition to important roles of herbaria and museums as collection of taxonomic types of biological samples are recognized within in the scientific community.

History of Codes of Nomenclature

Originally nomenclature of bacteria was initiated under the rules of the *International Code of Botanical Nomenclature* used for plants as well as for yeasts and fungi. However, bacterial nomenclature was separated from the plant one in 1958 and *The International Code of Nomenclature of Bacteria and Viruses* was established. Later, virus nomenclature was also separated in 1966, and bacteriological code was published as *The International Code of Nomenclature of Bacteria* (Editorial Board of the Judicial Commission of the International Committee on Nomenclature of Bacteria, 1966). The code stated that the type of a species or subspecies is preferably a designated type strain or in special cases it may be a description, a preserved specimen or preparation, or an illustration (Rule 9a). Furthermore, it also stated that the type of a species or subspecies is preferably a living strain maintained in a bacteriological laboratory, more particularly in one of the permanently established culture collections from which it would be available for study [Rule 9d(1)]. In the preface of 1975 edition (Lapage et al., 1975), the role of culture collection in bacterial taxonomy was stated as follows: the code recommends that, in the case of cultivable organisms, cultures of the type strains of newly named species and subspecies be deposited in culture collections from which they would be available. Until the 1990 revision, the exception of a living culture for the type strain was accepted by replacement of specimen or illustration for living culture. This was followed by the Bacteriological Code 1990 Revision (Lapage et al., 1992) and the following sentence was added: Rule 18a. The strain had to be maintained in pure culture and should agree closely in its characters with those in the original description." In addition, the recent edition of the *International Code of Nomenclature of Prokaryotes* (Parker et al., 2016) has prevented the possibility of the strain replacement with a statement that "as from 1 January 2001, a description, preserved (nonviable) specimen, or illustration may not serve as the type (Rule 18a)." In addition, as of 1 January 2001, the description of a new species, or new combinations previously represented by viable cultures must include the designation of a type strain, and a viable culture of that strain must be deposited in at least two publicly accessible culture collections in different countries from which subcultures must be available. The designations allotted to the strain by the culture collections should be quoted in the published description [Rule 30 (3) (b)]. Following the Prokaryotic Code, the maintenance of living culture for eukaryotic microorganisms is stated in the Botanical Code as "Whenever practicable a living culture should be prepared from the holotype material of the name of a newly described taxon of algae or fungi and deposited in at least two institutional culture or genetic resource collections (such action does not obviate the requirement for a holotype specimen under Art 8.4) (Recommendation 8B.1 of Article 8) (McNeill et al., 2006). Many of the new taxa or microorganisms recently proposed are from "hard-to-cultivate" microorganisms including extremophiles. Culture collections must therefore be equipped with expertize and advance instrumental techniques for the subsequent cultivation and characterization of these microorganisms.

Method of Preservation of Microbial Strains

Microbial strains must be maintained in the culture collections under good condition to preserve their original characteristics when they were deposited. One of the most apparent markers to evaluate the quality of preservation is viability. Culture collections, at which many and various kinds of microorganisms are preserved, common methods applicable for a wide variety of microorganisms should be employed for efficiency. Sample type and preservation characteristics are also important factors. Common methods used in culture collections are represented by freezing methods and drying methods where both methods aimed to stop the activities of cells for long-term preservation. In reality, they do not get more damages during preservation except external problems. Both methods have been in use and reliable for more than 50 years of effective storage. The freezing method under −80°C has been widely applied to most of bacteria, actinomycetes, yeasts, and spore-forming filamentous fungi only with some exception in fungi. This method is the first choice for a culture collection when it receives the cultures for deposit. The drying methods are less applicable to the variety of microorganisms than freezing methods and rather time-consuming. If either method cannot be applicable, the microbial strain must be maintained by a serial transfer or other methods. Viability is the ratio of colony forming units before and after preservation. Apart from viability or morphology, it is difficult to examine phenotypic characteristics changed (mostly lost) by the preservation. If the interest for the microorganism is a special characteristic, such as productivity of enzymes or secondary metabolites, the suitable method for preservation should be selected by tracing the characteristic which has of value to industry and it is not always correlated with viability.

Freezing Method

Freezing is one of the most essential method for long-term preservation. The cells cultivated in a suitable media are suspended in a solution and preserved in a freezer at −80°C or lower. This temperature is selected based on the phase transition temperature and freezing preservation higher than −50°C is not recommended.

For efficient preservation, the tubes and boxes in a freezer are recommended to be unified. For filamentous fungi, hyphae grown on an agar plate are cut together with the agar medium as 5 mm diameter plugs and placed into the preservation medium to avoid cell damage (Fig. 13.1).

Liquid-nitrogen tank is also an excellent container for the freezing preservation in which the samples are kept around −170°C in the gas phase at the upper part of the tank (Fig. 13.4). Liquid nitrogen tank has an advantage

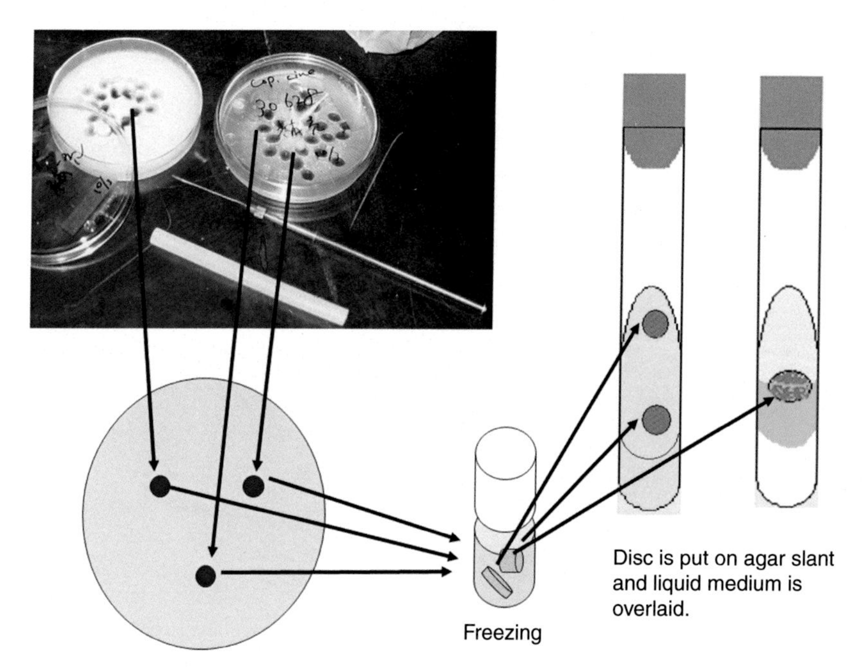

FIGURE 13.1 Freezing preservation of filamentous fungi. Agar plate on which filamentous fungi are grown is cut into 5 mm diameter discs using sterilized straw and put into cryoprotectant. The disc preserved is transferred on a fresh agar medium and incubate at appropriate temperature. *(By courtesy of A. Nakagiri.)*

to keep low temperature for several weeks without supply of electricity. Check points for freezing method is the cooling rate and the composition of the medium (cryoprotectant). Cooling rate have been generally recommended to be slow. However, it is also good to freeze cells quickly using liquid nitrogen.

The capacity of a deep freezer with 700 L is 27,000 tubes in 270 boxes each of which contains 100 of 2 mL cryotubes (Figs. 13.2–13.4).

Generally, prokaryotes, such as bacteria and actinomycetes which have small cells in size are not affected by the cooling rate in freezing. They can be frozen by putting the tubes in a freezer directly. However, filamentous fungi, which have relatively larger cells, are sensitive to the freezing process. Some fungi are recommended to be frozen by using programmed freezer providing 1°C/min cooling (Fig. 13.5). If this system is widely adapted, it will expand the range of filamentous fungi to be preserved by freezing. The medium used for cryoprotectant is usually 10%–20% (v/v) glycerol solution mostly in water. Sometimes it is necessary to modify the basal medium in consideration of the cultivation media, such as salt or sugar concentration. However, in case of the optimum cultivation medium is acidic or alkaline, the preservation medium should be neutral and the cells for

FIGURE 13.2 **Freezing preservation in a −80°C freezer.** The paper box specialized for freezing contains 100 × 2 mL cryotubes.

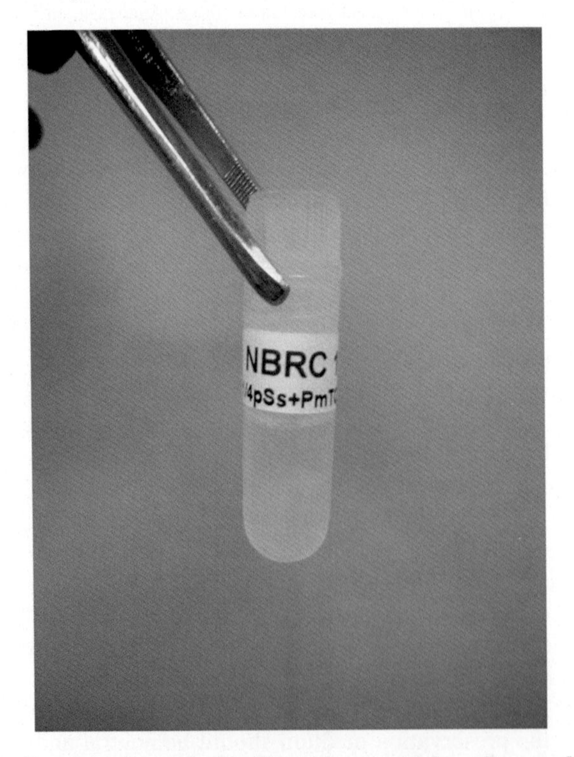

FIGURE 13.3 **Freezing tubes made of polypropylene containing cell suspended in cryoprotectant.** The tube is used for storage both in an electric freezer and in liquid nitrogen.

FIGURE 13.4 **Liquid nitrogen tank for freezing preservation (Capacity is 430 L).**

preservation should be washed with neutral buffer or with water. Freezing makes the pH of medium stronger by eliminating water as ice. Dimethylsulfoxide (DMSO) is an alternative cryoprotectant used for preservation medium. DMSO is used for the organisms which is more sensitive for possible cell disruption by freezing or those sensitive to organic compounds, such as chemolithoautotrophic bacteria. The concentration of DMSO is 5%–10% (v/v) in water. The procedure for freezing with DMSO is similar to that with glycerol. However, considering toxicity of DMSO, the cells are washed with water and centrifuge to remove DMSO prior to revival.

Other advantages of freezing preservation are the treatment of many samples at once and wide applicability. However, there is a possible risk caused by electric shortage or freezer equipment related problems, which can result in defrost of the samples. Accordingly, it is recommended to make back-up preservation for officially registered strains in a public culture collection. Freezing preservation in a liquid nitrogen tank is mostly used for this purpose.

Drying Methods

Drying preservation methods are largely divided into freeze-drying method and liquid (L) drying method. The former is a popular method that the cell

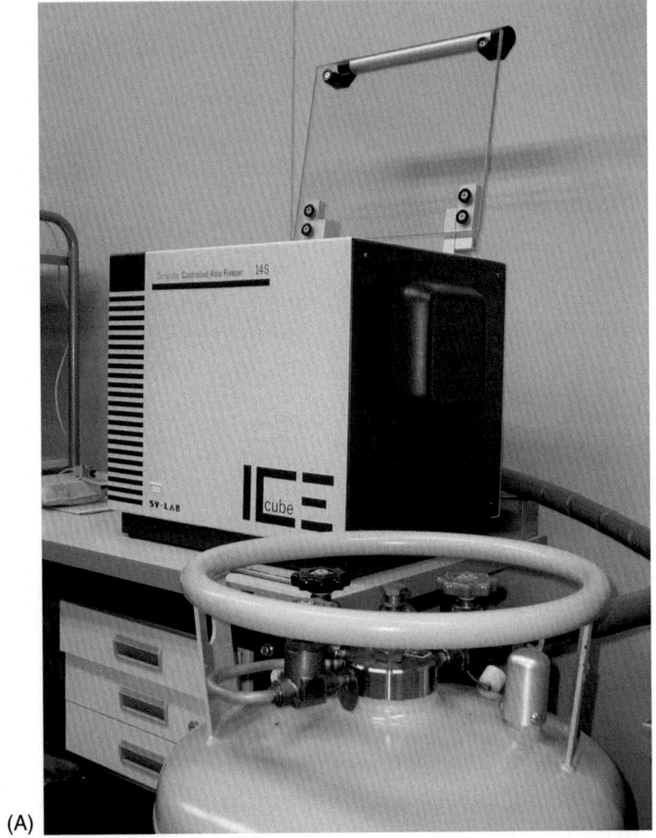

(A)

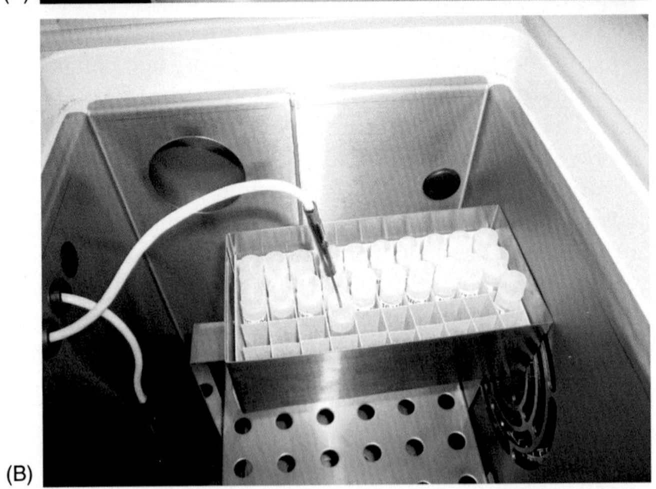

(B)

FIGURE 13.5 **Programming freezer.** (A) Body, (B) Inside.

suspension is dispensed into glass ampoules and then frozen in deep freezer prior to drying by freeze-drying machine (Fig. 13.6). Freeze-drying is carried out by evacuation in keeping sample frozen by sublimation. The temperature of samples is kept low to not to decompose the compounds in the cells. L-drying is the method to remove water from cell suspension without freezing to not to disrupt cells by freezing and this is said to be an advantage for Gram-negative

FIGURE 13.6 **Freeze-drying machine with manifold designed for ampoule preparation in a culture collection.**

bacteria. Freeze-drying method has an advantage in preparation to collect the ampoules containing cell suspension for freeze-drying in a freezer and freeze-dry at once.

On the other hand, L-drying requires soft evacuation not to make the samples frozen by vacuum and suitable drying time is rather short as 3 h. The essential procedure is shown in the Fig. 13.7. Over-drying leads low viability to some sensitive organisms. The medium used to suspend cells for drying are essentially 10% skim milk with 1% monosodium glutamate for freeze-drying and 0.1 M phosphate buffer with 3% monosodium glutamate (Table 13.1).

Glass ampoule is useful for keeping vacuum for long-term preservation by sealing completely with burner flame (Fig. 13.8A, B). Attention must be paid for decrease of viability in the early stage of long-term storage at 5°C. Some microorganisms suddenly lose viability in a few months to a few years even if they survive immediately after drying. There is a correlation of viability of L-drying between the results of 10 years of storage at 5°C and those of 2 weeks at 37°C using a wide variety of microorganisms. The acceleration test is useful for quality management of collection that L-dried ampoules are exposed at 37°C for 2 weeks to estimate the viability after 20 years (Fig. 13.9A, B). Drying in glass ampoules is useful not only for storage with low risk but also shipping without cooling in short time.

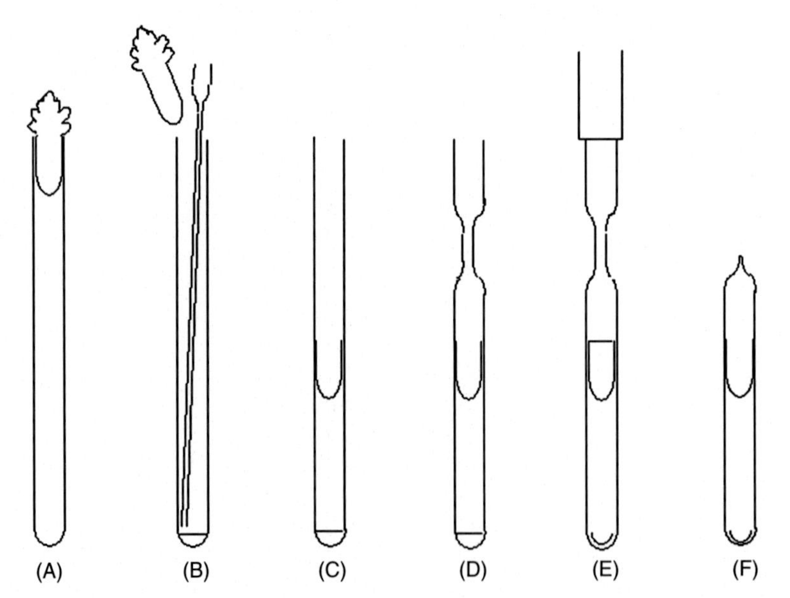

FIGURE 13.7 **Procedure of L-drying ampoule preparation.** (A) Sterile ampoule with cotton plug, (B) dispense cell suspension, (C) cut the plug and put into the center, (D) making constriction, (E) drying (F) seal by burner after dried.

TABLE 13.1 Medium for Preservation to Suspend Cells for Freezing and Drying

Method		Type of microorganisms	Medium composition[a]
Freezing		General	Peptone 0.5 g; glycerol 10 g; water 90 mL
Freeze drying		General	A: Skim milk 10 g; water 100 mL B: Monosodium glutamate 1 g; water 10 mL A and B. Mix after autoclave separately
L-Drying	SM1	General for bacteria	Monosodium glutamate 3 g; ribitol 1.5 g; L-cysteine 0.05 g; 0.1 M Potassium phosphate buffer (pH 7.0) 100 mL
	SM2	Marine bacteria	Monosodium glutamate 5 g; ribitol 1.5 g; sorbitol 1 g; 75% sea water 100 mL (pH 7.0)
	SM11	Yeasts	A: polyvinylpyrrolidone (K-90), 6 g; lactose, 5 g; water 75 mL. B: Monosodium glutamate, 3 g; 1 M potassium phosphate buffer (pH 7.0), 10 mL; water 15 mL. A and B. Mix after autoclave separately
	SM12	Filamentous fungi	Monosodium glutamate 3 g; 0.1 M Potassium phosphate buffer (pH 7.0) 100 mL

[a]*Sterilized by autoclaving.*

Maintenance in Active Culture

It is desirable for a culture collection to employ common preservation methods described earlier: freezing and/or drying. However, some microorganisms which are sensitive when any of these two methods is used they have to be maintained in active cultures. Active cultures are maintained by serial transfer may result in the loss of characteristics or reduction of activities. Selection of culture conditions including culture media, temperature, and storage temperature are of particular importance. To reduce the manpower for maintenance and spontaneous change of characteristics, long intervals during serial transfers are desirable. In addition, to avoid the possible death of a culture, at least two tubes must be maintained at different time of inoculation for each strain. When the culture collection receives the culture to preserve, the interval of transfer is extended step by step, such as 1 week for the first, 2 weeks for the second, 4 weeks for the third, etc. In the course of

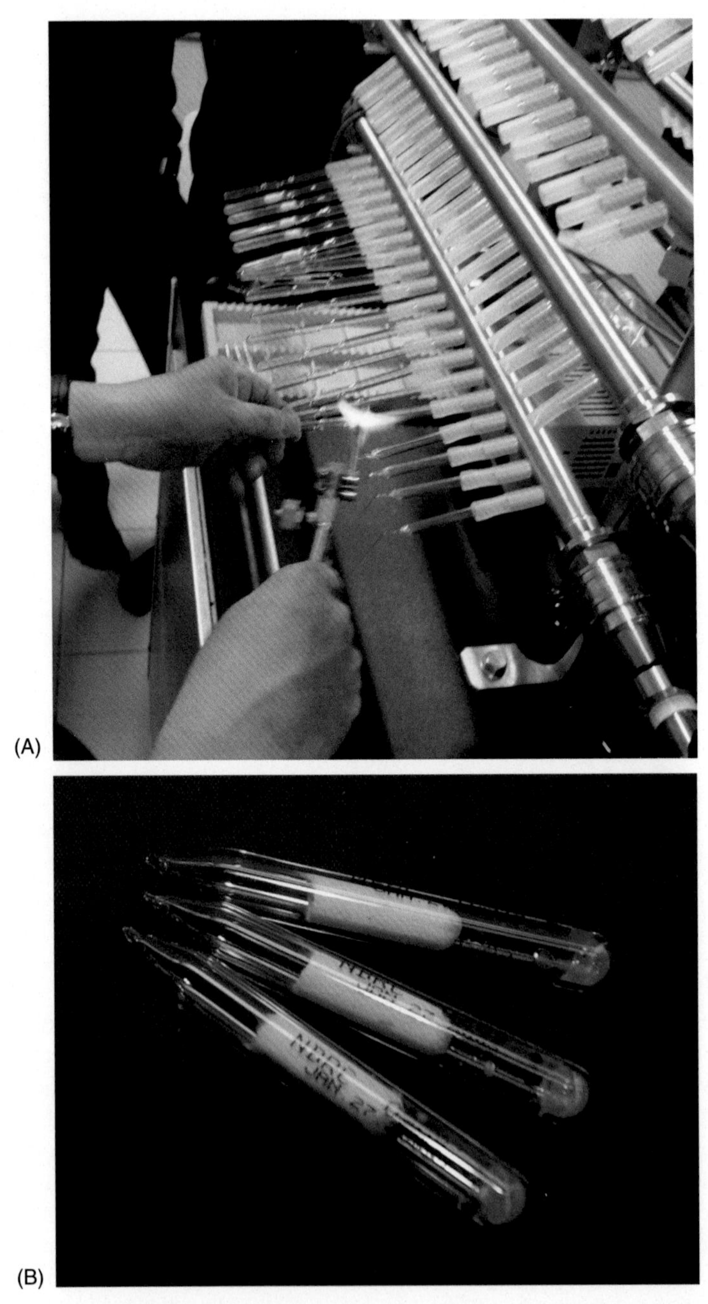

FIGURE 13.8 (A) Ampoule sealing by a gas burner after drying, (B) Ampoules of L-drying used for both storage and distribution.

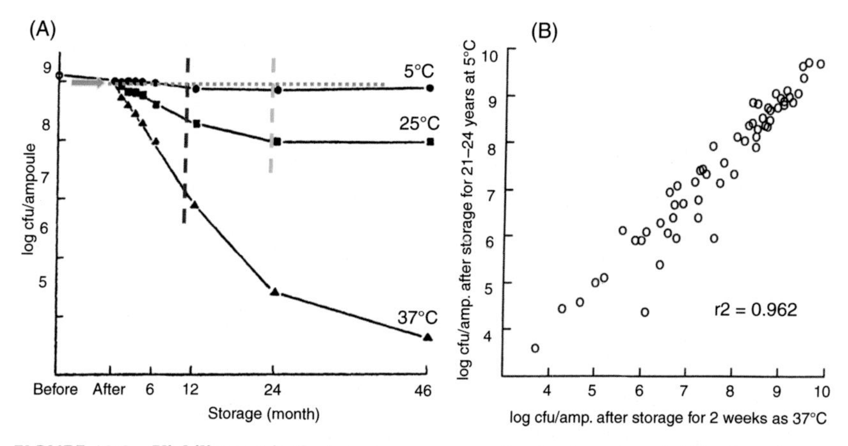

FIGURE 13.9 Viability test in the long-term preservation by L-drying. (A) Viable cell numbers of *Escherichia coli* in the long-term preservation stored in different temperatures (Banno and Sakane, 1981), (B) correlation of viability between long-term preservation at 5°C for 21–24 years and short-time acceleration test at 37°C for 2 weeks (Sakane and Kuroshima, 1997).

subculturing, attention should be paid to record the changes during maintenance, such as decrease of growth rate, losing color, etc. Optimum storage temperature also varies with the type of the strains. Some organisms are good at refrigerator temperature while some are at room temperature. Overlaying of mineral oil on the agar slope especially for yeasts and filamentous fungi is often useful (Fig. 13.10).

Quality Management of Microbial Strains

Essential qualities required to microbial strains in culture collections are: (1) viability, (2) purity, and (3) identity of the microbial strains. All must be rescored at the deposit of the strain and be confirmed with the depositors. Culture collections must manage the quality of bioresources in the collection by themselves according to their standard operation procedures (SOPs).

Viability

Viability is checked by using the most suitable growth medium which the depositor assigned for the strain. However, culture collections need to reduce the variety of the media for cultivation of the collection of microorganisms. It is desirable for the newly deposited strains grow on the similar ready-registered media in comparison with the depositor-specified medium. If the growth is not so much different, the recommended medium can be replaced. Viability includes the maintenance of activities of microorganisms indicated by the growth rate, spore or aerial mycelium formation, or pigment production, etc. on agar plates. When decrease of such growth activities is found, the culture is revived from the original or master stock suspension for comparison.

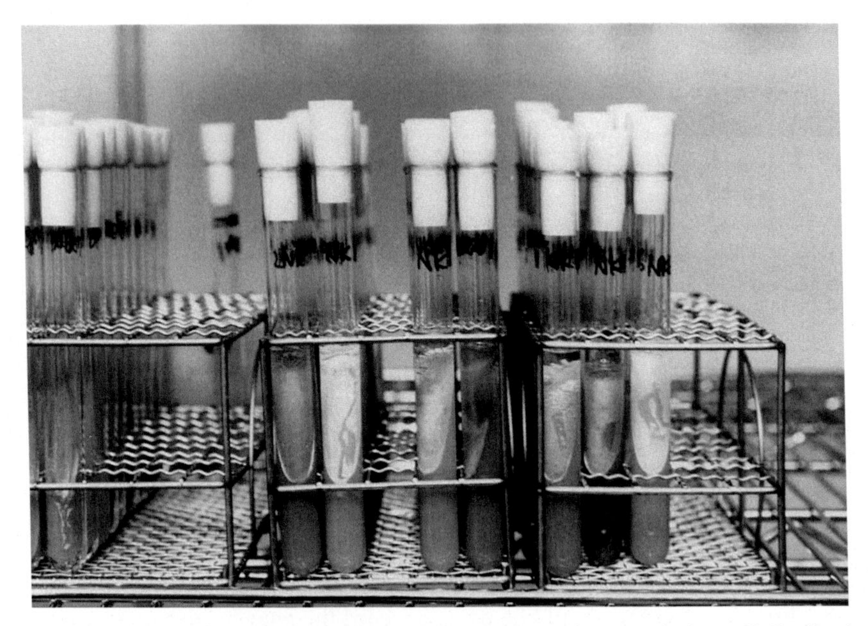

FIGURE 13.10 **Living cultures maintained by subcultures in a certain interval.** Sterilized mineral oil is overlaid on the slant cultures.

Purity

Purity test is essentially carried out by pure colony formation on agar plate. Attentions should be especially paid for some microorganisms, such as anaerobic bacteria which are generally cultivated in liquid medium for preparation of the reservation stocks. Colony formation test is desirable for purity check using cell suspension used for preservation. If this is difficult, microscopic, chemotaxonomic, or molecular techniques can be used. It also includes the negative control check that no growth is observed on the medium on which the organism must not grow. This check is important especially for a fastidious or autotrophic organism to detect contamination of faster growing heterotrophic organisms on richer medium. Some microorganisms show colony variant, such as rough-smooth types. The contamination and colony variant are checked by simple molecular technique, such as randomly amplified pleomorphic DNA (RAPD) technique.

Identity

Identity is important for culture collections for showing the data of the deposited strain in the catalog of the collection. When culture collections receive a microbial strain for deposit, the matching of the characteristics with the data sheet completed by the depositor must be confirmed prior to issuing the certificate of deposit. Not only the essential phenotypic characteristics, such as colony appearance, but also identity of the sequences of certain regions of ribosomal

RNA genes, mostly represented by 16S rRNA gene for the prokaryotes and gene of D1/D2 region of LSU rRNA genes for yeasts and fungi. The rRNA gene sequence is not perfect for taxonomic identification of the strain. However, it satisfies necessary requirement for the taxonomic placement of the microorganism and for finding replacement or mislabeling which often occur in the process of transfer of the strain between collection and depositor. Identity is important for use of the microorganism for further study or use based on the information from publication and databases. Recently, matrix-assisted laser desorption/ionization time of flight mass spectrometry (MALDI-TOF/MS) has been used for identification of microorganisms with commercially available databases by analyzing the mass spectra of ribosomal proteins. Considering the reproducibility of data and running cost, MALDI-TOF/MS is a suitable tool for culture collections to check the identity by the in-house database for quality management taking place of DNA sequencing to reduce the cost.

Other Quality Expected for the Strains in Culture Collections

In addition to the earlier three parameters, special characteristics of the microorganisms studied by the depositor must be checked and recorded. As such characteristics are unique and usually difficult to be checked by the collection. The culture collection requests the depositor to check it using the culture preserved in the collection and to certify it. Recent progress of molecular biology enables quality management of culture collections high. However, at the same time, the requirement to the quality is also becoming so high that the presence or sequence of genes must be held by the strain of which whole genome sequence is available from databases.

Quality and Certification of Culture Collections

Culture collections should possess standard operation procedures (SOPs) for their operation related to each preservation stage, such as: (1) accepting deposit, (2) preservation and the check methods at the deposit, (3) check methods during preservation, viability check in a certain interval, for distribution, for renewal preservation batch. Sometimes change of classification requires culture collections to examine additional characteristics to give names under the new taxonomic system. Both documents issued under the name of collections and records filed in the collections for in-house use are important for the evidence of deposit, identity, and distribution with quality confirmed under the SOP. Culture collections, especially large public culture collections are handling a wide variety of microorganisms for various purposes. The standard procedures are different between bacteria, yeasts, and fungi under different taxonomic methods. SOP is one of the most important manual to share the common standards applied among different sections of the collections. As an example, the ISO9001:2015 is an essential quality management system. However, this is certified by an external agent whether the organization is operated according to the procedure prepared by the organization and approved. There are no mandatory procedures specific to a microbial culture collection. As for reader's reference, World Federation for Culture Collections (WFCC) published the guidelines for operation of culture collections (WFCC, 2010). This will be helpful for the establishment of the quality management system of the collection. In 2007, Organization for Economic Co-Operation and Development (OECD) published OECD best practice guideline for biological resource centers (OECD, 2007). Since then, the certification system for biological resource centers which collect, preserve, and distribute bioresources has been discussion in various ways. Recent awareness in the needs of common standard for handling and transfer of biological resources makes the community to discuss the establishment of infrastructure including quality of materials and compliance with laws and regulations (see the next section).

Laws and Regulations Relevant to Handling and Transfer of Microorganisms

Microorganisms are very important biological materials to study essential biology and application to biotechnology. However, at the same time, they contain pathogenic microorganisms with significance for humans, animals, and plants. Therefore, they are controlled by so many laws and regulations nationally and internationally. In addition, the Convention on Biological Diversity (CBD) has been in force in 1993 and it states the sovereign right of the country on the biological resource which is originated from the country. This convention affects seriously the international sharing of biological resources within the scientific community as well as the bio-industry.

Laws and Regulations for Biosafety and Biosecurity

The laws for infectious diseases are aimed for human health and welfare. They are essentially domestic but mostly based on the guidance of WHO. Microorganisms are classified on the basis of the biosafety level. The guidance for transport of infectious substances is published by the World Health Organization (WHO) (WHO, 2012) in accordance with the United Nations (UN) recommendation (UN, 2013). Culture collections must use the packaging for deposit and distribution of microbial strains specified in the guidance especially when air freight is used (see Chapter 15 of this book).

Import of a plant pathogen is generally controlled by national law of individual countries in considerations of their circumstances. The Australia Group (AG) is an informal forum of countries which, through the harmonization of export controls, seeks to ensure that exports do not contribute to the development of chemical or biological weapons (Australia Group, 2016), and it has 41 participating countries as of June 2014. AG publishes the Common Control List Handbook containing the list of human and animal pathogens and toxins and plant pathogens for viruses, bacteria, and fungi (Australia Group, 2016). This list is used as a reference for the regulations of the member countries for the control of the transport of hazardous substances

The Convention on Biological Diversity *(CBD)* *and the Related Law and Regulations*

The Convention on Biological Diversity (CBD) has been in force in 1993 and it states the sovereign right of the country on the biological resource which is originated from the country (The Secretariat, 1992). Later, in October 2014, Nagoya protocol on access to genetic resources and the fair and equitable sharing of benefits arising from their utilization to the *Convention on Biological Diversity* (Nagoya Protocol) was in force (The Secretariat, 2011). *Nagoya Protocol* requires prior informed consent (PIC) and mutually agreed terms (MAT) for access to the biological material of the country. Based on the principle of the *Nagoya Protocol*, each country must set national law for access and use of the biological resources. Microbiologists must consider the laws and regulations of the both countries as the provider and the recipient when microbial strains are transferred from a country to another. Biological resource centers are expected to work for the microbiologists to transfer biological materials internationally in compliance with the laws and regulations of the both sides. Each country which ratified CBD and the *Nagoya Protocol* is recommended to set up simple and clear procedure to obtain PIC and MAT to access biological materials of the country to promote the use of the biological materials of the country (see Chapter 10 of this book).

Future Perspective for Biological Resource Centers

The scientific community needs biological materials to share internationally to confirm the validity of the results published in the scientific papers. On the other hand, CBD has made it difficult to transfer biological materials among the community. However, CBD is aimed to enhance the use of biological resources to develop application of life science and biotechnology as well as the basic science to produce benefit to share. Benefit is not only monetary benefit and it is important to develop capacity building through international cooperation. Microbial taxonomy is providing the tools and the knowledge to manage microbial materials in compliance with the laws relevant to the CBD. It is, therefore, expected for the countries to establish simple and clear scheme to access biological materials. Promotion of the importance of the use of microbial resources is a role of the microbial resource centers. Biological resource center is an infrastructure of providing diverse microorganisms, rich information, and document. Strengthening international network of biological resource center will enhance the use of overseas samples. The roles of biological resource centers are becoming important more than before from both scientific and social viewpoints.

ACKNOWLEDGMENTS

The author thanks Japan Collection of Microorganisms (JCM), RIKEN, Japan and NBRC, National Institute of Technology and Evaluation (NITE), Japan for giving the experience of microbial taxonomy and preservation in culture collections.

REFERENCES

Australia Group, 2016. Australia group common control list handbook Volume II: Biological weapons-related common control lists. Available from: http://www.australiagroup.net/en/controllist-handbooks.html

Banno, I., Sakane, T., 1981. Prediction of prospective viability of L-dried cultures of bacteria after long-term preservation. Inst. Ferm. Osaka Res. Comun. 10, 33–38.

Editorial Board of the Judicial Commission of the International Committee on Nomenclature of Bacteria, 1966. International code of nomenclature of bacteria. Int. J. Syst. Bacteriol. 16, 459–490.

Lapage, S.P., Sneath, P.H.A., Lessel, E.F., Seeliger, H.P.R., Clark, W.A. (Eds.), 1975. International Code of Nomenclature of Bacteria. American Society for Microbiology, Washington DC.

Lapage, S.P., Sneath, P.H.A., Lessel, E.F., Skerman, V.B.D., Seeliger, H.P.R., Clark, W.A. (Eds.), 1992. International Code of Nomenclature of Bacteria. American Society for Microbiology, Washington DC.

McNeill, J., Barrie, F.R., Burdet, H.M., Demoulin, V., Hawksworth, D.L., Marhold, K., Nicolson, D.H., Prado, J., Silva, P.C., Skog, J.E., Wiersema, J.H., Turlabd, N.J., 2006. International Code of Botanical Nomenclature (Vienna code). ARG Gentner Verlag KG, Regnum Vegetabile.

Organization for Economic Co-Operation and Development (OECD), 2007. OECD best practice guideline for biological resource centres. Available from: http://www.oecd.org/sti/biotech/oecdbestpracticeguidelinesforbiologicalresourcecentres.htm

Parker, C.T., Tindall, B.J. Garrity, G.M., 2015. International Code of Nomenclature of Prokaryotes. Int. J. Syst. Evol. Microbiol. [published ahead of print].

Sakane, T., Kuroshima, K., 1997. Viabilities of dried cultures of various bacteria after preservation for over 20 years and their prediction by the accelerated storage test. Microbiol. Cult. Coll. 13, 1–7.

The Secretariat of the Convention on Biological Diversity, Convention on Biological Diversity, (1992). Available from: https://www.cbd.int/convention/text/default.shtml

The Secretariat of the Convention on Biological Diversity, 2011. Nagoya protocol on access to genetic resources and the fair and equitable sharing of benefits arising from their utilization to the Convention on Biological Diversity. Available from: https://www.cbd.int/abs/text/default.shtml

United Nations, 2013. Recommendations on the transport of dangerous goods. Model regulations. Available from: http://www.unece.org/trans/danger/danger.html

World Federation for Culture Collections, 2010. World Federation for Culture Collections Guidelines for establishment and operation of collections of cultures of microorganism third ed. Available from: http://www.wfcc.info/guidelines/

World Health Organization, 2012. Guidance on regulations for the transport of infectious substances 2013-2014. Available from: (http://www.who.int/ihr/publications/who_hse_ihr_2012.12/en/)

Chapter 14

IP and The Budapest Treaty—Depositing Biological Material for Patent Purposes

Vera Bussas*, Avinash Sharma and Yogesh Shouche****

**Leibniz Institute DSMZ-German Collection of Microorganisms and Cell Cultures, Braunschweig, Germany; **Microbial Culture Collection, National Centre for Cell Science, Sutarwadi, Pashan Pune, Maharashtra, India*

INTELLECTUAL PROPERTY RIGHTS—WHY TO DEPOSIT BIOLOGICAL MATERIAL

Many discoveries in science fall in the category of "inventions" with commercial value and might result in patent applications. The invention of a product or a process is an intellectual property owned by the inventor. The term "intellectual property" refers to "creations of the human mind." According to the *Convention Establishing the World Intellectual Property Organization* (http://www.wipo.int/wipolex/en/treaties/text.jsp?file_id=283854) (WIPO, 1967) matters like literary, artistic and scientific works, performances of performing artists, phonograms, and broadcasts, inventions in all fields of human endeavor, scientific discoveries, industrial designs, trademarks, service marks, commercial names and designations, protection against unfair competition, and "all other rights resulting from intellectual activity in the industrial, scientific, literary or artistic fields" are protected by intellectual property rights.

Intellectual property rights are divided into two main categories, (1) the copyright and related rights and (2) industrial property rights.

- *Copyright and related rights* include rights of authors of books, paintings and computer programs, of actors, musicians etc. and grant protection for a period of at least 50 years after the death of the copyright owner. Thus authors, artists, singers etc. are encouraged and rewarded for their creative work.
- *Industrial property* comprises two main areas: trademarks and geographical indications (with the aim to stimulate and ensure fair competition and to protect consumers) on the one hand and industrial property on the

Microbial Resources. http://dx.doi.org/10.1016/B978-0-12-804765-1.00014-X

other hand which is supposed to stimulate innovation, design, and the creation of technology. The second one includes patent protection for inventions, industrial designs, and trade secrets. The protection serves as incentive for the necessary investments. The period of protection for patents is shorter than for a copyright and ceases after 20 years. Patent protection will guarantee the inventor some degree of exclusivity on the invention to assure a reasonable profit and to justify the risks of development. In return, the inventor discloses the invention by giving an exact written description of the process. The knowledge will be available to the public and others may build upon this knowledge. The disclosure gives the public the possibility to rework the invention for experimental purposes and, *after* expiration of the period of patent protection, for commercial purposes. The best-known examples are generic drugs which often are based on the patents that are expired. They are, for example, available for pain relief (Voltaren), antibiotics (Amoxil, Ampicillin), and Combivir (HIV treatment).

Most biological processes use live organisms, such as microorganisms, plant, or animal cells. This biological material could be naturally occurring, isolated from the environment or modified in the laboratory either by introducing mutations (improved strains) or genetically modified (recombinant strains). An example of former are antibiotic producing organisms like *Streptomyces venezuelae* which was cited as in the patent for the production of chloramphenicol. A well-known example of the later is the production of human insulin (humulin) using *Escherichia coli* as host. In such cases, even a detailed description of generating such cells may not ensure reproducibility of the process and this will necessitate deposition of such material at an independent place. Such material then becomes a "strain deposited for patent purposes" but is collicially known as a "patent strain." As patent offices do not have the facilities for the storage of microorganisms, independent and scientifically recognized culture collections have been entrusted with the handling of the biological material to be deposited. *The Budapest Treaty on the International Recognition of the Deposit of Microorganisms for the Purposes of Patent Procedure* (WIPO, 1989) regulates the process of patent deposits in an international framework. The most important feature of the treaty is that a patent deposit which has been made with one International Depositary Authority (IDA) (http://www.wipo.int/treaties/en/registration/budapest/) is sufficient and recognized by all member states of the Budapest Treaty (http://www.wipo.int/treaties/en/ShowResults.jsp?lang=en&treaty_id=7). Nonmember countries may also accept deposits according to the Budapest Treaty as per the norms of the country. To date (January 2017), 80 states worldwide and three intergovernmental industrial property organizations are party to the Treaty.

IDAs are impartial, objective, and available to any depositor on the same terms, and they comply with the requirements of secrecy. Information on sample deposited with an IDA is never disclosed to anyone, except to parties entitled to obtain a sample of the said organism. IDA tests the viability of the biological material promptly after receipt and stores the organisms for the minimum period of 30 years as specified by the treaty (Sharma and Shouche, 2014). The treaty also stipulates further storage of 5 years after the most recent request for the furnishing of a sample of the strain (Note 1). Samples of the deposited material are furnished for testing and verification to the entitled parties. At present (January 2017), 45 culture collections across the world have been given the status of IDA (Table 14.1).

TABLE 14.1 List of International Depositary Authorities (as of October 2016)

Institution	Country	Date status acquired
Advanced Biotechnology Center (ABC)	Italy	February 29, 1996
Agricultural Research Service Culture Collection (NRRL)	United States of America	January 31, 1981
American Type Culture Collection (ATCC)	United States of America	January 31, 1981
Banco Español de Algas (BEA)	Spain	October 28, 2005
Belgian Coordinated Collections of Microorganisms (BCCMTM)	Belgium	March 1, 1992
CABI Bioscience, UK Centre (IMI)	United Kingdom	March 31, 1983
Centraalbureau voor Schimmelcultures (CBS)	Netherlands	October 1, 1981
China Center for Type Culture Collection (CCTCC)	China	July 1, 1995
China General Microbiological Culture Collection Center (CGMCC)	China	July 1, 1995
Colección Chilena de Recursos Genéticos Microbianos (CChRGM)	Chile	March 26, 2012

(Continued)

TABLE 14.1 List of International Depositary Authorities (as of October 2016) (*cont.*)

Institution	Country	Date status acquired
Colección de Microorganismos del Centro Nacional de Recursos Genéticos (CM-CNRG)	Mexico	August 25, 2015
Colección Española de Cultivos Tipo (CECT)	Spain	May 31, 1992
Collection nationale de cultures de micro-organismes (CNCM)	France	August 31, 1984
Collection of Industrial Yeasts DBVPG	Italy	January 31, 1997
Culture Collection of Algae and Protozoa (CCAP)	United Kingdom	September 30, 1982
Culture Collection of Yeasts (CCY)	Slovakia	August 31, 1992
Czech Collection of Microorganisms (CCM)	Czech Republic	August 31, 1992
European Collection of Cell Cultures (ECACC)	United Kingdom	September 30, 1984
Guangdong Microbial Culture Collection Center (GDMCC)	China	January 1, 2016
IAFB Collection of Industrial Microorganisms	Poland	December 31, 2000
International Depositary Authority of Canada (IDAC)	Canada	November 30, 1998
International Patent Organism Depositary (IPOD), National Institute of Technology and Evaluation (NITE)	Japan	May 1, 1981
Istituto Zooprofilattico Sperimentale della Lombardia e dell'Emilia Romagna «Bruno Ubertini» (IZSLER)	Italy	February 9, 2015
Korean Agricultural Culture Collection (KACC)	Republic of Korea	May 1, 2015
Korean Cell Line Research Foundation (KCLRF)	Republic of Korea	August 31, 1993
Korean Collection for Type Cultures (KCTC)	Republic of Korea	June 30, 1990
Korean Culture Center of Microorganisms (KCCM)	Republic of Korea	June 30, 1990
Lady Mary Fairfax Cellbank Australia (CBA)	Australia	February 22, 2010

TABLE 14.1 List of International Depositary Authorities (as of October 2016) (*cont.*)

Institution	Country	Date status acquired
Leibniz-Institut DSMZ–Deutsche Sammlung von Mikroorganismen und Zellkulturen GmbH (DSMZ)	Germany	October 1, 1981
Microbial Culture Collection (MCC)	India	April 9, 2011
Microbial Strain Collection of Latvia (MSCL)	Latvia	May 31, 1997
Microbial Type Culture Collection and Gene Bank (MTCC)	India	October 4, 2002
National Bank for Industrial Microorganisms and Cell Cultures (NBIMCC)	Bulgaria	October 31, 1987
National Collection of Agricultural and Industrial Microorganisms (NCAIM)	Hungary	June 1, 1986
National Collection of Type Cultures (NCTC)	United Kingdom	August 31, 1982
National Collection of Yeast Cultures (NCYC)	United Kingdom	January 31, 1982
National Collections of Industrial, Food and Marine Bacteria (NCIMB)	United Kingdom	March 31, 1982
National Institute for Biological Standards and Control (NIBSC)	United Kingdom	December 16, 2004
National Institute of Technology and Evaluation, Patent Microorganisms Depositary (NPMD)	Japan	April 1, 2004
National Measurement Institute (NMI)	Australia	September 30, 1988
Polish Collection of Microorganisms (PCM)	Poland	December 31, 2000
Provasoli-Guillard National Center for Marine Algae and Microbiota (NCMA)	United States of America	April 26, 2013
Russian Collection of Microorganisms (VKM)	Russian Federation	August 31, 1987
Russian National Collection of Industrial Microorganisms (VKPM)	Russian Federation	August 31, 1987
VTT Culture Collection (VTTCC)	Finland	August 25, 2010

HOW TO DEPOSIT BIOLOGICAL MATERIAL—THE PRACTICAL PROCEDURE OF A PATENT DEPOSIT

Before filing a patent application, it is desirable to verify from a patent expert whether a patent deposit is necessary to ensure the reproducibility of the invention (Note 2). If the properties of the organism can be described in sufficient details in writing, for example, DNA sequence information etc., then a deposit may not be required. If, however, a sufficient written description is not possible, the material should be deposited. To be on the safer side, it might be advisable to deposit the biological material even in former case. If the biological material is already deposited, then it has to be ensured whether this deposit is deposited according to patent regulations. If the organism has already been deposited by the patent applicant according to the Budapest Treaty, the application can be filed immediately (Fig. 14.1).

A deposit which is not in accordance with the Budapest Treaty (a patent deposit performed before the collection was given the IDA status; a safe deposit) should be converted into a patent deposit as per Rule 6.4 (d) of the Treaty. The original accession date may—but not necessarily—remain valid but it may be easier to deposit the organism again according to the Treaty. If the organism has been deposited by a party different from the patent applicant it has to be checked whether the patent applicant is authorized to refer to the biological material in their application (Fig. 14.2). Deposits in the open part of a collection should not and type strains cannot be converted into deposits, according to the Budapest Treaty.

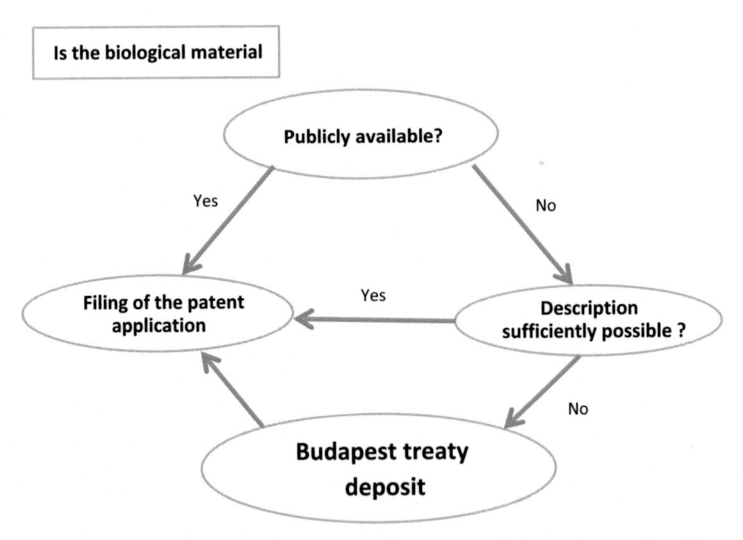

FIGURE 14.1 **Assistance key for decision making whether biological material has to be deposited or not.**

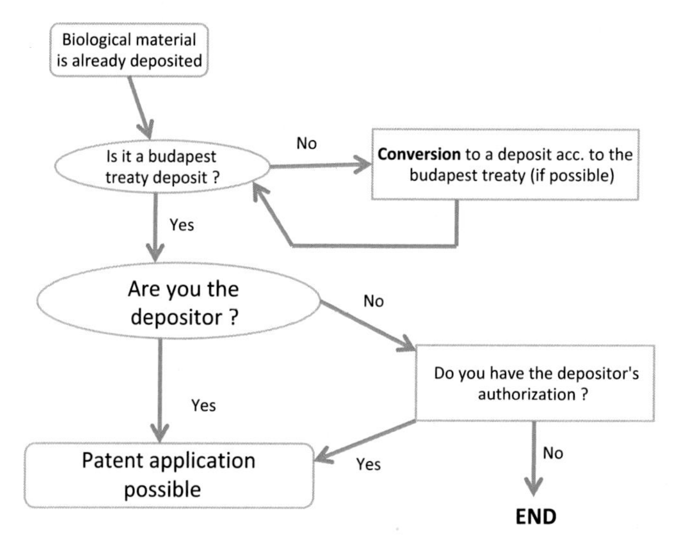

FIGURE 14.2 **Assistance key for decision making whether the correct kind of deposit has been chosen.**

Where to Deposit—Choice of an IDA

Once it is ensured that a patent deposit is advisable or necessary, the future depositor should first look for the suitable IDA as each IDA accepts different kinds of biological material. The Budapest Treaty does not offer a specific definition of the term "microorganism." It is interpreted in a broad sense: isolated DNA and cell cultures are regarded as microorganisms as well as bacteria or bacteriophages. The actual definition suitable for all kinds of biological material accepted as deposit for patent purposes states that the biological material must contain genetic information and must be self-replicating (e.g., bacteria, fungi) or of being reproduced in a biological system (e.g., plasmids, viruses) (European Parliament, 1998a). Considering the diverse nature of biological materials, the type of material accepted differs between the individual IDAs. There are some which accept only one or two kinds of organisms, whereas others accept a broader range of different kinds of biological material (Table 14.2).

Furthermore, an IDA decides about the Risk Group of the biological material it accepts. Biological material beyond these Risk Groups have to be refused and may not be accepted by the IDA. Therefore, material should at least be characterized to a certain minimal taxonomic level to enable its risk classification before it is sent to an IDA. Within the European Community, compliance to the Council Directive 2000/54/EEC (http://eur-lex.europa.eu/legal-content/EN/TXT/HTML/?uri=CELEX:32000L0054&from=EN) on the protection of workers from risks related to exposure to biological agents at work is mandatory

TABLE 14.2 Kinds of Biological Material Accepted by 12 out of 45 IDAs

IDAs	Kind of biological material accepted	Risk group
ABC	Animal and human cell lines	2
ATCC	Algae, bacteria, fungi, embryos, human, animal and plant cell cultures, bacteriophages, animal and plant viruses, seeds, DNA, RNA, Protozoa	4
BCCM	Bacteria, fungi, yeasts, plasmid DNA, RNA, animal and human cell lines	2
BNA	Microalgae, cyanobacteria and macroalgae	1
CBS	Bacteria, fungi, yeasts, bacteriophages, plasmid DNA	3, 2 for GEMs
CCAP	Algae, free-living protozoa, microorganisms	1
DSMZ	Bacteria, fungi, yeasts, bacteriophages, plasmid DNA, plant viruses, plant cell cultures, animal and human cell cultures	2
CABI Bioscience, UK Centre (IMI)	Fungi, bacteria, nematodes	2
MCC	Bacteria, fungi, yeasts, plasmid DNA	2
KCLRF	Animal, human and plant cell cultures	1
KCTC	Algae, animal and plant viruses, animal, human and plant cell cultures, bacteria, fungi, yeasts, bacteriophages, embryos, protozoa, seeds	1
VKM	Bacteria, fungi, yeast	1

(European Parliament, 2000). In the case of genetically modified organisms these must be classified accordingly (http://eur-lex.europa.eu/legal-content/EN/TXT/HTML/?uri=CELEX:32009L0041&from=EN) (European Parliament, 1998b).

As a result, the potential depositor and/or patent applicant first needs to classify the biological material that they wish to deposit before identifying the suitable IDA. The information about which kinds of biological material are accepted by an individual IDA can be obtained directly from the IDA (in most cases by consulting their respective websites) (Table 14.1) or by consulting the Guide to the Deposit of Microorganisms under the Budapest Treaty (http://www.wipo.int/treaties/en/registration/budapest/guide/index.html) (WIPO, 2016). The choice of the depositary should be made taking into account, inter alia: (1) the kind of biological material involved in the patent application; (2) the language(s) spoken and accepted by the IDA (to facilitate correspondence); and (3) the availability of the IDA considering possible import and export restrictions.

When to Deposit—Date of Patent Deposit

As the patent deposit can be interpreted as an essential part of a patent application it is extremely important that both, patent application and strain deposit, are performed before publication of any results (lecture, poster, publication). Otherwise, patent protection would not be grantable anymore as the invention would lack novelty. As it is desirable to file the patent application *including* the deposition statements, the deposit ideally should be performed in good time before the prospective filing date. Generally speaking, and depending on the respective patent legislation, the last date for depositing the biological material is the date of the filing of the patent application.

The Deposition Procedure Itself

While depositing a patent culture it should be accompanied by a deposit form which has been duly filled in and signed by an authorized official of the depositing organization. The depositor can obtain this form directly from the depositary. It is the deposition contract between the depositor and the IDA. The form should contain all information necessary for the IDA to handle the biological material and will allow the allocation of the material to the responsible IDA representative upon receipt at the culture collection. To allow the immediate processing of a deposit it is advisable to inform the IDA beforehand of an intended deposit and to specify which kind of biological material will be involved.

The culture should be supplied to the IDA in the amount and form requested by the culture collection in order to guarantee a smooth deposition procedure. The relevant information might be obtained directly from the IDA or the WIPO Guide to Deposit of Microorganisms under the Budapest Treaty (www.wipo.int/treaties/en/registration/budapest/guide). On the WIPO website, all necessary information is available. Depending on the IDA selected, different amounts of biological material might be requested:

The DSMZ (Leibniz Institute DSMZ-German Collection of Microorganisms and Cell Cultures) (https://www.dsmz.de/deposit/patent-deposit/biological-material-accepted-as-patent-deposit.html) as an example requests the following forms and amounts of biological material:

1. *Bacteria and fungi*—strains without plasmids or plasmid bearing strains—should, where possible, be deposited in the form of actively growing cultures (two separate preparations, preferably in the form of screw-capped test tubes). Freeze-dried and cryocultures are accepted as well.
2. *Plasmids* as isolated DNA preparations should be deposited in a minimum quantity of 2×20 μg (2 vials containing 20 μg each).
3. *Bacteriophages* should be deposited in minimum quantities of 2×5 mL (2 vials containing 5 mL each) having a minimal titer of 1×10^9 pfu/mL.

4. *Plant viruses* should be deposited in the form of dried or frozen plant material along with the host's seeds, unless the host is generally available. 100 μL of serum suitable for immunoelectron microscopy should be deposited in addition for the purity and identity test. When hybridomas for antibody testing of plants are deposited, the antigen (not pathogenic) necessary for the specificity test should be deposited at the same time.
5. *Plant cell cultures* should be deposited as active cultures in the form of a callus (4 Petri dishes), suspension (4 culture vessels) or—preferably—as frozen cultures in a dry shipper (25 ampoules).
6. *Human and animal cell cultures* should be deposited frozen in dry ice in a quantity of 12 cryoampoules containing 5×10^6 cells per ampoule, all prepared from the same batch of the culture.

The technical and administrative procedure of a deposition is as follows (see "checklist for depositors" in Fig. 14.3):

1. The depositor prepares the culture in the form and amount requested by the IDA.
2. The accession form is filled in and signed.

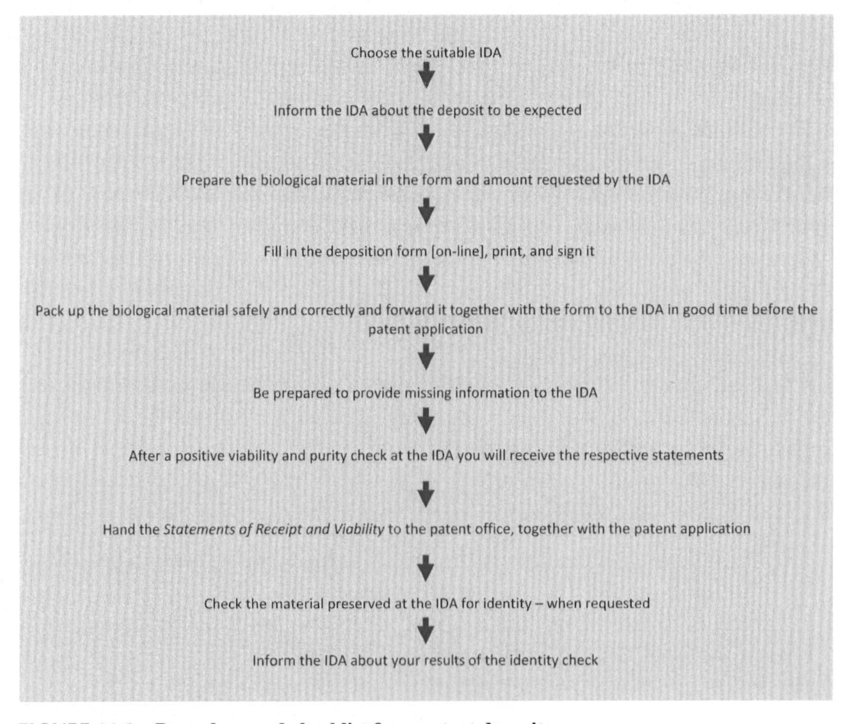

FIGURE 14.3 **Procedure and checklist for a patent deposit.**

3. The IDA is informed about the deposit to be expected (by phone, e-mail, fax, or letter).
4. The culture is packaged in a manner it will reach the IDA in a good condition. The relevant regulations for the shipping of biological material have to be taken into consideration (Rohde and Smith, 2013; Universal Postal Union, 2013; WHO, 1994).
5. Documents are checked for completeness at the IDA. If necessary, missing data will be requested from the depositor by the IDA.
6. After viability and purity of the culture is been checked at the IDA, the accession number will be assigned to the culture. Some collections assign provisional accession numbers when the culture reaches the IDA.
7. The depositor receives a statement of receipt and a statement of viability. These statements are important for the patenting procedure and should be submitted to the patent office involved when filing the application.
8. The biological material will be preserved and stored at the IDA.
9. After preservation at the IDA, the depositor will receive samples for identity checking (depending on the biological material deposited and the IDA chosen).
10. After checking the identity of the material, the depositor informs the IDA about the results of the check (material preserved at the IDA is identical/not identical).

According to established practice, an IDA will perform a viability test before assigning an accession number to the culture. Therefore, it must be in the interest of the depositor to send the cultures to the IDA as early as possible to provide sufficient time for the cultivation of the organism before the date of the filing of the patent application. This applies to all kinds of biological material. In general, the testing for viability and purity of bacterial and fungal cultures takes from one to several days. In the case of human and animal cell cultures viability testing, including the essential testing for contaminations with mycoplasmas, about 2 weeks might be required; the cultivation of plant cell cultures needs approximately 4 weeks. The individual time needed not only depends on the kind of biological material but may also vary considerably according to the species of organisms handled and often is even strain specific. A culture deposited for patent purposes cannot be claimed back after it has been deposited. Patent strains will be preserved by sufficient safety measures to minimize the risk of losing deposits. If, however, a strain deposited under the Budapest Treaty should die or be destroyed during the period of storage, it is the responsibility of the depositor to replace it by a viable culture of the same organism. Thus the patent can be kept alive. Therefore, it is advisable that the depositor also maintains the culture for the same period of time: in case the culture is for any reason no longer available from the Depositary Authority the stock may be replenished by the depositor.

HOW TO REQUEST THE FURNISHING OF SAMPLES OF PATENT ORGANISMS

As pointed out earlier the main reason for the deposit of biological material which is of relevance for a patent application is to render it available to entitled parties for verification, thus allowing the reproducibility of a protected invention. The accessibility to a deposited organism depends on the status of a patent. According to the Budapest Treaty, samples might be furnished either to interested industrial property offices (Rule 11.1), to the depositor and third parties who obtained the authorization of the depositor (Rule 11.2) or to parties legally entitled (Rule 11.3).

A request for the furnishing of a patent culture by a third party can be sent either to the IDA harboring the strain or, in certain cases, directly to the respective patent office involved. The IDA checks its files for information whether or under which conditions the culture is to be furnished. In most cases there is no information available whether the IDA is authorized to release a sample of the strain. The requestor will be asked to submit to the IDA a copy of the first page of the patent document in question and a copy of the page where the accession number of the collection of the strain in question is cited. After receipt of the requested documents the IDA can provide the relevant official forms (request for the release of a patent strain). In case of a national or a European patent the form must be filled in and sent to the relevant patent office for confirmation. The European Patent Office offers the necessary form 1140/1141 online (http://www.epo.org/applying/forms-fees/forms.html). If a US-Patent had been granted the strain is available without further examination by the patent office.

After receiving a pertinent confirmation from a patent office, the IDA will ship the sample taking into consideration of other national and international relevant regulations, for example, postal regulations (Universal Postal Union, 2013) or biological weapons conventions like the EU dual use directive (European Parliament, 1994) (Fig. 14.4) (Note 2). Please consult also (Rohde and Smith, 2013) and (WHO, 1994).

COMPLICATIONS ARISING DURING THE DEPOSITION PROCEDURE

There are a number of pitfalls in the course of a patent deposit which might cause serious problems to the patent applicant and therefore should be considered before the patent deposit:

Refusal of the IDA to Accept a Culture

An IDA may refuse to accept a culture for the following reasons:

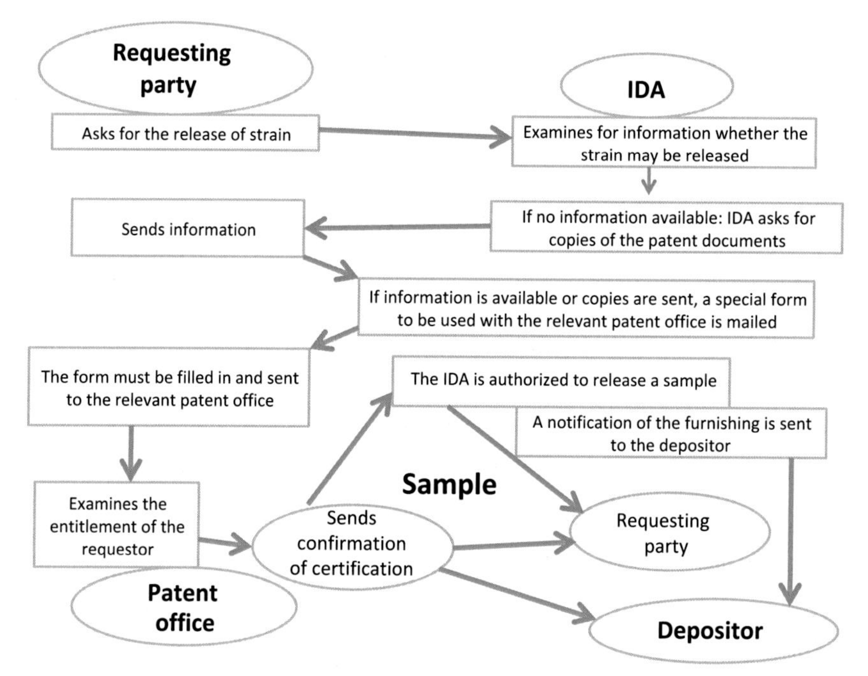

FIGURE 14.4 **Procedure for the furnishing of samples according to Rule 11.3 of the Budapest Treaty.**

- Where the biological material is not a kind of biological material that the IDA accepts (Rule 6.4 (a) i) (see section "Where to Deposit—Choice of an IDA").
- Where the conditions for cultivation are so exceptional that the IDA is technically not in a position to handle the organism. Here the depositary has the right to refuse an organism though it might belong to the kinds of organisms it gave assurances for (Rule 6.4 (a) ii).
- Where the organism is received by the IDA in a condition that clearly indicates that it is impossible to handle it properly (e.g. broken vials) (Rule 6.4 (a) iii).
- The most relevant reason to refuse a deposit, however, is if the organism proves to be nonviable at the first viability check. (Rule 10.2 (b)).

The Need of a New Deposit—the Reasons For

The viability of a deposited patent culture is tested from time to time at reasonable intervals. If an organism is no longer viable when tested at a later date the depositor within 3 months has the right to furnish the IDA with viable material of the organism originally deposited (Article 4(1)(a) i). The depositor has to confirm that the newly sent material is the same as the culture previously deposited (Article 4(1)(c)). In case the sending or receipt of samples of deposited organisms abroad is prevented by export or import restrictions the depositor has the right to make a new deposit of the organism with another IDA without losing the original deposit date (Article 4(1) (a) ii).

Transfer of a Deposited Organism

If an IDA temporarily, permanently, or partly ceases to carry out its functions as an IDA (e.g., restricts the kinds of biological material accepted) a transfer of the cultures in question to another IDA shall be done. (Article 4(1) (b) (i) (http://www.wipo.int/wipolex/en/treaties/text.jsp?file_id=283784#P74_5277).

Conversion of Deposits Made Outside the Budapest Treaty

If a patent deposit has been done before an IDA acquired the status of an IDA such a non-Budapest deposit might be converted into a deposit according to the Budapest Treaty since the patent depositary has been recognized as an IDA (Rule 6.4 (d)). A non-Budapest, nonpatent deposit with a recognized IDA might be converted into a valid Budapest Treaty deposit by the original depositor. This will depend on the culture collection's policy. The original deposition date might remain valid. The decision about the effective deposition date is up to national patent legislation. The date of receipt of the request for conversion is considered as conversion date. This date ideally should be before the date of the filing of the patent application, especially in the case of safe deposits (which have to be considered as not being available to the public at all). Thus the accessibility of the biological material in time can be assured.

Problems Which Can be Prevented

As patent deposits have to be regarded as part of the written manifestation, the date of deposit has to be before or latest the date of filing the patent application. It is advisable that deposits are made well in advance to avoid the following problems:

- The depositor might have to cope with unexpected delays due to the transport (e.g., postal strikes, accidents) or customs problems.
- If a package is damaged or the biological material is not in the form requested by the IDA or the culture is proven to be nonviable there would be no time left to send a replacement culture to the depositary.
- If a culture is not of the kind of organisms accepted by the IDA time would be required to find a suitable alternative IDA.

Therefore, proper communication between patent attorney, depositor, and IDA is required to ensure smooth deposition of culture well in advance before filing patent.

Type Strains as Patent Strains

A conflict might arise from the interest of a researcher to obtain both: patent protection and scientific merit. This conflict is based on the demand for the free availability of taxonomic reference strains which is in contrast to the patent practice of restricting the distribution of samples of patent deposits to entitled parties only. Many of the patented microorganisms isolated from unusual habitats are not only of industrial interest but are also new to systematists. Thus, the scientists might also be interested in describing a novel taxon and also having the name validly published. The strain *Paracoccus marcusii* MH1, for example, produces carotenoids as described in a US patent (Hirschberg and Harker, 1999), on the other hand being the type strain of the species (Hirschberg and Harker, 1999). Here, as a result, the situation arose where the type strain for this new taxon, DSM 11574 is under patent protection. From a scientific point of view, however, type strains should be readily accessible for the user and no additional restrictions should lengthen the request procedure. The problem can be solved by keeping both activities separate and depositing a parallel subculture of the patent strain under a different accession number in the public part of a collection. The deposit of type strains is governed by the International Code of Nomenclature of Prokaryotes (Parker et al., 2015). Rule 30 3 (b) is explicit in requiring:

"As of 1 January 2001, the description of a new species, or new combinations previously represented by viable cultures must include the designation of a type strain (see Rule 18a), and a viable culture of that strain must be deposited in at least two publicly accessible culture collections in different countries from which subcultures must be available. The designations allotted to the strain by the culture collections should be quoted in the published description. Evidence must be presented that the cultures are present, viable, and available at the time of publication."

While Rule 30 4 specifically forbids:

"Organisms deposited in such a fashion that access is restricted, such as safe deposits or strains deposited solely for current patent purposes, may not serve as type strains."

Other unacceptable practices include changing the status of a type strain following publication of the collection number restricting access as either a safe deposit or imposing restrictions required when a strain is deposited solely for patent purposes (Tindall and Garrity, 2008)

NOTES

1. Material deposited for patent purposes according to the Budapest Treaty is stored at the IDA for at least 30 years (plus additional 5 years after the most recent release). What will happen to the deposited patent organisms after the prescribed period of storage? Are they to be transferred into the

public section of the collection to guarantee the state of the art? Are they to be destroyed as they might represent a whole "factory" which is available so easily? Should they be handed back to the depositor who is the original owner of the material? As there is no jurisdiction concerning the fate of the organisms, some IDAs offer the possibility that the depositor himself decides what will happen to the organisms after 30 years (+ 5 years). The depositor's declaration will be valid until relevant jurisdiction comes in place.

2. Patent applicants might fear that third parties will abuse the biological material deposited, especially in case a patent has been withdrawn. To allay these fears the IDA promptly notifies the depositor in writing of each furnishing of a sample of the deposited organism. In addition, in Europe Rule 32 of the European Patent Convention (EPC) (European Patent Office, 2000) offers the option to choose the "expert solution" in connection with patent deposits. Following this rule, after publication until the date of granting the patent, samples of the deposited organism will only be provided to a recognized expert. Experts in the field of research are scientists nominated by the European Patent Office. Alternatively, a scientist might be recommended as expert by the requesting party. Here it would be up to the depositor to accept this person as independent expert. If a patent is withdrawn the expert solution is applicable for 20 years.
3. IDAs typically are service culture collections. It is recommended to follow the *OECD Best Practise Guidelines for Biological Resource Centres* (OECD, 2007). These culture collections are now referred to as Biological Resource Centres (BRC). The Organization for Economic Co-operation and Development (OECD) recognized the needs to harmonize the quality of collection management to better serve bioindustry. "Biological Resource Centre (BRC)" stands for user-driven public culture collections with an improved quality management system. Issues of public interest, such as biorisk, intellectual property rights and material transfer agreement are additional tasks required of BRCs. They are compliant with appropriate national law, regulations, and policies and fulfil many crucial roles, including the preservation and supply of biological resources for scientific, industrial, agricultural, environmental, and medical R&D and biotechnological processes, the performance of R&D on these biological resources, the conservation of biodiversity, and to serve as depositary of biological resources for protection of intellectual property.

CONCLUSIONS

International Depositary Authorities serve as permanent facilities for the preservation and storage of biological material under confidentiality for the purposes of patenting. They are recognized according to the Budapest Treaty. IDAs make these "patent strains" available to third party for research and reworkability.

The number of IDAs is constantly increasing. In 1981 five IDAs started their function, to-date 45 IDAs are in existence. The IDA network is not evenly distributed over the world. In the developing countries biotechnology starts budding and the establishment of IDAs in Africa and South America certainly will follow according to the needs in these areas.

ACKNOWLEDGMENTS

Authors are thankful to Ewald Glantschnig (WIPO), Kathrin Felsch, and Brian Tindall (both DSMZ) for their valuable input.

REFERENCES

European Parliament, 1994. Council Directive 3381/94 setting up a Community regime for the control of exports of goods with dual use. OJ No. L 367, pp. 1–12.

European Parliament, 1998a. Council Directive 98/44/EC on the legal protection of biotechnological inventions. OJ No. L 213, pp. 13–21.

European Parliament, 1998b. Council Directive 98/81/EC amending Directive 90/219/EEC on the contained use of genetically modified microorganisms. OJ No. L 330, pp. 13–31.

European Parliament, 2000. Council Directive 2000/54/EC on the protection of workers from risks related to exposure to biological agents at work. OJ No. L 262, pp. 21–45.

European Patent Office, 2000. Convention on the grant of European patents (European Patent Convention, EPC).

Hirschberg, J., Harker, M., 1999. Carotenoid-producing bacterial species and process for production of carotenoids using same. US Patent 5 935, 808.

OECD Best Practise Guidelines for Biological Resource Centres, 2007. Organisation for Economic Co-operation and Development, Geneva.

Parker, C.T., Tindall, B.J., Garrity, G.M., 2015. International code of nomenclature of prokaryotes. Int. J. Syst. Evol. Microbiol., Published Ahead of Print.

Rohde, C., Smith, D., 2013. European Biological Resource Centres Network Information Resource: International Regulations for Packaging and Shipping of Microorganisms, Available from: https://www.dsmz.de/deposit/deposit-in-the-open-collection/deposit-of-microorganisms/transport-requirements.html

Sharma, A., Shouche, Y., 2014. Microbial Culture Collection (MCC) and International Depositary Authority (IDA) at National Centre for Cell Science Pune. Indian J. Microbiol. 54 (2), 129–133.

Tindall, B.J., Garrity, G.M., 2008. Proposals to clarify how type strains are deposited and made available to the scientific community for the purpose of systematic research. Int. J. Syst. Evol. Microbiol. 58, 1987–1990.

Universal Postal Union, 2013. Official compendium of information of general interest concerning the implementation of the Letter Post Services (2013 Berne Convention and Letter Post Regulations).

World Health Organization, 1994. Guidance on Regulations for the Transport of Infectious Substances, WHO/HSE/GCR/2012.12, Geneva, 2011, European Parliament (1994) Council Directive 3381/94 setting up a Community regime for the control of exports of goods with dual use. OJ No. L 367, pp. 1–12.

World Intellectual Property Organization, 1967. Convention establishing the World Intellectual Property Organization (1967) as amended 1979, Geneva; 1979.

World Intellectual Property Organization, 1989. Budapest Treaty on the international recognition of the deposit of microorganisms for the purposes of patent procedure. (1977), as amended in 1980, and Regulations (as in force since 1981), Geneva; 1989.

World Intellectual Property Organization, 2016. Guide to the deposit of microorganisms under the Budapest Treaty., Geneva, updated regularly, see www.wipo.int/treaties/en/registration/budapest/guide.

FURTHER READING

Harker, M., Hirschberg, J., Oren, A., 1998. Paracoccus marcusii sp. nov., an orange Gram-negative coccus. Int. J. Syst. Evol. Microbiol. 48, 543–548.

Chapter 15

Biosafety, Transport and Related Legislation Concerning Microbial Resources—An Overview

Vera Bussas
Leibniz Institute DSMZ-German Collection of Microorganisms and Cell Cultures, Braunschweig, Germany

SAMPLING OF MICROORGANISMS—CONVENTION ON BIOLOGICAL DIVERSITY AND ITS NAGOYA PROTOCOL

The route to utilization of a given microorganism starts with taking samples from its habitat. In the past there were no clear-cut regulations, and scientists performed sampling wherever they wanted, often ignoring the (at least ethical) rights of the landowner or the country of origin, thus bioprospecting sometimes ended up in biopiracy. The term biopiracy is defined as the commercial use of bioactive compounds or genetic sequences by companies from an industrialized, technologically advanced country without having sought the consent from the providing country and without giving fair compensation to the landowner or the nation from whose territory the materials were collected. Thus, the sampling has been essentially illegal in most cases, with similar consequences associated with the resulting patent.

Infamous examples of early microbial biopiracy include the production of the antibiotic erythromycin from bacteria isolated from soil collected in the Philippines (Bunch and McGuire, 1952) as well as apramycin originating from samples from India (Tomita et al., 1977). Today, the very first steps to be considered before sampling are to obtain the necessary access rights and sampling agreements regulating possible benefit sharing. Internationally the *Convention on Biological Diversity* (CBD) aims at the conservation of the biological diversity, the sustainable use of its components, and the fair and equitable sharing of the benefits arising out of the utilization of genetic resources (United Nations, 1992). This includes the appropriate access to genetic resources and the appropriate transfer of relevant technologies, taking into account all rights

Microbial Resources. http://dx.doi.org/10.1016/B978-0-12-804765-1.00015-1

over these resources. The CBD was signed in 1992 in Rio de Janeiro/Brazil. To date (January 2017) there are 196 parties; the CBD is part of the United Nations Environment Programme (UNEP) (www.unep.org/).

The CBD's key elements of access and benefit sharing (ABS) have been codified in the *Nagoya Protocol on Access to Genetic Resources and the Fair and Equitable Sharing of Benefits Arising from their Utilization to the Convention on Biological Diversity* (United Nations, 2011) (www.cbd.int/convention/text/). The intention of this international agreement is clearly defined, and entered into force on October 12, 2014: any sampling of biological material (plants, animals, microorganisms, DNA) needs the prior informed consent (PIC) (national/provincial/local) in the provider country and mutually agreed terms (MAT) to clarify fair and equitable ABS. The exact date of ratification depends on regional/national legislation; in the European Union it is exactly this date as laid down in the Regulation 511/2014 of the European Parliament and of the Council on compliance measures for users from the *Nagoya Protocol on Access to Genetic Resources and the Fair and Equitable Sharing of Benefits Arising from their Utilization in the Union* (European Parliament and Council, 2014).

ISOLATION OF MICROORGANISMS—BIOSAFETY CONSIDERATIONS

With the samples from the collection area in hand, the scientist enters the laboratory and starts isolation procedures. What has to be considered before getting started? The researcher should be aware of the fact that the sample might carry infectious microorganisms as well as mostly harmless ones, accordingly biosafety aspects have to be considered. Biosafety aims at the reduction or elimination of exposure to and removal of the risk of potential infections and the contamination of the environment by potentially hazardous biological agents (World Health Organization, 2004a). This goal can be achieved by laboratory control and containment measures as access restrictions, personnel expertise and training, suitable equipment, and safe methods. Biosafety is based on risk assessment, personnel responsibility, and material transfer documentation.

WHO Hazard Classification System

Microorganisms have different hazard potentials and are classified according to the harm they might cause to humans or animals. "Harm"—as defined by the *International Standard Organization* (www.iso.org/iso/home.html) (International Organization for Standardization, 2014)—is the injury or damage to the health of people, or damage to property or the environment, whereas the term "hazard" is used to define a potential source of harm (like fire hazard, cutting hazard, or biological hazard). The combination of the probability of occurrence of harm and the severity of that harm finally is called "risk." The classification of microorganisms according to the risk they impose is used as the basis for existing restrictions on who is allowed to import, export, or work with which organisms. To allow a scientist to allot his isolated microorganism correctly to Risk Groups, the organisms must have been identified taxonomically and allocated to an already existing systematic taxon, at least at the species level. The World Health Organization (2004a) established a classification system of four Risk Groups for microorganisms to be applied for the work in a laboratory. The organisms are allocated to the Risk Groups according to the increasing risk they impose for human beings or animals. With the WHO system as backbone, national and international laws, directives, guidelines, and administrative provisions have been established worldwide. The definitions of Risk Groups and examples are largely concordant and have been compiled here from selected international and national legal works:

International	WHO Laboratory Biosafety Manual (World Health Organization, 2004a, 2006)
Regional	Europe: Council Directives 2000/54/EC on the Protection of Workers from Risks Related to Exposure to Biological Agents at Work (European Parliament, 2000)
	Australia/New Zealand: Australian/New Zealand Standard™ AS/NZS 2243.3:2010, Safety in Laboratories, Part 3: Microbiological Safety and Containment (Australian/New Zealand Standard™, 2010)
National	Canada: Canadian Biosafety Standard, Public Health of Canada (Canadian Biosafety Standard, 2015)
	Germany: Ordinance on Safety and Health Protection at Workplaces Involving Biological Agents (Biological Agents Ordinance-BioStoffV) (German Ordinance, 2013, 2013) (see Appendix)
	United Kingdom: The Approved List of Biological Agents Advisory Committee on Dangerous Pathogens (Advisory Committee on Dangerous Pathogens, 2013)
	United States of America: U.S. Department of Health and Human Services Public Health Service, CDC-NIH—Biosafety in Microbiological and Biomedical Laboratories (U.S. Department of Health and Human Services, 2009)

Organisms known to be without any hazardous potential based on long-term experience will be placed in Risk Group 1. Potentially hazardous organisms, rated according to their rising pathogenic potential, are allocated to Risk Groups 2, 3, or 4.

(A) Nonhazardous Organisms

Risk Group 1: organisms and viruses with no risk or which are most unlikely to cause diseases in man and vertebrates.

Examples are as follows:

- *Lactobacillus helveticus*—used for cheese fermentation (Parmesan, Romano, Provolone, mozzarella)
- *Bacillus subtilis*—ubiquitous (water, air, soil)

- *Aspergillus oryzae*—used for the production of miso or soya sauce
- obligately psychrophilic organisms (growth only below 20°C)
- obligately thermophilic organisms (growth only above 50°C)
- obligately phototrophic organisms (no growth without light)
- obligately chemolithotrophic organisms (no growth with organic matter)
- saprophytically living organisms; no infectious potential for humans or animals
- individual strains of organisms of Risk Group 2 or higher which definitely lost their pathogenicity (e.g., *Escherichia coli* K12, *Salmonella typhi* 21a) or those having been in use for a long period of time (e.g., for production purposes) without having ever caused diseases (e.g., certain strains of *Lactobacillus rhamnosus*)

(B) Potentially Hazardous Organisms

Risk Group 2: organisms and viruses which may cause diseases in animals and employees, but are unlikely to spread in the community; the risk for the laboratory worker is low to medium and weak for the general population as prophylactic and/or therapeutic measures exist.

Examples are as follows:

- *Acinetobacter baumannii*: multidrug resistant, one of the most difficult antimicrobial-resistant gram-negative microorganisms to control and treat; survives for prolonged periods under a wide range of environmental conditions; causes outbreaks of infection and health care–associated infections, including bacteremia, pneumonia, meningitis, urinary tract infection, and wound infection (Eliopoulos et al., 2008).
- *Streptococcus mutans*: lives on the mucous membrane of the human mouth and is involved in the development of caries; facultatively a causing agent for endocarditis.
- *Vibrio cholerae*: causing agent of cholera; high dose needed for infection; infection mainly via drinking water spoilt with feces; infection by air excluded; cholera is an acute diarrheal disease that can kill within hours if left untreated [World Health Organization, 2004b, Fact Sheets, 2014 (#107)].
- *Candida albicans*: lives on the mucous membrane of the human mouth and intestines; causes mucosal mycosis.
- Herpes simplex virus type 1 (HSV-1): highly contagious infection, common and endemic throughout the world. The virus causes lifelong infection, and there is no cure, although treatment can reduce symptoms [World Health Organization, 2004b, Fact Sheets, 2016 (#400)].
- Rabies virus: invades the central nervous system; active vaccination after infection possible; incubation time between 1 week and 1 year; lethal without medical attention; immunoprophylaxis possible. Infection almost always fatal following the onset of clinical signs; virus transmitted by domestic dogs (bites or scratches, usually via saliva) [World Health Organization, 2004b, Fact Sheets, 2016 (#99)].

Risk Group 3: organisms and viruses which may cause severe and life-threatening diseases; regarding infectiousness, pathogenicity, and the existence of prophylactic and/or therapeutic measures, the risk is medium to high for the laboratory worker, whereas the risk for the community is weak to medium.

Examples are as follows:

- *Mycobacterium tuberculosis*: causing agent of tuberculosis; extremely infectious, spreading by air; severe disease with lengthy therapy; protection by vaccination possible; curable and preventable; spread from person to person. About one-third of the world's population has latent tuberculosis; people infected have a 10% lifetime risk of falling ill with TB [World Health Organization, 2004b, Fact Sheets, 2016 (#104)].
- *Yersinia pestis*: causing agent of pestilence/plague; as little as five cells are sufficient for an infection; acutely life-threatening disease; therapy with antibiotics successful only directly after infection; protection by vaccination possible; transmission between animals and humans by the bite of infected fleas, direct contact, and inhalations; fatality ratio of 30%–60% if left untreated. The "Black Death" during the 14th century caused an estimated 50 million deaths [World Health Organization, 2004b, Fact Sheets, 2014 (#267)].
- *Coccidioides immitis*: causes severe inner mycosis, initial infection of the lungs, spreads to other organs of the body; existing medicines do not guarantee complete recovery.
- Human immunodeficiency virus (HIV): causing agent of AIDS; virus extremely unstable; targets the immune system and weakens people's defense systems against infections and some types of cancer. Infected individuals gradually become immune-deficient. The most advanced stage of HIV infection is Acquired Immunodeficiency Syndrome (AIDS), which can take from 2 to 15 years to develop depending on the individual. AIDS is defined by the development of certain cancers, infections, or other severe clinical manifestations. There is no cure for HIV infection. However, effective antiretroviral (ARV) drugs can control the virus and help prevent transmission, so that people with HIV, and those at substantial risk, can enjoy healthy and productive lives [World Health Organization, 2004b, Fact Sheets, 2016 (#360)].

Risk Group 4: biological agents causing severe diseases and representing a serious risk to employees and the general population; regarding infectiousness, pathogenicity, and the usually lacking existence of effective prophylactic and/or therapeutic measures, the risk to the laboratory worker and the community is high.

To date only certain viruses have been allocated to Risk Group 4.

Examples are as follows:

- Herpes B virus: acute encephalomyelitis, resulting in death or severe neurologic impairment. Respiratory involvement and death can occur 1 day to 3 weeks after symptom onset. B virus infection in humans usually occurs as

a result of bites or scratches from macaques—a genus of Old World monkeys. Laboratory infections known; spreading by air; no therapy or prophylactic measures possible; up to 100% lethality. The virus can be present in the saliva, feces, urine, or nervous tissue of infected monkeys and may be harbored in cell cultures derived from infected monkeys (Hilliard, 2007).
- Lassa fever virus: systemic disease with inflamed and ulcerated throat, fever and convulsions; death within 10–14 days; acute hemorrhagic illness in West Africa; transmission to humans via contact with food or household items contaminated with rodent urine or feces. Person-to-person infections and laboratory transmission also possible. Overall case fatality rate is 1%. Early supportive care with rehydration and symptomatic treatment improves survival; disease especially severe late in pregnancy, with maternal death and/or fetal loss in more than 80% of cases during the third trimester [World Health Organization, 2004b, Fact Sheets, 2016 (#179)].
- Pox virus (Variola): causing agent of pocks; vaccination led to possible eradication worldwide.
- Ebola virus: causes Ebola virus disease (EVD), formerly known as Ebola hemorrhagic fever; severe, often fatal illness in humans (fatality 50%); transmission to people from wild animals, spread in the human population through human-to-human transmission; no licensed treatment [World Health Organization, 2004b, Fact Sheets, 2016 (#103)].

Interestingly, some countries like China, Japan, and Russia decided to use a different classification system where Risk Group 4 is foreseen for the harmless organisms, whereas Risk Group 1 contains the most dangerous ones (Qian, 2014). Such classification might result in confusion, for example, when sending biological material of a Risk Group 1 organism from a country following WHO classification (harmless according to WHO) to a country with the opposite classification system where Risk Group 1 is of the most dangerous category. Serious problems might be encountered during importation of the biological material. The more so as the laboratory safety level in most countries may follow the WHO system where the laboratory safety levels BSL 1 to BSL4 (Risk Group 1 to 4 organisms) are defined according to increasing risk of the organisms. As a result further confusion may arise creating significant risk to workers. There are numerous text books available giving guidance how laboratories need to be equipped to handle biological material according to the necessary laboratory containment level. The one most commonly referred to is the World Health Organization's Laboratory Safety Manual (World Health Organization, 2004a) where an excellent compilation of the relationship between the Risk Groups and biosafety levels are illustrated (Table 15.1). The manual also provides practical guidance for the work to be conducted in a microbiology laboratory. One of the key features, the *Good Microbiological Techniques* (GMT), includes below listed measures:

- The access to a laboratory must be restricted.
- Laboratory coats have to be worn.

TABLE 15.1 Relation of Risk Groups to Biosafety Levels, Practices, and Equipment (World Health Organization, 2004a)

Risk group	Biosafety level	Type of laboratory	Laboratory practices	Safety equipment
1	Level 1	Basic teaching, research	GMT	none
2	Level 2	Primary health and diagnostic services; research	GMT + protective clothing, biohazard sign	BSC
3	Level 3	Special diagnostic services; research	Level 2 + special clothing, controlled access, directional airflow	BSC and/or other primary devices for all activities
4	Maximum containment Level 4	Dangerous pathogen unit	Level 3 + airlock entry, shower exit, special waste disposal	Class III BSC, or positive pressure suits in conjunction with Class II BSC, double-ended autoclave (through the wall), filtered air

BSC, Biological safety cabinet; GMT, good microbiological techniques.

- As a matter of course, no eating, no drinking, no chewing gums, or application of cosmetics is tolerated.
- Doors and windows have to be closed during work in a laboratory.
- Aerosols should be prevented by necessary and possible measures (e.g., laminar air flow; aerosol tight centrifuge beakers).
- Safety glasses and gloves should be worn whenever personal protection is necessary (e.g., when working with glass, liquid nitrogen, always in the case of skin injuries, etc.).
- Personal hygiene is a must: before leaving the laboratory hands have to be cleaned thoroughly by disinfection followed by washing, drying, and skin care.

Necessary Permits and Allowances

A purpose specific and adequately equipped laboratory is indispensable for the work with microbial isolates. In most countries, it will be necessary to care for the registration of the working area, be it for the handling of

- Human pathogens
- Animal pathogens
- Plant pathogens
- Genetically modified organisms

In Germany, as an example, the respective laws are the acts dealing with the prevention and control of infectious diseases in humans (German Federal Law, 2000), the infectious diseases of animals enactment (German Federal Law, 1985), the German plant protection act (German Federal Law, 2012), and the law regulating genetic engineering (German Federal Law, 1993). In addition, in accordance with the German Social Security System, the Employer's Liability Insurance Association (www.eubusiness.com/europe/germany/staff-welfare/) needs to be informed. This system is specific for Germany and is to protect employees from liabilities arising from disease, fatality, or injury resulting from work.

Besides the registration of the facilities, *personal* allowances might be necessary. These will require personnel expertise and training. Thus, scientists will take over personal responsibility for their work and collaborators. These personal allowances again will be granted for the work with human, animal, and/or plant pathogens or genetically modified organisms.

VALUABLE MICROBIAL ISOLATES—BIOSECURITY ISSUES

Any isolated bioactive microorganism with bioactivity and commercialization potential might also be the producer of toxic compounds such as neurotoxins, not only dangerous to humans but also to animals. Such properties leads us to the field of biosecurity. The term "biosecurity" can have different meanings in different contexts. In ecology, it stands for the protection of natural resources from biological invasion and threats; in agriculture, it describes the precautions to be taken to minimize the risk of introducing an infectious agent into a population. In food supply, it deals with the protection of a nation's food supply and agricultural resources from accidental contamination and bioterrorism attacks.

Finally, in 2006 the WHO in collaboration with the Food and Agriculture Organization (FAO) and the World Organization for Animal Health (OIE) coined the expression "Laboratory Biosecurity" which is restricted only to *laboratory* environments (World Health Organization, 2006). Laboratory biosafety (as already explained earlier) and biosecurity are related, but conceptually not identical. Biosafety aims at the reduction or elimination of exposure of individuals and the environment to potentially hazardous biological agents, whereas biosecurity primarily wants to prevent unauthorized access, loss, theft, misuse, diversion, or intentional release of biological materials. Moreover, notably biosecurity does not only refer to potentially hazardous materials or dual-use goods. Dual-use goods are those which can be used in military spheres leading to harmful misuse as well as for civilian purposes and peaceful activities. *The Biological and Toxin Weapons Convention* [Convention on the Prohibition of the Development, Production and Stockpiling of Bacteriological (Biological) and Toxin Weapons and on their Destruction] (BTWC) (http://disarmament.un.org/treaties/t/bwc/text) (United Nations Office for Disarmament Affairs, 1975) bans the development, production, stockpiling, acquisition, and retention of microbial or other biological agents or toxins without justification for prophylactic, protective, or other peaceful purposes. The Convention was signed in 1972 and came into force in 1975, has 175 member states (September 2016), and is supported by the United Nations Office for Disarmament Affairs (UNODA). To strengthen the Convention, confidence-building measures (CBM) have been established with annual reports and an implementation support unit (ISU) to assist its state parties in implementing the Convention.

In case of dual-use goods, the aim of the biosecurity is to protect the human beings from their potential to cause harm. And also other materials have to be protected from unauthorized access, loss or theft! Here we have valuable biological materials requiring administrative oversight and control to protect either their economic or historical (archival) value. Thus, valuable biological materials (VBM) could very well include pathogens and toxins, and as well nonpathogenic organisms, vaccine strains, foods, genetically modified organisms (GMOs), extraterrestrial samples, and patent strains. The main focus of biosecurity therefore lies in the limitation of access to the facilities, the materials themselves as well as the related information. Biosecurity measures are embedded in biosafety levels, offering complementary measures with common components. Scientists have to take over the responsibility for the microorganisms they work with, for their use in a nonharmful way, also by protecting them "physically." This responsibility of scientists is also reflected in codes which have been established in the last years. Some examples are as follows.

Codes

There is a "universal" code in existence to guide the conduct of those involved in the life sciences. It has been established by the United Nations Educational, Scientific and Cultural Organization (UNESCO) and is called "Universal Declaration on Bioethics and Human Rights" (UNESCO, 2006). It addresses ethical issues related to medicine, life sciences, and associated technologies as applied to human beings, taking into account their social, legal, and environmental dimensions. Some examples of codes with a tighter scope from different sources that are currently in use are as follows.

Focus "Bioweapons"

1. WHO—World Health Organization, 194 countries and 2 associate members: Public Health Response to biological and chemical weapons: WHO Guidance (1970, 2004) (www.who.int/csr/delibepidemics/biochemguide/en/).
2. IUMS—International Union of Microbiological Societies, representing over 100 countries, is one of 31 Scientific Unions of the International Council of Science (ICSU): Code of Ethics against Misuse of Scientific Knowledge, Research and Resources (2008) (www.iums.org/index.php/code-of-ethics)

(IUMS, 2008)—"IUMS also strives to promote ethical conduct of research and training in the areas of biosecurity and biosafety so as to prevent use of microorganisms as biological weapons and therefore to protect the public's health and to promote world peace."

3. ASM—American Society for Microbiology—world's largest scientific society of individuals interested in the microbiological sciences with more than 47,000 members: Code of Ethics (2005) (www.asm.org/index.php/governance/code-of-ethics) (ASM, 2005)—"ASM members are obligated to discourage any use of microbiology contrary to the welfare of humankind, including the use of microorganisms as biological weapons and not to engage knowingly in research for the production or promotion of biological warfare agents."
4. Leibniz Association—connects 91 German independent research institutions ranging from the natural, engineering, and environmental sciences via economics, spatial, and social sciences to the humanities: Biosecurity Code (Leibniz Association, 2012)—a globally applicable code of conduct specifically dedicated to biosecurity; addresses the regulations governing potential dual use of biological materials, associated information and technologies to reduce the potential for their malicious use. It aims to raise awareness of regulatory needs and to protect researchers, their facilities, and stakeholders. It specifically addresses the work of public service culture collections and describes the issues of importance and the controls or practices that should be in place. The code was introduced to the Seventh Review Conference to the Biological and Toxin Weapons Convention (BTWC), United Nations, Geneva, 2011; the delegates to the States' parties recommended that this code of conduct be broadly applied in the life sciences among microbiologists.

Focus "Gene Technology"

1. FAO/WHO—Codex Alimentarius (since 1963—Guidelines for the Conduct of Food Safety Assessment of Foods Derived from Recombinant-DNA Plants/Animals/using Recombinant DNA Microorganisms Foods derived from modern biotechnology, 2009) (www.fao.org/fao-who-codexalimentarius/en/) (FAO/WHO, 1992).

Focus "Medicine"

1. WMA—World Medical Association (since 1947), membership of more than 100 national medical associations representing some 10 million physicians—Declaration of Helsinki (DoH) (1964; 2013) (www.wma.net/en/30publications/10policies/b3/ (World Medical Association, 2013)—set of ethical principles for the medical community regarding human experimentation.

There are also codes in existence proposed by individual persons. One example is the code developed by Somerville and Atlas—*Code of Ethics for the*

Life Sciences—"All persons and institutions engaged in any aspect of the life sciences must ensure that their discoveries and knowledge do no harm by refusing to engage in any research intended to facilitate or of being used to facilitate bioterrorism or biowarfare; work for ethical and beneficent advancement, development, and use of scientific knowledge. .... seek to restrict dissemination of dual-use information and knowledge ..." (Atlas and Somerville, 2004).

In 2002, the Stockholm International Peace Research Institute (SIPRI) (www.sipri.org/) conducted a survey to get to know how many scientific organizations worldwide had some kind of code of ethics at that time. Only 11% out of 71 international scientific organizations and 12% of 267 national or regional scientific organizations had a code in place at that time (Reyes, 2003). As a consequence of the increased fear of terrorist attacks, the number of codes—as you can see from the dates of drawing up of the codes presented earlier—nowadays is higher.

SHIPPING OF MICROORGANISMS—NO TRIVIAL ISSUE

Once the isolate of interest is defined in one laboratory, it will become an isolate of interest for other researchers as well due to its unique properties. The scientist might intend to send it to other colleagues for further research and cooperation, to a company plant, or to a microbial resource center, be it for the deposit in the open collection to make it available to the scientific community, be it for patent purposes, be it as a safe deposit. To do so the scientist will have to follow a series of required steps. The samples of the strain, representing valuable biological material, should arrive at the collection in a good condition—viable and uncontaminated. Along the transport chain there will be secretaries, postal, customs, and other employees who might get in touch with the material during transport in an inadvertent manner. Regulations concerning transport have been drawn up not to endanger people who are not involved in the project and to guarantee that the microorganism will reach its destination in the original status. Again the scientist here takes over the responsibility, for the correct classification, packaging, labeling, and marking of the samples. Shippers of infectious biological substances of Risk Group 2 and higher will need training offered by airlines and courier services.

The demands on the export, packaging, and shipping of biological material are manifold and the numbers of possible mistakes are many. Before a living microorganism is offered for dispatch, whether it is a transborder supply or supply within a country, it has to be made sure that the organism does not fall into wrong hands and that the dispatch is legitimate.

Export and Import of Biological Material and Dispatch Within a State

As a general rule, biological material—at least pathogens or otherwise hazardous biological material—should never be sent to private persons. Only adequately trained staff in sufficiently and adequately equipped laboratories should receive microorganisms to guarantee safe handling and proper use. A second very important restriction is that before every single dispatch of a microorganism the addressee (person and/or company) has to be checked against applicable sanctions lists following antiterrorism laws (e.g., EEC/881/2002 in the European Union) (European Parliament, 2016).

Export

The export of microorganisms is usually governed by national law. And also international or regional (e.g., in Europe) legislation exists aiming at the harmonization on the lists of microorganisms affected. With few exemptions, organisms classified in Risk Group 1 are admitted for export without restrictions. A total embargo for a given state, however, would exclude even such organisms from export. Sanctions by the United Nations may be in place and have to be observed. More general restrictions are imposed by those laws and directives controlling the export of goods with a possible dual use, which means they could be used for civil or military purposes. A regulative example is EEC Council Directive 3381/94/EEC setting up a Community regime for the control of exports of goods with dual use (European Parliament, 1994). Organisms categorized as potential biological weapons include the causative agents of anthrax (*Bacillus anthracis*, Risk Group 3), pestilence (*Y. pestis*, Risk Group 3), or toxin producing strains of organisms like *Aspergillus fumigatus*, *Clostridium botulinum*, and *Staphylococcus aureus* (all Risk Group 2).

Other examples are certain plant (crop) pathogenic species and all those genetically modified microorganisms that meet the definition of the respective paragraph in the EU list of dual-use goods (European Community, 2000). Another impediment can be the *Convention on International Trade in Endangered Species of Wild Fauna and Flora* (CITES) (www.cites.org/eng/disc/text.php) (International Union for Conservation of Nature, 1973) with the aim to ensure that international trade in specimens of wild animals and plants does not threaten their survival. This convention was signed in 1973 in Washington and has 183 parties to date (January 2017); it is part of the UNEP (www.unep.org). Biological products (blood, tissue, hair, feathers, cell and tissue cultures, DNA) of endangered species are subject to CITES restrictions.

Import

The import of biological material might be subject to the quarantine regulations of a given recipient country. These restrictions often depend on the special ecological or climatic situation of a certain country: it is free of a certain pathogen and it might be expected that an introduced microorganism could spread excessively. Any individual case of import of microorganisms to, for example, Australia, Brazil, Canada, and New Zealand requires an import permit.

The European Community Commission Directive 2000/29/EC (European Community, 2000) restricts the import of organisms harmful to plants and plant products and their spread within the Community. In Germany, comparable restrictions concerning, for example, *Synchytrium endobioticum*, the causative agent of potato cancer or *Erwinia amylovora* causing fire blight in fruit trees are imposed according to the German Plant Protection Act (German Federal Law, 2012). Similarly, the German Infectious Diseases of Animals Imports Enactment (German Federal Law, 1982, 2015) restricts the import of *Chlamydia psittaci* causing psittacosis, or *Zymonema farcinimosus* responsible for lymphatic vessel inflammation of one-hoofed animals. In other countries similar laws exist.

Dispatch Within a State

Handling of certain kinds of biological material might be restricted to certain persons or registered laboratories having acquired the necessary permissions or registrations. Pathogens or otherwise hazardous biological material should only be shipped after having verified that the recipient has the respective permissions to handle the material, such as human pathogens, animal pathogens, plant pathogens, or genetically modified organisms. It should be noted that also the international dispatch of certain agents listed as potential dual-use material might be regulated: a signed end-user certificate is possibly required. The best sources of information are the respective national export offices. Table 15.2 demonstrates which issues have to be clarified before any dispatch of biological material.

The Transport Itself

After having clarified that no restrictions impede the dispatch the sample can be prepared for shipment.

Packaging of Biological Material

The principle of the safe packaging of biological material relies on the system of triple packaging. This system which specifically has been designed for the transport of infectious substances serves as the only protection for people, animals, and the environment. Nonhazardous biological substances (allocated to Risk Group 1) are not regulated by dangerous goods transport regulations and usually postal regulations and requirements as laid down by the Universal Postal Union (UPU) (International Bureau of the Universal Postal Union, 2013) (www.upu.int/uploads/tx.../actInFourVolumesLetterPostManualEn.pdf) and by the National Postal Authorities (e.g., Germany: www.dhl.de/de/express/versenden/versandberatung/gefahrgut.html) apply. Nevertheless, they of course should reach their destination in a viable and undamaged state, as

TABLE 15.2 Issues to be Clarified Before the Dispatch of Biological Material

Risk group	Destination	Notice	Action/proof	Kind of dispatch
1	Domestic	National laws like: – Genetic Engineering Act – Plant Protection Act – Infectious Diseases of Animals Act	Permission/ Registration/ Authorization	Mail
1	International	– Plant Protection Act – Air mail might not be permitted – Dual Use Act – Total embargo?	 Only for civil use! No dispatch	Mail/air mail Air freight
2	Domestic	National laws like: – Infectious Diseases Act – Infectious Diseases of Animals Act – Plant Protection Act – Genetic Engineering Act – Dual Use Acts	Permission/ Registration/ Authorization/ Registered laboratory Only for civil use!	Private carriers by road or air
2	International	– Dual Use Acts – Import or quarantine restrictions – Plant Protection Act	Only for civil use! Import permit Permission	Private carriers by road or air
2	Non-OECD country	– Embargo – Act dealing with foreign trade	No dispatch! No dispatch!	
3 and 4	*See* Risk Group 2	*See* Risk Group 2	*See* Risk Group 2	Private carrier

fast as possible and should be packaged accordingly. Material allocated to Risk Group 2 and higher falls under the "UN Model Regulations on the Transport of Dangerous Goods" affecting all modes of transport (road, air, rail, waterways) (United Nations, 2015).

The triple packaging is composed of as follows: the packaging must consist of (1) a watertight primary receptacle, (2) a watertight secondary packaging, and (3) an outer packaging. Absorbent material has to be placed between the primary and the secondary packaging. Several primary receptacles should be wrapped individually. There should be sufficient absorbent material present (e.g., cotton wool) to absorb the entire liquid if a leakage occurs. The outer packaging bears all the labels and documentation (World Health Organization, 2015) (Fig. 15.1A–B).

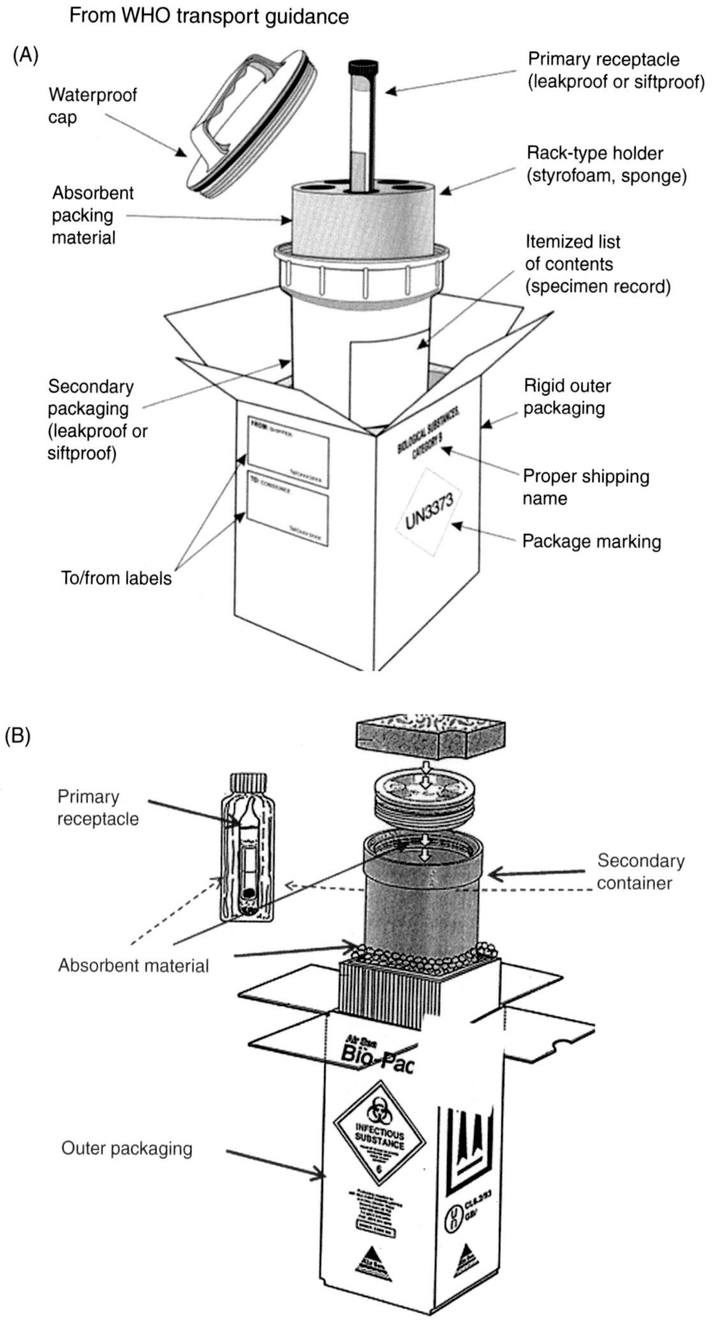

From WHO transport guidance

(A)

Waterproof cap

Primary receptacle (leakproof or siftproof)

Rack-type holder (styrofoam, sponge)

Absorbent packing material

Itemized list of contents (specimen record)

Secondary packaging (leakproof or siftproof)

Rigid outer packaging

Proper shipping name

Package marking

To/from labels

UN3373

(B)

Primary receptacle

Secondary container

Absorbent material

Outer packaging

Bio-Pac

INFECTIOUS SUBSTANCE

FIGURE 15.1 **Triple packaging for infectious substances according to UN Packing Instruction (A) PI620 and (B) PI650.**

Transport by Mail

National postal services are merged in the UPU. In the Letter Post Manual (International Bureau of the Universal Postal Union, 2013), guidelines are laid down for the postal transport and packaging of biological substances. It is differentiated between noninfectious organisms (Risk Group 1) and infectious organisms of category B (mainly Risk Group 2).

Generally, the exchange of biological substances by mail is restricted to countries whose postal administrations agree to accept biological material. The national postal services of several countries do not allow shipment (receipt or dispatch) of biological substances by mail at all, being by surface or air. All postal administrations prohibit infectious substances of category A like all other kinds of dangerous goods in the mail.

It has to be indicated which organisms are shipped (e.g., on a delivery slip between the outer and secondary container. Shipments of biological material into foreign countries might be subject to customs inspection. For this purpose they must carry the green customs declaration CN22. Indicate number and type of material (e.g., three bacterial cultures, two plasmid preparations). The note "No commercial value" should be added for the free exchange between laboratories or any other supply free of charge. If infectious substances (Category B) are shipped by air mail, the International Air Transport Association (IATA) Dangerous Goods Regulations (www.iata.org/publications/dgr/Pages/index.aspx) (IATA, 2017) have to be followed (see section "Transport as Freight").

Transport as Freight

If shipments with biological material are not accepted in (air) mail, shipping by using private carrier service and (air) freight has to be chosen. The carrier will decide on the fastest mode of transport (road, rail, air, waterways). The United Nations Committee of Experts for the Transport of Dangerous Goods established a system for the classification of all kinds of dangerous substances called "Orange Book" (United Nations, 2015), with the aim to ensure safe shipping for all modes of transport. The nine different classes include explosives, gases, flammable liquids, flammable solids, oxidizing substances, organic peroxides, toxic and infectious substances, radioactive material, corrosives, and miscellaneous dangerous goods. The UN number, proper shipping name, class, hazard label, packing group (not relevant for biological substances), and packing instruction (Table 15.3) are important for shipping procedures. Infectious substances are classified in Class 6, Division 6.2. It is differentiated between:

- *noninfectious material*: no dangerous goods, labeled as "exempt biological specimen without pathogens"—not subject to Dangerous Goods Regulations
- *infectious substances, category A*: exposure may cause permanent disability, life-threatening or fatal disease (Risk Group 3 and 4 organisms, some

TABLE 15.3 Relevant Extract From the IATA Dangerous Goods List

UN-No.	Name and description	Class	Hazard label	Packing instruction	Max. net quantity per package	
					Passenger aircraft	Cargo aircraft
1845	Carbon dioxide, solid (dry ice)	9	Miscellaneous	954	200 kg	200 kg
2814	Infectious substance, affecting humans (solid or liquid)	6.2	Infectious substance	620	50 g or mL	4 kg or L
2900	Infectious substance, affecting animals (solid or liquid)	6.2	Infectious substance	620	50 g or mL	4 kg or L
3245	Genetically modified microorganisms *(acc. 3.9.2.5.2)*	9	None	959	No limit	No limit
3373	Biological Substance, category B	6.2	None	650	4 kg or L	4 kg or L

Risk Group 2 organisms like *C. botulinum*, polio virus, *Mycoplasma mycoides*), labeled as: "infectious substance, affecting humans" or "infectious substance, affecting animals"

- *infectious substances, category B*: infectious, but not meeting the criteria of category A (mainly Risk Group 2)

The International Civil Aviation Organization (ICAO) established guidelines for the transport of dangerous goods by air, implemented by the IATA. Air transport is playing a major role today, at least in case of international exchange of biological material. In this context, the IATA Infectious Substances Shipping Guidelines (IATA, 2017) are especially concerned with the shipping of infectious biological material, whereas the transport of noninfectious biological material is not included. When sending infectious substances of category A by air, a Shipper's Declaration for Dangerous Goods has to be filled in (in English). The Shipper's Declaration is a legal document and has to be signed by a trained person: general or job-specific training of a person involved in packing and shipping of dangerous goods is a prerequisite according to IATA DGR, and certified training courses have to be attended by shippers of dangerous goods. The shipper is responsible for the correct documentation.

The UN numbers for infectious substances of category A affecting humans and those affecting animals are UN 2814 and UN 2900, respectively. Those of category B are assigned to UN3373. All consignments containing infectious substances of category A are to be accompanied by the transport emergency card (available from courier services) and must be packed in UN certified packagings. UN certified combination packaging and labels are commercially available, labels might already be printed on ready-to-use packaging.

In case the packaging contains infectious substances as the only dangerous good, the following labeling is necessary: (1) UN number plus technical (scientific) name of the infectious substance (UN 2814, UN 2900, or UN3373, respectively) (www.iata.org/publications/dgr/Pages/index.aspx); (2) name, address, and telephone number of sender; (3) name, address, and telephone number of recipient; (4) hazard label (not for UN3373); and (5) packaging orientation label (if it contains liquid material). Advance arrangements prior to dispatch are indispensable: the sender is urgently advised to contact the courier service when preparing a shipment. Also, the consignee has to be informed of the intended dispatch of infectious substances.

For transport of infectious substances by road or by rail, the ADR (*Accord Européen relatif au transport international des marchandises dangereuses par routes*) (www.unece.org/trans/danger/publi/adr/adr2015/15contentse.html) (United Nations Economic Commission for Europe, 2016) and COTIF/RID (regulations concerning the international carriage of dangerous goods by rail) (www.cit-rail.org/en/rail-transport-law/cotif/) (Intergovernmental Organization for International Carriage by Rail, 2015), respectively, apply in Europe. Rules for the international traffic of dangerous goods by inland waterways (ADN) (www.unece.org/trans/danger/publi/adn/adn2015/15files_e.html) (United Nations Economic Commission for Europe, 2016) have been developed, and the carriage of dangerous goods by sea is internationally regulated by the International Maritime Dangerous Goods Code (IMDG) (www.imo.org/en/Publications/IMDGCode/Pages/Default.aspx) (International Maritime Organization, 2016).

Transport of Genetically Modified Organisms

When the microorganisms to be shipped have been genetically modified in a way that does not occur naturally, they can be allocated to one of the following three groups for air transport:

1. Organisms being infectious to humans and/or animals of category A. They are to be shipped as infectious substances according to IATA DGR, Class 6, Division 6.2, UN 2814 or UN 2900, following Packing Instruction 620.
2. Organism being allocated to category B. Transport has to be performed as biological substances, category B, UN3373, applying Packing Instruction 650.
3. Organisms capable of altering animals, plants, or microbiological substances in a way that does not occur in nature but which do not meet the definition

of an infectious substance. They are allocated to Class 9 (Miscellaneous dangerous substances and articles), UN 3245. The applicable Packing Instruction is PI 959.

In the context of transborder movement of genetically modified organisms, the *Cartagena Protocol on Biosafety to the Convention on Biological Diversity* (Secretariat of the Convention on Biological Diversity, 2000) has to be followed. It is an international agreement in the framework of the CBD (United Nations, 1992) which aims to ensure the safe handling, transport, and use of living modified organisms (LMOs) resulting from modern biotechnology and that may have adverse effects on biological diversity. It entered into force on September 2003. Its vision is to make biological diversity adequately protected from any adverse effects of living modified organisms. Thus, the mission is to strengthen global, regional, and national action and capacity to ensure an adequate level of protection in the field of the safe transfer, handling, and use of living modified organisms with possible adverse effects on the conservation and sustainable use of biological diversity. Risks to human health are also taken into account and the focus specifically lies on transboundary movements.

Transport of Patient Specimens

Patient specimens must be assigned to the most appropriate UN number, be it UN2814 or UN2900 for human or animal material of category A, be it UN 3373 for those of category B and as "exempt human/animal specimen" if there is only a minimal likelihood that pathogens are present.

Use of Dry Ice

If dry ice (carbon dioxide, solid) is used as a refrigerant to keep the biological material cold during transport, a special containment has to be used allowing the constant release of carbon dioxide gas and thus preventing the build up of pressure that may cause rupturing of the packaging. There are special combination containments for infectious substances on the market that can keep the low temperature up to 3 days. Dry ice is a dangerous good (Class 9, miscellaneous dangerous goods, UN 1845, Packing Instruction 954) (www.iata.org/publications/dgr/Pages/index.aspx). It has to be included in the documentation that accompanies the consignment and the latter has to be labeled accordingly.

CONCLUSIONS

As it has been shown, legislation concerning the handling of biological material is diverse and complex, numerous steps have to be done before biological material can be handled and shipped in compliance with the law. Note that this chapter provides an overall picture on the complex topic of biolegislation. All these regulations and especially those concerning dispatch and transport of infectious substances may underlie constant changes. Therefore, it is necessary to keep ones self informed about recent changes when working with biological material as well as referring to local level rules and regulations which might differ in different countries. Currently, significant activities are taking place related to the CBD and its NP which result in the implementation of regional and national laws and regulations.

ABBREVIATIONS

ABS access and benefit sharing
ADN Accord Européen relatif au transport international des marchandises dangereuses par voies de navigation intérieures (European agreement concerning the international carriage of dangerous goods by inland waterways)
ADR Accord Européen relatif au transport international des marchandises dangereuses par routes (European agreement concerning the international carriage of dangerous goods by road)
BTWC Biological and Toxin Weapons Convention
CBD Convention on Biological Diversity
COTIF Convention concerning International Carriage by Rail
DGR Dangerous Goods Regulations
GMT Good Microbiological Techniques
IATA International Air Transport Association
ICAO International Civil Aviation Organization
IMDG International Maritime Dangerous Goods Code
MAT mutually agreed terms
NP Nagoya Protocol
PI Packing Instruction
PIC prior informed consent
RID Regulations concerning the International Carriage of Dangerous Goods by Rail
UN United Nations
UPU Universal Postal Union
WHO World Health Organization

ACKNOWLEDGMENTS

The author thanks Kathrin Felsch and Brian Tindall (DSMZ) for proofreading.

REFERENCES

Advisory Committee on Dangerous Pathogens, 2013. The Approved List of Biological Agents, third ed. Public Health and Safety Executive 08/13 Misc208 (rev2).

American Society for Microbiology (ASM). Code of Ethics, 2005. Available from: www.asm.org/index.php/component/content/article/114-unknown/unknown/4089-code-of-ethics.

Atlas, R.M., Somerville, M.A., 2004. Ethics: a weapon to counter bioterrorism. Science 307 (5717), 1881–1882.

Australian/New Zealand Standard™ AS/NZS 2243.3, 2010. Safety in laboratories, part 3 microbiological safety and containment, ISBN 978 0 7337 6996 2.

Bunch, R.L., McGuire, J.M., 1952. Erythromycin, its salts, and method of preparation. US patent 2,653,899.

Canadian Biosafety Standard, 2015. Public Health Agency of Canada, second ed., ISBN 978-1-100-25771-6, publ. no. 140467.

Eliopoulos, G.M., Maragakis, L.L., Perl, T.M., 2008. *Acinetobacter baumannii*: epidemiology, antimicrobial resistance, and treatment options. Clin. Infect. Dis. 46 (8), 1254–1263.

European Community, 2000. Commission Directive 2000/29/EC on protective measures against the introduction into the Community of organisms harmful to plants or plant products and against their spread within the Community. OJ no. L 169: 1-112.

European Parliament, 1994. Council Directive 3381/94 setting up a Community regime for the control of exports of goods with dual use. OJ no. L 367: 1-12.

European Parliament, 2000. Directive 2000/54/EC on the protection of workers from risks related to exposure to biological agents at work. OJ 2000, no. L262: 21-45.

European Parliament, 2016. Council Regulation 881/2002 imposing certain specific restrictive measures directed against certain persons and entities associated with Usama bin Laden, the Al-Qaida network and the Taliban, and repealing Council Regulation (EC) No. 467/2001 prohibiting the export of certain goods and services to Afghanistan, strengthening the flight ban and extending the freeze of funds and other financial resources in respect of the Taliban of Afghanistan. OJ 2002, no. L139: 9-22, last updated 2016. OJ no. L12: 42.

European Parliament and Council, 2014. Regulation (EU) No 511/2014 on compliance measures for users from the Nagoya Protocol on Access to Genetic Resources and the Fair and Equitable Sharing of Benefits Arising from their Utilization in the Union. OJ L 150: 59-71.

German infectious diseases of animals' imports enactment, 1982. (BGBl. I, p. 1728), last changed in 2015 (BGBl. I, p.1474).

German infectious diseases of animals' enactment, 1985. German Federal Law Gazette (BGBl.) Part I, p. 2123, last changed in 2014 (Federal Law Gazette.(BGBl.) Part I, p. 388).

German law regulating genetic engineering, 1993. (German Federal Law Gazette (BGBl.) Part I, p. 2066), last changed in 2015 (Federal Law Gazette.(BGBl.) Part I, p. 1474).

German act dealing with the prevention and control of infectious diseases in man, 2000. Federal Law Gazette (BGBl.) Part I, p. 1045, last changed in 2015 (German Federal Law Gazette (BGBl.) Part I, p. 2229).

German plant protection act (Pflanzenschutzgesetz), 2012. German Federal Law Gazette (BGBl.) Part I, p. 148, 1281, last changed in 2015 (German Federal Law Gazette.(BGBl.) Part I, p. 1474).

German Ordinance on Safety and Health Protection at Workplaces Involving Biological Agents (Biological Agents Ordinance—BioStoffV), 2013. Federal Law Gazette (BGBl.) Part I, p. 2514. Available from: http://www.gesetze-im-internet.de/englisch_biostoffv/.

Hilliard, J., 2007. Monkey B virus. In: Arvin, A., Campadelli-Fiume, G., Mocarski, E. et al., (Eds.), Human Herpesviruses: Biology, Therapy, and Immunoprophylaxis. Cambridge University Press, Cambridge, pp. 1031–1042.

IATA, 2017. Dangerous Goods Regulations, fifty-eighth ed. Montreal, Geneva, ISBN 978-92-9252-916-1.

Intergovernmental Organization for International Carriage by Rail, 2015. Regulations concerning the international carriage of dangerous goods by rail (RID 2015—Règlement concernant le transport international ferroviaire des marchandises dangereuses), TSO, London, ISBN 9788086206424.

International Bureau of the Universal Postal Union, 2013. Letter Post Manual, Berne.

International Maritime Organization, 2016. International Maritime Dangerous Goods Code (IMDG 2016). London, ISBN 978-92-8011-6366.

International Organization for Standardization, 2014. ISO/IEC, Guide 51:2014. Safety aspects—guidelines for their inclusion in standards.

International Union for Conservation of Nature, 1973. Convention on International Trade in Endangered Species of Wild Fauna and Flora (CITES), last amended in 1983.

International Union of Microbiological Societies (IUMS), 2008. Code of Ethics against Misuse of Scientific Knowledge, Research and Resources. Available from: www.iums.org/index.php/code-of-ethics.

Joint FAO/WHO Codex Alimentarius Commission, 1992. Codex Alimentarius. Food and Agriculture Organization of the United Nations, Rome.

Leibniz Association. Biosecurity Code, 2012. Available from: www.leibniz-gemeinschaft.de/medien/aktuelles/news-details/article/leibniz_gemeinschaft_verabschiedet_verhaltenskodex_zur_biosicherheit_100000547/.

Qian, W. 2014. Efforts to Strengthen Biosafety and Biosecurity in China. Lecture Notes. Available from: https://www.nonproliferation.org/wp-content/uploads/2014/02/070917_wang.pdf.

Reyes, D., 2003. The ethics of biowarfare. Available from: www.actionbioscience.org/biotechnology/reyes.html.

Secretariat of the Convention on Biological Diversity, 2000. Cartagena Protocol on Biosafety to the Convention on Biological Diversity: Text and Annexes. Montréal, ISBN 92-807-1924-6.

Tomita, K., Tsukiura, H., Kawaguchi, H., 1977. Fermentation process for producing apramycin and nebramycin factor V'. US patent 4,032,404.

United Nations, 1992. Convention on Biological Diversity. UNEP/CBD.

United Nations, 2011. Nagoya Protocol on Access to Genetic Resources and the Fair and Equitable Sharing of Benefits Arising from their Utilization to the Convention on Biological Diversity, ISBN 92-9225-306-9. European Parliament and Council, 2014. Regulation (EU) No 511/2014 on compliance measures for users from the Nagoya Protocol on Access to Genetic Resources and the Fair and Equitable Sharing of Benefits Arising from their Utilization in the Union. OJ L 150: 59-71.

United Nations, 2015. Recommendations on the Transport of Dangerous Goods. Model Regulations ('Orange Book'), nineteenth ed. New York and Geneva, ISBN 978-92-1-139155-8.

United Nations, Economic Commission for Europe, 2016. European agreement concerning the international carriage of dangerous goods by inland waterways (ADN 2017—Accord Européen relatif au transport international des marchandises dangereuses par voies de navigation intérieures). New York and Geneva, ECE/TRANS/258, ISBN 978-92-1-139157-2.

United Nations, Economic Commission for Europe, 2016. European agreement concerning the international carriage of dangerous goods by road (ADR 2017—Accord Européen relatif au transport international des marchandises dangereuses par routes), New York and Geneva, ECE/TRANS/257, ISBN 978-92-1-139156-5.

United Nations Educational, Scientific and Cultural Organisation (UNESCO), 2006. Universal Declaration on Bioethics and Human Rights. SHS/EST/BIO/06/1.

United Nations Office for Disarmament Affairs, 1975. Convention on the Prohibition of the Development, Production and Stockpiling of Bacteriological (Biological) and Toxin Weapons and on their Destruction.

U.S. Department of Health and Human Services, Public Health Service Centers for Disease Control and Prevention, National Institutes of Health, 2009. Biosafety in Microbiological and Biomedical Laboratories, fifth ed. HHS Publ. No. (CDC) 21-1112.

World Health Organization, 2004a. Laboratory Biosafety Manual, third ed. World Health Organization, Geneva.

World Health Organization. 2004b. Public health response to biological and chemical weapons: WHO guidance—2nd ed. World Health Organization, Geneva, ISBN 92 4 154615 8.

World Health Organization, 2006. Biorisk management. Laboratory biosecurity guidance. WHO/CDS/EPR/2006.6.

World Health Organization, 2015. Guidance on Regulations for the Transport of Infectious 2015-2016. WHO/HSE/GCR/2015.2.

World Medical Association, 2013. In: World Medical Association Declaration of Helsinki: ethical principles for medical research involving human subjects. JAMA 310 (20), 2191-2194.

FURTHER READING

World Health Organization, 2014. Media Centre, Fact Sheets: Infectious Diseases. Available from: www.who.int/topics/infectious_diseases/factsheets/en/.

World Health Organization, 2016. Media Centre, Fact Sheets: Infectious Diseases. Available from: www.who.int/topics/infectious_diseases/factsheets/en/.

Index

B

C

V

W

X

Y

Z

Printed in the United States
By Bookmasters